Interactive
Textbook

HOLT, RINEHART AND WINSTON
A Harcourt Education Company
Orlando • Austin • New York • San Diego • London

Contents

CHAPTER 9 It's Alive!! Or Is It?

CHAPTER 10 Cells: The Basic Units of Life

CHAPTER 11 Classification

CHAPTER 12 Bacteria and Viruses

CHAPTER 13 Protists and Fungi

CHAPTER 14 Introduction to Plants

CHAPTER 15 Introduction to Animals

CHAPTER 16 Interactions of Living Things

CHAPTER 17 Cycles in Nature

CHAPTER 18 Properties and States of Matter

CHAPTER 24 Heat and Heat Technology

CHAPTER 1 Science in Our World

SECTION 1 | # Science and Scientists

BEFORE YOU READ

After you read this section, you should be able to answer these questions:

• What is science?

• How does science affect our lives?

• Where do scientists work?

What Is Science?

You probably do not know it, but there are ways you think and act like a scientist. You think like a scientist any time you ask a question about the world around you. For example, you might wonder why your reflection in a spoon is upside down.

You act like a scientist any time you investigate a question you ask. For example, you may investigate how your reflection changes in different mirrors. Asking questions and searching for answers are part of a process of gathering knowledge about the natural world. This process is called **science**. ☑

The world around you is full of amazing things that can lead you to ask questions and search for answers. The girl shown below is thinking like a scientist because she is asking questions about things she has observed in the world around her.

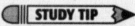

STUDY TIP

Outline As you read, make an outline of this section. Use the headings from the section in your outline.

READING CHECK

1. Define What is science?

_____

_____

_____

_____

Why does the mirror fog when I shower?

How are a frog and a lizard different?

How do birds know where to go when they migrate?

OBSERVATION

Anywhere you look you may see things that make you wonder. You could look in your home, neighborhood, city, or a nearby forest or beach. You could even look beyond Earth to the moon, sun, and other objects in our universe. Obviously, there are many things that can cause people to think of a question.

Once you ask a question, you can look for an answer. There are several different methods you can use to start your search for an answer. One way you may find the answer is by looking around and making observations.

For example, you may wonder whether clouds and weather are related. To find the answer, you could observe the type of clouds you see each day and the day's weather. You may find that stormy weather is associated with one type of cloud, while nice weather is associated with another type of cloud.

RESEARCH

In addition to making observations, you can also research what scientists already know about your question. There are many places to do research and find information. The girl shown below is working in a library, a good place to conduct research.

At a library, you can search for information in books, in magazines, and on the Internet. Always be sure to use information that comes from a reliable source, whether it is a book, magazine article, or a site on the World Wide Web. Ask your teacher or the librarian to find out if a source is reliable. ☑

Critical Thinking

2. Describe Give an example of how making observations helped you answer a question.

_____

_____

_____

_____

☑ READING CHECK

3. Identify List three places to research information in a library.

_____

_____

EXPERIMENTATION

You may also answer your question by doing an experiment. Many experiments can be simple to perform and use materials you can easily obtain at home. However, other experiments are more difficult to perform and may require materials found only in a science laboratory.

All experiments must be carefully planned. You should research similar experiments that have been done in the past. Follow safety precautions. During an experiment, make careful observations and write them down. You may observe something unexpected. You may need to research more, repeat the experiment, or perform a different experiment. ☑

Finding an answer to a question may involve several steps, including making observations, doing research, and performing an experiment.

☑ READING CHECK

4. Describe What could you do if you observed something unexpected during an experiment?

_____

_____

_____

TAKE A LOOK
5. Describe What safety precaution is the girl in this experiment using?

_____

_____

What Role Does Science Play in Our Lives?

Scientists cannot answer every question that they ask. Yet, they continue to ask more questions and search for answers. The answers they find often affect people in their daily lives. For example, scientists are always searching for new medicines, ways to make machines more efficient, and more accurate ways to predict the weather. Their discoveries have resulted in saving lives, resources, and the environment.

SECTION 1 Science and Scientists *continued*

SAVING LIVES

One question scientists have asked is, "How can a person be safer while riding a bicycle?" This question has led to several answers. One answer is to make better helmets. Science has helped people develop new materials that make stronger helmets. A strong helmet helps protect a rider's head if it hits the ground. A strong helmet can prevent injuries that may cause brain damage or death. In this way, science is helping to save lives.

SAVING RESOURCES

Another question scientists ask is "How can we conserve natural resources such as iron and trees?" One answer is to recycle. Scientists have developed efficient ways to recycle many things such as steel, aluminum, paper, glass, and even some plastics. In this way, science helps make resources last longer. The figure below shows the resources saved by recycling one metric ton of paper.

Compared with making the paper originally, recycling 1 metric ton (1.1 tons) of paper:

 produces 30 kg (66 lb) less air pollution

 uses 2.5 m³ (3.3 yd³) less landfill space

 uses 18.7 fewer trees

 uses 4,500 kWh less energy

 uses 29,100 L (7,700 gal) less water

 uses 1,800 L (470 gal) less oil

Critical Thinking

6. Predict What might happen to our natural resources if we could not recycle them?

_____

_____

_____

SAVING THE ENVIRONMENT

Still another question scientists have asked is, "How can we protect the environment from pollution?" Pollution in the air, water, and soil can harm our health. Pollution is also harmful to plants and animals. ☑

One way scientists have helped reduce pollution is by finding ways that cars can produce less exhaust. By inventing lighter materials, scientists can build lighter cars. Lighter cars use less fuel and make less exhaust. Scientists also develop new types of cars that are better for the environment, such as the one on the following page.

READING CHECK

7. List Pollution can be found in what three things?

_____

_____

SECTION 1 Science and Scientists *continued*

This car does not pollute the air directly because it runs on batteries rather than gasoline.

TAKE A LOOK
8. Explain How is a car that runs on batteries better for the city the car is driven in?

_____

_____

_____

_____

Where Can You Find a Scientist at Work?

Scientists work in many different places including laboratories, forests, offices, oceans, and space. The following table summarizes some jobs that use science.

Job title	Description
Cartographer	person who makes maps of the surface of Earth
Engineer	person who uses scientific knowledge in order to build things
Environmental scientist	person who studies how humans interact with their environment
Science educator	person who teaches other people about science
Zoologist	person who studies animals

TAKE A LOOK
9. Explain How does an experiment like this one help protect the environment?

_____

_____

_____

_____

These environmental scientists are testing water quality.

Section 1 Review

SECTION VOCABULARY

science the knowledge obtained by observing natural events and conditions in order to discover facts and formulate laws or principles that can be verified or tested	

1. Define In your own words, write a definition for the term *science*.

_____

_____

_____

2. Identify List three processes that can be used to answer a science question.

_____

_____

_____

3. Describe How might an engineer use science in order to save lives?

_____

_____

_____

4. Calculate Complete the following table.

Resources saved	Amount saved by recycling 1 metric ton of paper	Amount saved by recycling 3 metric tons of paper
Trees	18.7 trees	
Water		87,300 L
Energy	4,500 kWh	

5. Explain Name one way that science has played a role in your life today.

_____

_____

_____

CHAPTER 1 Science in Our World
SECTION 3 **Scientific Models**

BEFORE YOU READ

After you read this section, you should be able to answer these questions:

• What are the three types of scientific models?

• How do scientists use models to help them understand scientific information?

What Is a Scientific Model?

How can baking soda, vinegar, and clay help you understand the way a volcano erupts? You might not think these things could help you. However, you could use them to build a model of a volcano. The model may help you understand how a real volcano erupts.

A **model** is something that scientists use to represent an object or process. For example, models of human body systems help you learn how the body works. Models can also help you learn about the past or predict the future. However, a model cannot tell you everything about the thing it represents. A model is always slightly different than what it represents.

There are three kinds of scientific models: physical models, mathematical models, and conceptual models. The model volcano shown below is an example of a physical model.

A model volcano can help you understand what happens when a real volcano erupts.

STUDY TIP

Compare After you read this section, make a chart comparing scientific theories and scientific laws.

Critical Thinking

1. Compare What are some differences you might find between a model and the thing it represents?

_____

_____

_____

TAKE A LOOK

2. Describe How is a model volcano useful for learning about the way a real volcano erupts? Give two ways.

_____

_____

_____

SECTION 3 Scientific Models *continued*

Critical Thinking

3. Describe How could a physical model help you understand an invisible sound wave?

_____

_____

_____

_____

PHYSICAL MODELS

A *physical model* is a model you can see and touch. Some physical models can help you study things that are too small to see. For example, a ball-and-stick model can show how atoms are arranged in a molecule. Other physical models can help you study things that are too large to see all at once. For example, you can't see the entire Earth or solar system. A model can help you picture them in your mind.

A physical model can also help you understand a concept. Launching a model of a space shuttle can help you understand how a real space shuttle is launched.

MATHEMATICAL MODELS

A *mathematical model* is made of mathematical equations and data. You can't see a mathematical model the way you can see a physical model. However, you can use it to understand systems and make predictions. For example, meteorologists use models to predict the weather.

Some mathematical models are simple. Others are very complicated and are operated by computers. Scientists used a mathematical model run on a computer to predict how fast the number of people on Earth will grow. The graph below shows the model's prediction.

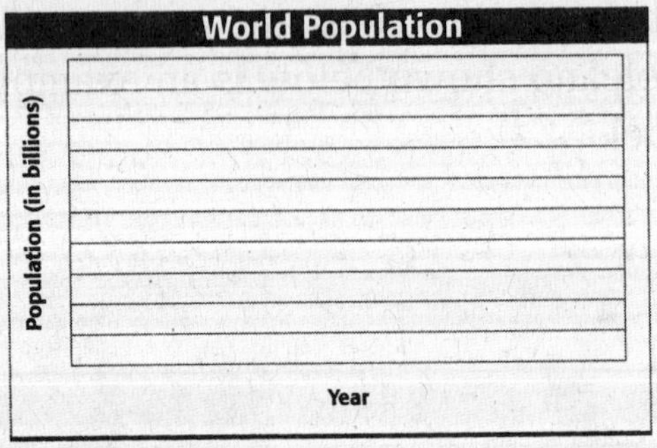

World Population

Population (in billions)

Year

Math Focus

4. Calculate According to the graph, how much did the global population grow between 1950 and 2000?

CONCEPTUAL MODELS

Conceptual models are used to help explain ideas. Conceptual models may be based on systems of ideas or data from experiments. Some conceptual models explain natural events or processes that cannot be observed.

One example of a conceptual model is the system that scientists use to classify living things. By using a system of ideas, scientists can put organisms into groups based on what the organisms have in common.

How Do Models Help Build Scientific Knowledge?

Scientists use models to help them look for new information. Models are also used to help explain scientific theories. In science, a **theory** is an explanation for many hypotheses and observations. Theories are supported by many tests and observations. A theory can explain why something happens and can predict what will happen in the future. ☑

When a model gives scientists new information, that information can support a theory or show it is wrong. As models give new information and scientists make new observations, new theories are developed over time. New models are constructed to demonstrate new theories. The figure below shows an old model of Earth and the current model that replaced it.

SCIENTIFIC LAWS

In science, a **law** is a summary of many experimental results and observations. It tells you how things work. A law is different from a theory. A law tells you only what happens, while a theory tells you why it happens. ☑

An example of a scientific law is the *law of conservation of mass*. It states that the total mass of a substance does not change in a chemical change. The law does not explain why this is true or make any predictions. It tells you only what happens during any chemical change.

☑ **READING CHECK**

5. Define What is a scientific theory?

_____

_____

_____

TAKE A LOOK

6. Explain Why might scientists come up with a new theory?

_____

_____

_____

☑ **READING CHECK**

7. Define What is a scientific law?

_____

_____

Section 3 Review

SECTION VOCABULARY

law a summary of many experimental results and observations; a law tells how things work **model** a pattern, plan, representation, or description designed to show the structure or workings of an object, system, or concept	**theory** an explanation that ties together many hypotheses and observations

1. Identify What are three types of models used by scientists?

_____

2. Identify Fill in the blanks in the table to tell whether the example is a physical, mathematical, or conceptual model.

Model	Type of Model
Computer prediction of the path of a hurricane	
Ball-and-stick model of a molecule	physical
Plastic skeleton	
Classification of living things	

3. Identify A model of Earth can help show you what the inside of Earth looks like. What are two ways that this model is different from the object it represents?

_____

_____

4. Compare What is the difference between a scientific theory and a scientific law?

_____

_____

_____

5. Describe Which kind of model would you use to represent a human heart? Explain the reason for your choice.

_____

_____

_____

CHAPTER 1 Science in Our World
SECTION 4 Tools, Measurement, and Safety

BEFORE YOU READ

After you read this section, you should be able to answer these questions:

• What tools do scientists use to make measurements?

• What is the International System of Units?

• How can you stay safe in science class?

What Kinds of Tools Do Scientists Use?

Would you use a hammer to tighten a bolt on a bicycle? If you did, you probably wouldn't be very successful in your task. To be successful, you need the right tools. Scientists use many tools. A *tool* is anything that helps you do a task.

TOOLS FOR SEEING

If you observe a jar of pond water, you may see a few creatures swimming around. However, you might not be able to see them very well. A microscope, like the one in the figure below, can help you see things that are too small to see with just your eyes. A microscope has special lenses that make things appear larger than they are. When you place something in front of the lens, it is magnified. That way, you can see things that are very small.

STUDY TIP

Outline As you read, underline the main ideas in each paragraph. When you finish reading, make an outline of the section using the ideas you underlined.

Critical Thinking

1. Identify What other tools make the size of an object appear larger?

_____

_____

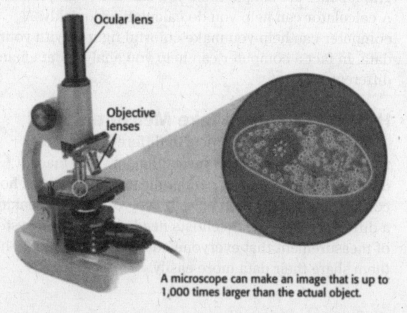

A microscope can make an image that is up to 1,000 times larger than the actual object.

TOOLS FOR MEASURING

Remember that one way to collect data during an experiment is to take measurements. To take the most accurate measurements, you need the correct tools. The figure below shows some of the tools scientists use to take measurements.

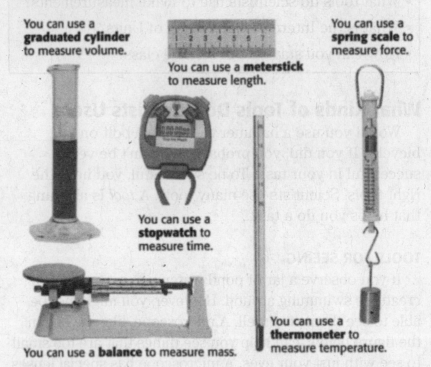

You can use a **graduated cylinder** to measure volume.

You can use a **meterstick** to measure length.

You can use a **spring scale** to measure force.

You can use a **stopwatch** to measure time.

You can use a **balance** to measure mass.

You can use a **thermometer** to measure temperature.

TAKE A LOOK
2. Identify Which tool can you use to measure mass?

_____

_____

TOOLS FOR ANALYZING

Some tools help you analyze your data. A pencil and graph paper are simple tools you can use to graph data. A calculator can help you do calculations quickly. A computer can help you make colorful figures with your data. In fact a computer can help you analyze data many different ways. ☑

READING CHECK
3. Identify What type of tool can help you analyze data in many different ways?

_____

_____

How Do Scientists Take Measurements?

Many years ago, scientists in different countries used different systems of measurement. This made it difficult for scientists to communicate. Just imagine how confusing it would be in class if everyone were speaking a different language! Scientists needed a standard system of measurement that everyone could use. This would help them share their data more easily.

SECTION 4 Tools, Measurement, and Safety *continued*

THE INTERNATIONAL SYSTEM OF UNITS

In the late 1700s, the French Academy of Sciences created a system of measurement that would be used by scientists throughout the world. Today, this system is called the *International System of Units*, abbreviated as SI. You might know it as the metric system.

All SI units are written in multiples of 10. This makes it easy to change from one unit to another. Some common SI units are shown in the table below. ☑

Common SI Units		
Length	meter (m)	
	kilometer (km)	1 km = 1,000 m
	decimeter (dm)	1 dm = 0.1 m
	centimeter (cm)	1 cm = 0.01 m
	millimeter (mm)	1 mm = 0.001 m
	micrometer (μm)	1 μm = 0.000 001 m
	nanometer (nm)	1 nm = 0.000 000 001 m
Area	square meter (m²)	
	square centimeter (cm²)	1 cm2 = 0.0001 m²
Volume	cubic meter (m³)	
	cubic centimeter (cm³)	1 cm³ = 0.000 001 m³
	liter (L)	1 L = 1 dm³ = 0.001 m³
	milliliter (mL)	1 mL = 0.001 L = 1 cm³
Mass	kilogram (kg)	
	gram (g)	1 g = 0.001 kg
	milligram (mg)	1 mg = 0.000001 kg
Temperature	Kelvin (K)	0°C = 273 K
	Celsius (°C)	100°C = 373 K

LENGTH AND AREA

The **meter** (m) is the basic SI unit for length. An Olympic-sized swimming pool, for example, is 50 m long. Knowing the length of an object can help you calculate its area. Area is a measure of the surface of an object. **Area** is based on two measurements: length and width. Objects with different shapes can have the same surface area. The units for area are called square units, such as square kilometers (km²) and square meters (m²). The equation below describes how to calculate the area of a square or rectangle.

$$area = length \times width$$

✓ READING CHECK

4. Describe Why is it easy to change from one SI unit to another?

_____

_____

Math Focus

5. Calculate How many meters are in 50 kilometers?

_____

Critical Thinking

6. Explain How can two differently shaped objects have the same area?

_____

_____

_____

Math Focus

7. Calculate What is the area of a rectangle that is 6 cm long and 4 cm wide? Show your work.

MASS

Mass is a measure of how much matter is in an object. The *kilogram* (kg) is the basic SI unit for mass. The kilogram is used to describe the mass of large objects, such as sacks of grain. *Grams* (g) are used to describe the mass of smaller objects, such as apples. A medium-sized apple, for example, has a mass of about 100 g. One thousand grams equals one kilogram. ☑

READING CHECK

8. Define What is mass?

_____

_____

VOLUME

Volume is the amount of space that something takes up. Volume is based on three measurements: length, width, and height. The following equation shows how the volume of a box is calculated:

$$volume = length \times width \times height$$

The units for volume of an object are called cubic units, such as cubic meters (m^3) and cubic centimeters (cm^3). The volume of a liquid, however, is measured in *liters* (L). One cubic meter equals 1,000 L. One cubic centimeter equals one milliliter (mL).

Many objects have an irregular shape. The figure below shows how to measure the volume of an irregularly shaped object, such as a rock. Notice in the left figure that the volume of the water in the graduated cylinder is 70 mL. Notice in the right figure that adding the rock to the water increases the volume to 80 mL. This increase of 10 mL represents the volume of the rock. Because 1 mL = 1 cm^3, the volume of the rock is 10 cm^3.

TAKE A LOOK

9. Explain Why is it necessary to use a different method to measure the volume of an irregularly shaped object?

_____

_____

_____

_____

TEMPERATURE

Temperature is a measure of how hot or cold something is. You probably use degrees Fahrenheit (°F) to describe temperature. Scientists often use degrees Celsius (°C). However, the SI unit for temperature is the kelvin (K, without a degree sign). The figure below shows the relationship between the Celsius scale and the Fahrenheit scale. ☑

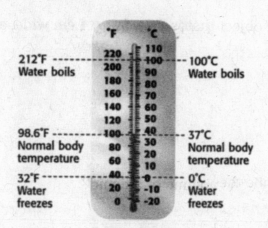

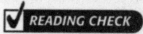

READING CHECK

10. Define What is the SI unit for temperature?

_____

_____

TAKE A LOOK

11. Calculate The equation to convert from degrees Celsius to Kelvins is K = °C + 273. What temperature does water boil on the Kelvin scale?

_____

_____

How Can You Be Safe in Science Class?

Science is exciting and fun, but it can be dangerous. The following rules will help keep you safe while doing experiments.

• Always get permission before performing an experiment.
• Always listen to your teacher's instructions.
• Read directions carefully, especially safety information. ☑

The safety symbols in the figure below are important. Learn them so that you and others will be safe in the lab.

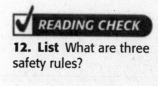

READING CHECK

12. List What are three safety rules?

_____

_____

_____

Section 4 Review

SECTION VOCABULARY

area a measure of the size of a surface or a region	**temperature** a measure of how hot (or cold) something is; specifically, a measure of the average kinetic energy of the particles in an object
mass a measure of the amount of matter in an object; a fundamental property of an object that is not affected by the forces that act on the object, such as the gravitational force	**volume** a measure of the size of a body or region in three-dimensional space
meter the basic unit of length in the SI (symbol, m)	

1. **Calculate** What is the volume of an object that is 3 cm long, 4 cm wide, and 6 cm high? Show your work

2. **Identify** What unit is used to describe the volume of a liquid?

_____

3. **Identify** Which of the safety symbols shown below would you see in a lab procedure that asks you to pour acid into a beaker? Explain the reason for your choice(s).

A

B

C

_____
_____
_____

4. **Describe** How would you measure the volume of a strawberry?

_____
_____
_____
_____
_____

CHAPTER 2 | Weathering and Soil Formation
SECTION 1 | **Weathering**

National Science
Education Standards
ES 1c, 1d, 1k

BEFORE YOU READ

After you read this section, you should be able to answer these questions:

• What is weathering?
• What causes mechanical weathering?
• What causes chemical weathering?

What Is Weathering?

How do large rocks turn into smaller rocks? **Weathering** is the process in which rocks break down. There are two main kinds of weathering: mechanical weathering and chemical weathering.

What Is Mechanical Weathering?

Mechanical weathering happens when rocks are broken into pieces by physical means. There are many *agents*, or causes, of mechanical weathering. ☑

ICE

Ice is one agent of mechanical weathering. Cycles of freezing and thawing can cause *ice wedging*, which can break rock into pieces.

The cycle of ice wedging starts when water seeps into cracks in a rock. When the water freezes, it expands. The ice pushes against the cracks. This causes the cracks to widen. When the ice melts, the water seeps further into the cracks. As the cycle repeats, the cracks get bigger. Finally, the rock breaks apart.

STUDY TIP

Compare Make a chart showing the ways that mechanical weathering and chemical weathering can happen.

☑ **READING CHECK**

1. Define What is mechanical weathering?

_____

_____

Critical Thinking

2. Infer Would ice wedging happen if water did not expand as it froze? Explain your answer.

_____

_____

_____

_____

Ice Wedging

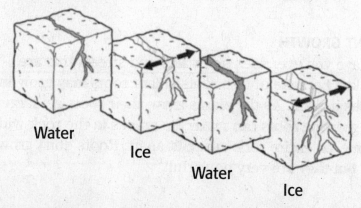

Water

Ice

Water

Ice

WIND, WATER, AND GRAVITY

As you scrape a large block of chalk against a board, tiny pieces of the chalk rub off on the board. The large piece of chalk wears down and becomes smaller. The same process happens with rocks. **Abrasion** is a kind of mechanical weathering that happens when rocks are worn away by contact with other rocks. Abrasion happens whenever one rock hits another. Water, wind, and gravity can cause abrasion.

Water can cause abrasion by moving rocks and making them hit each other. The rocks in this river are rounded because of abrasion.

Wind can cause abrasion when it blows sand against rocks. This rock has been shaped by blowing sand.

Gravity can cause abrasion by making rocks rub against each other as they slide downhill. As the rocks grind against each other, they are broken into smaller pieces.

TAKE A LOOK
3. Explain How does running water cause abrasion?

_____

_____

_____

_____

PLANT GROWTH

Have you ever seen sidewalks and streets that are cracked because of tree roots? Plant roots may grow into cracks in rock. As the plants grow, their roots get larger. The growing roots can make the cracks in the rock wider. In time, an entire rock can split apart. Roots don't grow fast, but they are very powerful!

ANIMALS

Did you know that earthworms cause a lot of weathering? They tunnel through the soil and move pieces of rock around. This motion breaks some of the rocks into smaller pieces. It also exposes more rock surfaces to other agents of weathering.

Any animal that burrows in the soil causes mechanical weathering. Ants, worms, mice, coyotes, and rabbits are just a few of the animals that can cause weathering. The mixing and digging that animals do can also cause chemical weathering, another kind of weathering.

What Is Chemical Weathering?

In addition to physical weathering, rocks can be broken down by chemical means. **Chemical weathering** happens when rocks break down because of chemical reactions.

Water, acids, and air are all agents of chemical weathering. They react with the chemicals in the rock. The reactions can break the bonds in the minerals that make up the rock. When the bonds in the minerals are broken, the rock can be worn away. ☑

WATER

If you drop a sugar cube into a glass of water, the sugar cube will dissolve after a few minutes. In a similar way, water can dissolve some of the chemicals that make up rocks. Even very hard rocks, such as granite, can be broken down by water. However, this process may take thousands of years or more.

Chemical Weathering in Granite

Granite is made of many different minerals. Rain and air can contain chemicals that react with the minerals.

Eventually, many of the minerals in the granite will be broken down. The small pieces of minerals that are left are called sediment.

The chemicals in rain and air can break down the bonds in the minerals. Rain can dissolve some of the minerals in the rock and wash them away.

STANDARDS CHECK

ES 1k Living organisms have played many roles in the Earth system, including affecting the composition of the atmosphere, producing some types of rocks, and contributing to the weathering of rocks.

Word Help: role
a part or function; purpose

Word Help: affect
to change; to act upon

4. Describe How can earthworms cause weathering?

_____
_____
_____
_____
_____
_____

☑ **READING CHECK**

5. List What are three agents of chemical weathering?

_____
_____
_____

TAKE A LOOK
6. Infer What do you think is the reason it takes a very long time for granite to break down?

_____
_____
_____
_____

ACID PRECIPITATION

Precipitation, such as rain and snow, always contains a little bit of acid. However, sometimes precipitation contains more acid than normal. Rain, sleet, or snow that contains more acid than normal is called **acid precipitation**.

Acid precipitation forms when small amounts of certain gases mix with water in the atmosphere. The gases come from natural sources, such as active volcanoes. They are also produced when people burn fossil fuels, such as coal and oil. ☑

The acids in the atmosphere fall back to the ground in rain and snow. Acids can dissolve materials faster than plain water can. Therefore, acid precipitation can cause very rapid weathering of rock.

ACIDS IN GROUNDWATER

In some places, water flows through rock underground. This water, called *groundwater*, may contain weak acids. When the groundwater touches some kinds of rock, a chemical reaction happens. The chemical reaction dissolves the rock. Over a long period of time, huge caves can form where rock has been dissolved.

This cave formed when acids in groundwater dissolved the rock.

✔️ **READING CHECK**

7. Identify What are two sources of the gases that produce acid precipitation?

_____

_____

TAKE A LOOK

8. Explain Caves like the one in the picture are not found everywhere. What do you think controls where a cave forms?

_____

_____

_____

ACIDS FROM LIVING THINGS

All living things make weak acids in their bodies. When the living things touch rock, some of these acids are transferred to the surface of the rock.

The acids react with chemicals in the rock and weaken it. The different kinds of mechanical weathering can more easily remove rock in these weakened areas. ☑

The rock may also crack in the weakened areas. Even the smallest crack can expose more of the rock to both mechanical weathering and chemical weathering.

AIR

Have you ever seen a rusted car or building? Rusty metal is an example of chemical weathering. Metal reacted with something to produce rust. What did the metal react with? In most cases, the answer is air.

The oxygen in the air can react with many metals. These reactions are a kind of chemical weathering called *oxidation*. Rust is a common example of oxidation. Rocks can rust if they have a lot of iron in them.

Many people think that rust forms only when metal gets wet. In fact, oxidation can happen even without any water around. However, when water is present, oxidation happens much more quickly.

Oxidation can cause rocks to weaken. Oxidation changes the metals in rocks into different chemicals. These chemicals can be broken down more easily than the metals that were there before.

READING CHECK

9. Explain How can acids from living things cause weathering?

_____

_____

_____

_____

_____

Factor	How does it cause chemical weathering?
Water	
Acid precipitation	
Acids in groundwater	
Acids from living things	
Air	

TAKE A LOOK

10. Describe Fill in the blank spaces in the table to describe how different factors cause chemical weathering.

Name _____ Class _____ Date _____

Section 1 Review

SECTION VOCABULARY

abrasion the grinding and wearing away of rock surfaces through the mechanical action of other rock or sand particles	**mechanical weathering** the process by which rocks break down into smaller pieces by physical means
acid precipitation rain, sleet, or snow that contains a high concentration of acids	**weathering** the natural process by which atmospheric and environmental agents, such as wind, rain, and temperature changes, disintegrate and decompose rocks
chemical weathering the process by which rocks break down as a result of chemical reactions	

1. List What are three things that can cause abrasion?

_____

_____

_____

2. Explain Fill in the spaces to show the steps in the cycle of ice wedging.

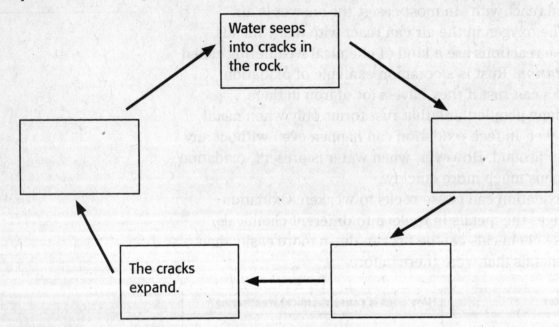

3. Identify How can acids cause chemical weathering?

_____

_____

_____

4. Compare How is mechanical weathering caused by ice wedging similar to mechanical weathering caused by plant roots?

_____

_____

CHAPTER 2 | Weathering and Soil Formation

SECTION 2 | Rates of Weathering

BEFORE YOU READ

After you read this section, you should be able to answer these questions:

• What is differential weathering?

• What factors affect how fast rock weathers?

National Science Education Standards
ES 1c, 1d

What Is Differential Weathering?

Hard rocks, such as granite, weather more slowly than softer rocks, such as limestone. **Differential weathering** happens when softer rock weathers away and leaves harder, more resistant rock behind. The figures below show an example of how differential weathering can shape the landscape.

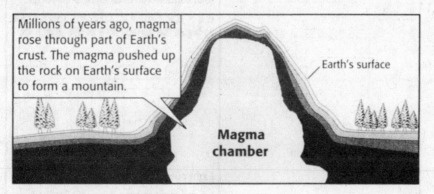
Millions of years ago, magma rose through part of Earth's crust. The magma pushed up the rock on Earth's surface to form a mountain.
Earth's surface
Magma chamber

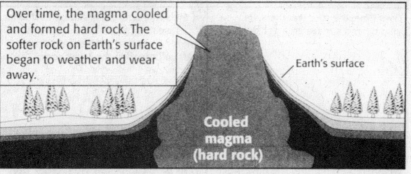
Over time, the magma cooled and formed hard rock. The softer rock on Earth's surface began to weather and wear away.
Earth's surface
Cooled magma (hard rock)

Today, all of the soft rock of the mountain has weathered away. The only thing that is left is the hard rock from inside the mountain. This hard rock forms the unusual structure called Devils Tower, in Wyoming.

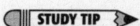**STUDY TIP**

Apply Concepts After you read this section, think about how the different factors that control weathering can affect objects that you see every day. Discuss your ideas in a small group.

**Say It**

Discuss In a small group, share your ideas about how some different landscape features formed because of differential weathering.

TAKE A LOOK

1. Infer Imagine that the rock on the outside of the mountain was much harder than the cooled magma. How might Devils Tower look today? Explain your answer.

_____

_____

_____

_____

SECTION 2 Rates of Weathering *continued*

How Does a Rock's Surface Area Affect Weathering?

Most chemical weathering takes place only on the outer surface of a rock. Therefore, rocks with a lot of surface area weather faster than rocks with little surface area. However, if a rock has a large volume as well as a large surface area, it takes longer for the rock to wear down. The figure below shows how the surface area and volume of a rock affect how fast it wears down.

Critical Thinking

2. Explain Why does most chemical weathering happen only on the outer surface of a rock?

_____

_____

_____

_____

_____

Math Focus
3. Calculate Determine the surface areas of the large cube and the eight smaller cubes. Write these values in the blank lines on the figure.

Imagine a rock in the shape of a cube. Each side of the rock is 4 m long. The volume of the cube is 4 m × 4 m × 4 m = 64 m³. Each face of the rock has an area of 4 m × 4 m = 16 m². Because there are six faces on the rock, it has a surface area of 6 × 16 m² = _____ m². 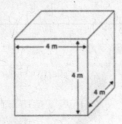 Cube with 4 m sides: volume = 64 m³, surface area = _____ m²	Now imagine eight smaller rocks that are shaped like cubes. Each small rock's side is 2 m long. The volume of each small rock is 2 m × 2 m × 2 m = 8 m³. The total volume of all 8 rocks is 8 × 8 m³ = 64 m³, the same as the large rock. Each face of each small rock has an area of 2 m × 2 m = 4 m². Each small rock has six faces, and there are eight rocks total. Therefore, the total surface area of all eight small rocks is 8 × 6 × 4 m² = _____ m². This is twice as big as the surface area of the large rock. 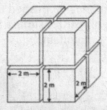 Eight cubes with 2 m sides: volume = 64 m³, surface area = _____ m²
Over time, the rock weathers. Its volume and surface area get smaller. 	In the same amount of time, the smaller rocks weather more and become much smaller. They lose a larger fraction of their volume than the larger rock.
More time passes. The large rock is weathered even more. It is now much smaller than it was before it was weathered. 	In the same amount of time, the small rocks have completely worn away. They took much less time to wear away than the large rock. Even though their volume was the same, they had more surface area than the large rock. The large surface area allowed them to wear away more quickly.

TAKE A LOOK
4. Apply Concepts Why do the edges and corners of the cubes weather faster than the faces? (Hint: remember that objects with large surface areas weather quickly.)

_____

_____

_____

How Does Climate Affect Weathering?

The rate of weathering in an area is affected by the climate of that area. *Climate* is the average weather conditions of an area over a long period of time. Some features of climate that affect weathering are temperature, moisture, elevation, and slope.

TEMPERATURE

Temperature is a major factor in both chemical and mechanical weathering. Cycles of freezing and thawing increase the chance that ice wedging will take place. Areas in the world that have many freezes and thaws have faster mechanical weathering than other regions do.

High temperatures can also speed up weathering. Many chemical reactions happen faster at higher temperatures. These reactions can break down rock quickly in warm areas. ☑

MOISTURE

Water can interact with rock as precipitation, as running water, or as water vapor in the air. Water can speed up many chemical reactions. For example, oxidation can happen faster when water is present.

Water is also important in many kinds of mechanical weathering. For example, ice wedging cannot happen without water. Abrasion also happens faster when water is present.

ELEVATION AND SLOPE

Elevation and slope can affect how fast weathering occurs. *Elevation* is a measure of how high a place is above sea level. *Slope* is a measure of how steep the sides of a mountain or hill are. The table below shows how elevation and slope affect weathering.

Factor	How the factor affects weathering
Elevation	Rocks at high elevations are exposed to low temperatures and high winds. They can weather very quickly. Rocks at sea level can be weathered by ocean waves.
Slope	The steep sides of mountains and hills make water flow down them faster. Fast-moving water has more energy to break down rock than slow-moving water. Therefore, rocks on steep slopes can weather faster than rocks on level ground.

✓ READING CHECK

5. Identify What is one way that temperature affects mechanical weathering?

_____

_____

_____

TAKE A LOOK
6. Explain Why do rocks on the sides of mountains weather faster than rocks on level ground?

_____

_____

_____

_____

Section 2 Review

SECTION VOCABULARY

differential weathering the process by which softer, less weather resistant rocks wear away at a faster rate than harder, more weather resistant rocks do	

1. Identify What are four factors that affect how fast weathering happens?

_____

_____

_____

2. Explain Why does it take less time for small rocks to wear away than it does for large rocks to wear away?

_____

_____

_____

3. Describe Imagine a rock on a beach and a rock on the side of a mountain. How would the factors that control weathering be different for these two rocks?

_____

_____

_____

4. Apply Concepts Two rivers run into the ocean. One river is very long. The other is very short. Which river probably drops the smallest rock pieces near the ocean? Explain your answer.

_____

_____

_____

_____

CHAPTER 2 | Weathering and Soil Formation
SECTION 3 | From Bedrock to Soil

National Science Education Standards
ES 1c, 1e, 1g, 1k

BEFORE YOU READ

After you read this section, you should be able to answer these questions:

- What is soil?
- How do the features of soil affect the plants that grow in it?
- What is the effect of climate on soil?

Where Does Soil Come From?

What do you think of when you think of soil? Most people think of dirt. However, soil is more than just dirt. **Soil** is a loose mixture of small mineral pieces, organic material, water, and air. All of these things help to make soil a good place for plants to grow.

Soil is made from weathered rocks. The rock that breaks down and forms a soil is called the soil's **parent rock**. Different parent rocks are made of different chemicals. Therefore, the soils that form from these rocks are also made of different chemicals. ☑

Bedrock is the layer of rock beneath soil. Because the material in soil is easily moved, the bedrock may not be the same as the soil's parent rock. Soil that has been moved away from its parent rock is called *transported soil*.

In some cases, the bedrock is the same as the parent rock. In these cases, the soil remains in place above its parent rock. Soil that remains above its parent rock is called *residual soil*.

STUDY TIP
Summarize in Pairs Read this section quietly to yourself. Then, talk about the material with a partner. Together, try to figure out the parts that you didn't understand.

READING CHECK
1. Explain Why are different soils made of different chemicals?

_____

_____

The soil is weathered from bedrock.	The soil is carried in from another place.
The bedrock is the same as the parent rock.	The bedrock is not the same as the parent rock.

_____   _____

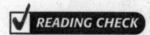
TAKE A LOOK
2. Identify Fill in the blanks with the terms *residual soil* and *transported soil*.

What Are the Properties of Soil?

Some soils are great for growing plants. However, plants cannot grow in some other soils. Why is this? To better understand how plants can grow in soil, you must know about the properties of soil. These properties include soil texture, soil structure, and soil fertility.

SOIL TEXTURE

Soil is made of particles of different sizes. Some particles, such as sand, are fairly large. Other particles are so small that they cannot be seen without a microscope. **Soil texture** describes the amounts of soil particles of different sizes that a soil contains. ☑

Soil texture affects the consistency of soil and how easily water can move into the soil. *Soil consistency* describes how easily a soil can be broken up for farming. For example, soil that contains a lot of clay can be hard, which makes breaking up the soil difficult. Most plants grow best in soils that can be broken up easily.

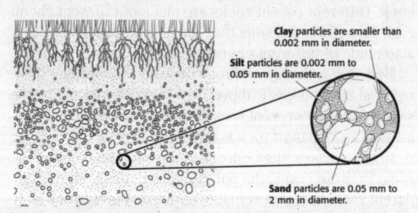

Clay particles are smaller than 0.002 mm in diameter.

Silt particles are 0.002 mm to 0.05 mm in diameter.

Sand particles are 0.05 mm to 2 mm in diameter.

Soil contains particles of many different sizes. However, all of the particles are smaller than 2 mm in diameter.

SOIL STRUCTURE

The particles in soil are not always evenly mixed. Sometimes, particles of a certain kind of material will form clumps in the soil. **Soil structure** describes the arrangement of particles in a soil.

✓ READING CHECK

3. Define What is soil texture?

_____

_____

_____

Math Focus

4. Calculate How many times larger is the biggest silt particle than the biggest clay particle?

_____

SECTION 3 From Bedrock to Soil *continued*

SOIL FERTILITY

Plants need to get nutrients from soil in order to grow. Some soils are rich in nutrients. Other soils may have few nutrients or may be unable to give the nutrients to plants. The ability of soil to hold nutrients and to supply nutrients to plants is called *soil fertility*.

Some of the nutrients in a soil come from its parent rock. Other nutrients come from **humus**. Humus is the organic material that forms in soil from the remains of plants and animals. These remains are broken down into nutrients by decomposers, such as bacteria and fungi. It is humus that gives dark-colored soils their color. ☑

What Are the Different Layers in Soil?

Most soil forms in layers. The layers are horizontal, so soil scientists call them *horizons*.

Horizon name	Description
O	The O horizon is made of decaying material from dead organisms. It is found in some areas, such as forests, but not in others.
A	The A horizon is made of topsoil. Topsoil contains more humus than any other soil horizon does.
E	The E horizon is a layer of sediment with very few nutrients in it. The nutrients in the E horizon have been removed by water.
B	The B horizon is very rich in nutrients. The nutrients that were washed out of other horizons collect in the B horizon.
C	The C horizon is made of partly weathered bedrock or of sediments from other locations.
R	The R horizon is made of bedrock that has not been weathered very much.

Water dissolves and removes nutrients as it passes through the soil. This is called **leaching**.

READING CHECK

5. Define What is humus?

_____

_____

_____

STANDARDS CHECK

ES 1e Soil consists of weathered rocks and decomposed organic material from dead plants, animals, and bacteria. Soils are often found in <u>layers</u>, with each having a different chemical composition and texture.

Word Help: <u>layer</u> a separate or distinct portion of matter that has thickness

6. Identify Which three soil horizons probably contain the most nutrients?

_____

_____

_____

SECTION 3 From Bedrock to Soil *continued*

Why Is the pH of a Soil Important?

The *pH scale* is used to measure how acidic or basic something is. The scale ranges from 0 to 14. A pH of 7 is a *neutral* pH. Soil that has a pH below 7 is *acidic*. Soil that has a pH above 7 is *basic*.

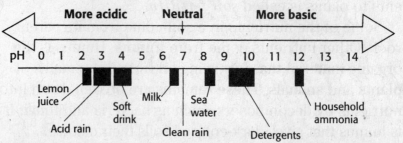

TAKE A LOOK
7. Identify Which is more acidic, lemon juice or a soft drink?

_____

_____

The pH of a soil affects how nutrients dissolve in the soil. Many plants are unable to get certain nutrients from soils that are very acidic or basic. The pH of a soil therefore has a strong effect on soil fertility. Most plants grow best in soil with a pH of 5.5 to 7.0. A few plants grow best in soils with higher or lower pH.

Soil pH is determined partly by the soil's parent rock. Soil pH is also affected by the acidity of rainwater, the use of fertilizers, and the amount of chemical weathering. ☑

8. List What are three things that affect soil pH?

_____

_____

_____

How Does Climate Affect Soil?

Soil types vary from place to place. The kinds of soils that develop in an area depend on its climate. The different features of these soils affect the number and kinds of organisms that can survive in different areas.

TROPICAL CLIMATES

Tropical rain forests receive a lot of direct sunlight and rain. Because of these factors, plants grow year-round. The heat and moisture also cause dead organisms to decay easily. This decay produces a lot of rich humus in the soil.

Even though a lot of humus can be produced in tropical rain forests, their soils are often poor in nutrients. One reason for this is that tropical rain forests have heavy rains. The heavy rains in this climate zone can leach nutrients from the topsoil. The rainwater carries the nutrients deep into the soil, where the plants can't reach them. In addition, the many plants that grow in tropical climates can use up the nutrients in the soil. ☑

9. Explain Why are many tropical soils poor in nutrients?

_____

_____

SECTION 3 From Bedrock to Soil *continued*

DESERTS AND ARCTIC CLIMATES

Deserts and arctic climates receive little rainfall. Therefore, the nutrients in the soil are not leached by rainwater. However, the small amount of rain in these climates makes weathering happen more slowly. As a result, soil forms slowly.

Few plants and animals live in deserts and arctic climates. Therefore, most soils there contain very little humus.

Sometimes, desert soils can become harmful, even to desert plants. Groundwater can seep into the desert soil. The groundwater often contains salt. When the water evaporates, the salt is left in the soil. The salt can build up in the soil and harm plants.

TEMPERATE FORESTS AND GRASSLANDS

Most of the continental United States has a temperate climate. Because the temperature changes often, mechanical weathering happens quickly in temperate climates. Thick layers of soil can build up.

Temperate areas get a medium amount of rain. The rain is enough to weather rock quickly, but it is not enough to leach many nutrients from the soil.

Many different kinds of plants can grow in temperate soils. Therefore, they contain a lot of humus. The large amount of humus makes the soils very rich in nutrients. The most fertile soils in the world are found in temperate climates. For example, the Midwestern part of the United States is often called the United States' "breadbasket" because of the many crops that grow there.

Critical Thinking

10. Apply Concepts As in deserts, groundwater in arctic climates can contain salt. Salt does not build up in arctic soils as quickly as in desert soils. What do you think is the reason for this?

_____

_____

_____

Type of climate	Description of climate	Features of the soil in this climate
Tropical climates	warm temperatures a lot of rain many living things	
Deserts and arctic climates		has little humus poor in nutrients
	medium amount of rain temperature changes often	

TAKE A LOOK
11. Describe Fill in the chart to show the features of soils in different climates.

Section 3 Review

SECTION VOCABULARY

bedrock the layer of rock beneath soil	**soil** a loose mixture of rock fragments, organic material, water, and air that can support the growth of vegetation
humus dark, organic material formed in soil from the decayed remains of plants and animals	**soil structure** the arrangement of soil particles
leaching the removal of substances that can be dissolved from rock, ore, or layers of soil due to the passing of water	**soil texture** the soil quality that is based on the proportions of soil particles
parent rock a rock formation that is the source of soil	

1. Summarize What are three properties of soil?

_____

_____

_____

2. Compare What climate feature do arctic climates and desert climates share that makes their soils similar?

_____

_____

3. Analyze How can flowing water affect the fertility of soils?

_____

_____

4. Identify How does soil pH affect plant growth?

_____

_____

_____

5. Explain What determines a soil's texture?

_____

_____

6. Identify Name three things that are found in soils.

_____

_____

_____

SECTION 4 | Soil Conservation

BEFORE YOU READ

After you read this section, you should be able to answer these questions:

• Why is soil important?

• How can farmers conserve soil?

Why Is Soil Important?

You have probably heard about endangered plants and animals. Did you know that soil can be endangered, too? Soil can take many years to form. It is not easy to replace. Therefore, soil is considered a nonrenewable resource.

Soil is important for many reasons. Soil provides nutrients for plants. If the soil loses its nutrients, plants will not be able to grow. Soil also helps to support plant roots so the plants can grow well.

Animals get their energy from plants. The animals get energy either by eating plants or by eating animals that have eaten plants. If plants are unhealthy because the soil has few nutrients, then animals will be unhealthy, too.

Soil also provides a home, or *habitat*, for many living things. Bacteria, insects, mushrooms, and many other organisms live in soil. If the soil disappears, so does the habitat for these living things. ☑

Soil is also very important for storing water. It holds water that plants can use. Soil also helps to prevent floods. When rain falls, the soil can soak it up. The water is less likely to cause floods.

What does soil provide?	Why is it important?
Nutrients	
Habitat	
Water storage	

If we do not take care of soils, they could become unusable. In order to keep our soils usable, we need to conserve them. **Soil conservation** means protecting soils from erosion and nutrient loss. Soil conservation can help to keep soils fertile and healthy.

STUDY TIP

Compare Create a chart that shows the similarities and differences in the ways that farmers can help conserve soil.

READING CHECK

1. Explain Why is soil important for animals?

_____

_____

_____

TAKE A LOOK

2. Identify In the table, fill in the reasons that nutrients, habitat, and water storage are important.

How Can Soil Be Lost?

Soil loss is a major problem around the world. One cause of soil loss is soil damage. Soil can be damaged if it is overused. Overused soil can lose its nutrients and become infertile. Plants can't grow in infertile soil.

Plants help to hold water in the soil. If plants can't grow somewhere because the soil is infertile, the area can become a desert. This process is known as *desertification*.

EROSION

Another cause of soil loss is erosion. **Erosion** happens when wind, water, or gravity transports soil and sediment from one place to another. If soil is not protected, it can be exposed to erosion.

Plant roots help to keep soil in place. They prevent water and wind from carrying the soil away. If there are no plants, soil can be carried away through erosion. ☑

How Can Farmers Help to Conserve Soil?

Farming can cause soil damage. However, farmers can prevent soil damage if they use certain methods when they plow, plant, and harvest their fields.

CONTOUR PLOWING

Water that runs straight down a hill can carry away a lot of soil. Farmers can plow their fields in special ways to help slow the water down. When the water moves more slowly down a hill, it carries away less soil. *Contour plowing* means plowing a field in rows that run across the slope of a hill.

Contour plowing helps water to run more slowly down hills. This reduces erosion because _____

✔ **READING CHECK**

3. Describe How do plant roots prevent erosion?

_____

_____

_____

_____

TAKE A LOOK

4. Identify Fill in the blank line in the figure to explain how contour plowing reduces erosion.

SECTION 4 Soil Conservation *continued*

TERRACES

On very steep hills, farmers can use terraces to prevent soil erosion. *Terraces* change one very steep field into many smaller, flatter fields.

Terraces keep water from running downhill very quickly.

NO-TILL FARMING

In *no-till farming*, farmers leave the stalks from old crops lying on the field while the newer crops grow. The old stalks protect the soil from rain and help reduce erosion.

The stalks left behind in no-till farming reduce erosion by protecting the soil from rain.

COVER CROPS

Cover crops are crops that are planted between harvests of a main crop. Cover crops can help to replace nutrients in the soil. They can also prevent erosion by providing cover from wind and rain.

CROP ROTATION

If the same crop is grown year after year in the same field, the soil can lose certain nutrients. To slow this process, a farmer can plant different crops in the field every year. Different crops use different nutrients from the soil. Some crops used sin crop rotation can replace soil nutrients.

Critical Thinking

5. Infer What do you think is the reason farmers use terraces only on very steep hills?

_____

_____

_____

_____

Critical Thinking

6. Apply Concepts How can crop rotation affect the number of plants that soil can support?

_____

_____

_____

_____

_____

_____

Section 4 Review

SECTION VOCABULARY

erosion the process by which wind, water, ice, or gravity transports soil and sediment from one location to another	**soil conservation** a method to maintain the fertility of soil by protecting the soil from erosion and nutrient loss

1. Define Write your own definition for soil conservation.

_____

_____

2. Identify Give three things that soil provides to living things.

_____

_____

_____

3. Compare How is weathering different from erosion?

_____

_____

_____

4. Identify What are two causes of soil loss?

_____

_____

5. List What are five ways that farmers can help to conserve soil?

_____

_____

_____

_____

_____

6. Explain How does no-till farming help to reduce erosion?

_____

_____

_____

_____

CHAPTER 3 | The Flow of Fresh Water

SECTION 1 | The Active River

National Science Education Standards
ES 1c, 1f

BEFORE YOU READ

After you read this section, you should be able to answer these questions:

• How does moving water change the surface of Earth?

• What is the water cycle?

• What factors affect the rate of stream erosion?

What Is Erosion?

Six million years ago, the Colorado River began carving through rock to form the Grand Canyon. Today, the river has carved through 1.6 km (about 1 mi) of rock!

Before the Grand Canyon was formed, the land was flat. Then the rock in the area began to lift upward because of plate tectonics. As Earth's crust lifted upward, water began to run downhill. The moving water cut into the rock and started forming the Grand Canyon.

Over millions of years, water cut into rock through the process of erosion. During **erosion**, wind, water, ice, and gravity move soil and rock from one place to another. Water is the main force in forming the Grand Canyon and in changing the Earth's landscape. ☑

STUDY TIP

Describe As you read, make a list of the different ways in which water can change the landscape of Earth.

READING CHECK

1. Identify What formed the Grand Canyon?

_____

_____

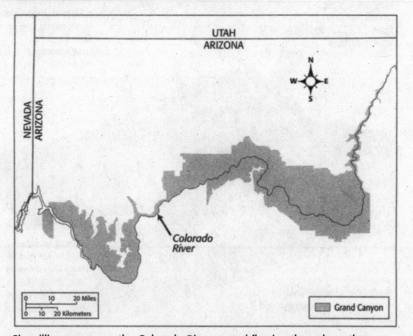

Six million years ago, the Colorado River started flowing through northern Arizona. Today, it has carved the Grand Canyon, which is about 1.6 km deep and 446 km long.

Math Focus
2. Calculate How long is the Grand Canyon in miles? Show your work.

1 km = 0.62 mi

_____

_____

_____

How Does the Water Cycle Work?

Have you ever wondered where the water in rivers comes from? It is part of the water cycle. The **water cycle** is the nonstop movement of water between the air, the land, and the oceans. The major source of energy that drives the water cycle is the sun.

In the water cycle, water comes to Earth's surface from the clouds as rain, snow, sleet, or hail. The water moves downward through the soil or flows over the land. Water that flows over the land collects in streams and rivers and flows to the oceans.

Energy from the sun changes the water on Earth's surface into a gas that rises up to form clouds. The gas is called *water vapor*. The water vapor in clouds moves through the atmosphere until it falls to Earth's surface again.

3. Define What is the water cycle?

_____

_____

_____

_____

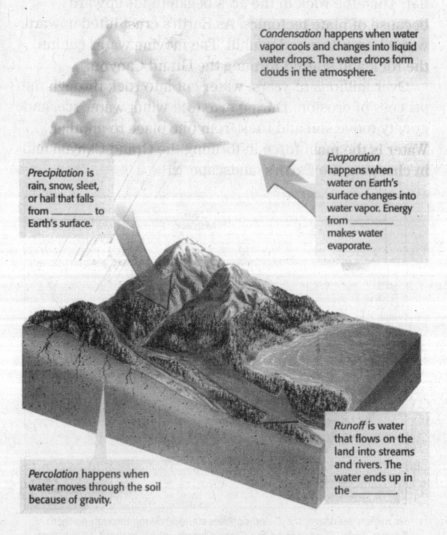

Condensation happens when water vapor cools and changes into liquid water drops. The water drops form clouds in the atmosphere.

Precipitation is rain, snow, sleet, or hail that falls from _____ to Earth's surface.

Evaporation happens when water on Earth's surface changes into water vapor. Energy from _____ makes water evaporate.

Runoff is water that flows on the land into streams and rivers. The water ends up in the _____.

Percolation happens when water moves through the soil because of gravity.

TAKE A LOOK
4. Identify In the figure, fill in the blank lines with the correct words.

What Is a River System?

What happens when you turn on the shower in your family's bathroom? When water hits the shower floor, the individual drops of water join together to form small streams. The small streams join together to form larger streams. The larger streams carry the water down the drain.

The water in your shower is like the water in a river system. A *river* is a stream that has many tributaries. A *river system* is a group of streams and rivers that drain an area of land. A **tributary** is a stream that flows into a lake or a larger stream. ☑

How Do River Systems Work?

River systems are divided into areas called watersheds. A **watershed** is the land that is drained by a river system. Many tributaries join together to form the rivers in a watershed.

The largest watershed in the United States is the Mississippi River watershed. It covers over one-third of the United States. It has hundreds of tributaries. The Mississippi River watershed drains into the Gulf of Mexico.

Watersheds are separated from each other by an area of higher ground called a **divide**. All of the rivers on one side of a divide flow away from it in one direction. All of the rivers on the other side of the divide flow away from it in the opposite direction. The Continental Divide separates the Mississippi River watershed from the watersheds in the western United States.

> **✔ READING CHECK**
>
> **5. Define** What is a river system?
>
> _____
> _____
> _____

TAKE A LOOK
6. List Name three rivers that are tributaries to the Mississippi River.

_____
_____
_____

SECTION 1 The Active River *continued*

What Is a Stream Channel?

A stream forms as water wears away soil and rock to make a channel. A **channel** is the path that a stream follows. As more soil and rock are washed away, the channel gets wider and deeper.

Over time, tributaries flow into the main channel of a river. The main channel has more water in it than the tributaries. The larger amount of water makes the main channel longer and wider. ☑

What Causes Stream Erosion?

Gradient is a measure of the change in the height of a stream over a certain distance. Gradient can be used to measure how steep a stream is. The left-hand picture below shows a stream with a high gradient. The water in this stream is moving very fast. The fast-moving water easily washes away rock and soil.

A river that is flat, as in the right-hand picture, has a low gradient and flows slowly. The slow water washes away less rock and soil.

This stream has a large gradient. It flows very fast.

This stream has a small gradient. It flows very slowly.

The amount of water that flows in a stream during a certain amount of time is called the *discharge*. The discharge of a stream can change. A large rainfall or a lot of melted snow can increase the stream's discharge.

When the discharge of a stream gets bigger, the stream can carry more sediment. The larger amount of water will flow fast and erode more land.

✓ **READING CHECK**

7. Explain Why is the main channel of a river longer and wider than the channels of its tributaries?

_____

_____

_____

TAKE A LOOK

8. Explain How does the gradient of a stream affect how much erosion it causes?

_____

_____

_____

_____

How Does a Stream Carry Sediment?

A stream's **load** is the material carried in the stream's water. A fast-moving stream can carry large rocks. The large rocks can cause rapid erosion by knocking away more rock and soil.

A slow-moving stream carries smaller rocks in its load. The smaller particles erode less rock and soil. The stream also carries material that is dissolved in the water.

Large rocks that bounce along the bottom of the stream are called *bed load*.

Materials that are floating in the water are called *suspended load*. They often make the stream look muddy or cloudy.

Tiny particles that are dissolved in the water are called *dissolved load*.

STANDARDS CHECK

ES 1c Land forms are the result of a combination of constructive and destructive forces. Constructive forces include crustal deformation, volcanic eruption, and deposition of sediment, while destructive forces include weathering and erosion.

9. Identify How does a stream carry material from one place to another? Give three ways.

_____

_____

_____

How Do Scientists Describe Rivers?

All rivers have different features. These features can change with time. Many factors, such as weather, surroundings, gradient, and load, control the changes in a river. Scientists use special terms to describe rivers with certain features.

YOUTHFUL RIVERS

Youthful rivers are fast-flowing rivers with high gradients. Many of them flow over rapids and waterfalls. Youthful rivers make narrow, deep channels for the water to flow in. The picture below shows a youthful river.

TAKE A LOOK
10. Describe What features of this river tell you that it is a youthful river?

_____

_____

_____

MATURE RIVERS

Mature rivers erode rock and soil to make wide channels. Many tributaries flow into a mature river, so mature rivers carry large amounts of water. The picture below shows a mature river bending and curving through the land. The curves and bends are called *meanders*. ☑

☑ **READING CHECK**

11. Explain Why do mature rivers carry a lot of water?

_____

_____

TAKE A LOOK
12. Identify Label the meanders on this picture of a mature river.

SECTION 1 The Active River *continued*

REJUVENATED RIVERS

Rejuvenated rivers form where land has been raised up by plate tectonics. This gives a river a steep gradient. Therefore, rejuvenated rivers flow fast and have deep channels. As shown in the picture below, steplike gradients called *terraces* may form along the sides of rejuvenated rivers.

TAKE A LOOK

13. Identify Label the terraces on this picture of a rejuvenated river.

OLD RIVERS

Old rivers have very low gradients. Instead of widening and deepening its channel, an old river deposits soil and rock along its channel. Since very few tributaries flow into an old river, the river does not quickly erode land. Old rivers have wide, flat floodplains and many meanders. In the picture below, a bend in an old river's channel has eroded into a lake. This is called an *oxbow lake*.

Old river

Oxbow lake

Critical Thinking

14. Infer Very few tributaries flow into an old river. Do you think it will have a large or a small discharge?

_____

_____

Section 1 Review

SECTION VOCABULARY

channel the path that a stream follows	**load** the materials carried by a stream
divide the boundary between drainage areas that have streams that flow in opposite directions	**tributary** a stream that flows into a lake or into a larger stream
erosion the process by which wind, water, ice, or gravity transports soil and sediment from one location to another	**water cycle** the continuous movement of water between the atmosphere, the land, and the oceans
	watershed the area of land that is drained by a river system

1. Explain Why do most rivers have wider channels than most streams?

_____

_____

2. Show a Sequence Fill in the Process Chart to show what happens in the water cycle.

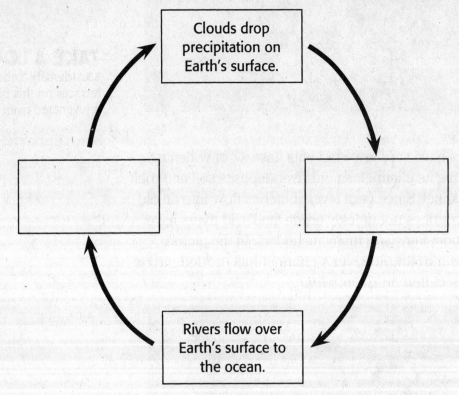

Clouds drop precipitation on Earth's surface.

Rivers flow over Earth's surface to the ocean.

3. Identify What is the main source of energy for the water cycle?

_____

4. Describe How do rivers change Earth's surface?

_____

_____

CHAPTER 3 The Flow of Fresh Water

SECTION 2 Stream and River Deposits

National Science Education Standards
ES 1c

BEFORE YOU READ

After you read this section, you should be able to answer these questions:

- What types of deposits are caused by streams?
- Why do floods happen?
- How do floods affect humans?

How Do Rivers and Streams Rebuild Land?

Earlier, you learned that rivers and streams erode rock and soil. However, rivers also carry this loose rock and soil to new places. Rivers can help to form new land through deposition. **Deposition** is the process in which material is laid down, or dropped.

The rock and soil that a river erodes from the land move downstream in the river's load. When the flow of water slows down, the river deposits, or drops, some of its load. Material, such as rock and soil, that is deposited by rivers and streams is called *sediment*. ☑

In a river, erosion happens on the outside of a bend, where the water flows quickly. Deposition happens on the inside of a bend, where the water flows more slowly.

STUDY TIP

Compare In your notebook, create a table to compare the different landforms created by stream deposits.

✓ **READING CHECK**

1. Define What is sediment?

_____

_____

_____

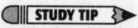

Erosion happens along the outside of a bend, where water is flowing _____.

Deposition happens along the inside of a bend, where water is flowing _____.

TAKE A LOOK
2. Identify Fill in the blanks in the figure with the correct words.

PLACER DEPOSITS

Heavy minerals, such as gold, can be carried by very fast-moving rivers. In places where the rivers slow down, the heavy minerals may be deposited. This kind of sediment is called a *placer deposit*.

What Are Deltas and Alluvial Fans?

Rivers deposit their loads of rocks and soil when their flow of water slows down. When a river enters the ocean, it flows much more slowly. Therefore, it deposits its load into the ocean.

When the river enters the ocean, it deposits its load under the water in a fan-shaped pattern called a **delta**. The river deposits can build up above the water's surface to form new land and build new coastline.

Rivers and streams can also deposit their loads on dry land. When a fast-moving stream flows from a mountain onto flat land, the stream slows down quickly. As it slows down, the stream deposits its rocks and soil in a fan-shaped pattern known as an **alluvial fan**.

Why Do Rivers and Streams Flood?

Rivers and streams are always changing. They may have different amounts of water in them in different seasons. If there is a lot of rain or melting snow, a stream will have a lot of water in it.

Floods are natural events that can happen with the change of seasons. A *flood* happens when there is too much water for the channel to hold. The stream flows over the sides of its channel. ☑

During a flood, the land along the sides of the stream is covered in water. The stream drops its sediment on this land. This area is called a **floodplain**. Floodplains are good areas for farming because flooding brings new soil to the land.

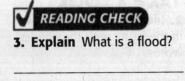

Investigate Learn about a place in the world where people live on a delta. Give a short talk to your class about the place you studied.

Say It

READING CHECK
3. Explain What is a flood?

_____

_____

_____

_____

SECTION 2 Stream and River Deposits *continued*

What Are the Effects of Floods?

Floods are powerful and can cause a lot of damage. They can ruin homes and buildings. They can also wash away land where animals and people live.

In 1993, when the Mississippi River flooded, farms and towns in nine states were damaged. Floods can cover roads and carry cars and people downstream. They can also drown people and animals. ☑

Floods sometimes happen very fast. After a very bad rainstorm, water can rush over the land and cause a *flash flood*. Flash floods can be hard to predict and are dangerous.

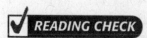

Floodwater can flow strongly enough to move cars. These people were trapped in their car in a flash flood.

Humans try to control flooding by building barriers around streams. One kind of barrier is called a dam. A *dam* is a barrier that can guide floodwater to a reservoir, such as a lake or pond. A dam can prevent flooding in one area and create an artificial lake in another area. The water in the artificial lake can be used for farming, drinking, or producing electricity.

Another barrier that people build is called a levee. A *levee* is a wall of sediment on the side of a river. The barrier helps keep the river from flooding the nearby land. Many levees form naturally from sediment that is deposited by the river. People can use sandbags to create artificial levees. ☑

✓ **READING CHECK**

4. List List three ways that floods can be harmful to people and animals.

_____
_____
_____

TAKE A LOOK

5. Explain Why can a flash flood be dangerous to people driving in cars?

_____
_____
_____
_____

✓ **READING CHECK**

6. Identify What do people do to try to protect themselves from a flood?

_____
_____
_____
_____

Section 2 Review

SECTION VOCABULARY

alluvial fan a fan-shaped mass of material deposited by a stream when the slope of the land decreases sharply	**deposition** the process in which material is laid down
delta a fan-shaped mass of material deposited at the mouth of a stream	**floodplain** an area along a river that forms from sediments deposited when the river overflows its banks

1. Explain Why do floods happen?

_____

_____

_____

2. Compare Complete the table to describe the features of different kinds of stream deposits.

Type of Deposit	How is it formed?	Where is it formed?
Alluvial fan		
Delta		
Floodplain		

3. Explain How can a flood be both helpful and harmful to people?

_____

_____

_____

_____

4. List Give two kinds of barriers that people use to control floodwater.

_____

_____

CHAPTER 3 The Flow of Fresh Water

SECTION 3 Water Underground

National Science
Education Standards
ES 1c

BEFORE YOU READ

After you read this section, you should be able to answer these questions:

• What is a water table?

• What is an aquifer?

• What is the difference between a spring and a well?

Where Is Fresh Water Found?

Some of the Earth's fresh water is found in streams and lakes. However, a large amount of water is also found underground. Rainwater and water from streams move through the soil and into the spaces between rocks underground. This underground area is divided into two zones. The *zone of aeration* is the area that rainwater passes through. The spaces between particles in the zone of aeration contain both water and air.

The *zone of saturation* is the area where water collects. The spaces between particles in the zone of saturation are filled with only water. The water found inside underground rocks is called *groundwater*. ☑

The zone of aeration and the zone of saturation meet at a boundary called the **water table**. The depth of the water table is not the same all the time or in all places. The water table can move closer to the surface during wet seasons and farther from the surface during dry seasons. In wet regions, the water table may be just below the surface. In dry regions, the water table may be hundreds of meters below the surface.

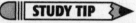

STUDY TIP

Summarize As you read, underline the important ideas in this section. When you are finished reading, write a one- or two-paragraph summary of the section, using the underlined ideas.

READING CHECK

1. Define What is groundwater?

_____

_____

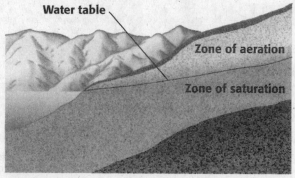

The water table is the boundary between the zone of aeration and the zone of saturation.

TAKE A LOOK

2. Infer If a region receives a lot of rainfall, will the water table in the region probably rise or fall?

_____

_____

SECTION 3 Water Underground *continued*

How Can Water Pass Through Rock?

A layer of rock or sediment that stores groundwater is called an **aquifer**. Most aquifers are made of sedimentary rock. There can be many *pores*, or open spaces, between the particles in an aquifer. The more open spaces there are, the more water the aquifer can hold. The fraction of a rock's volume that is taken up by pores is called the rock's **porosity**. ☑

Imagine filling a jar with large pebbles. The pebbles cannot fill all of the space in the jar, so there will be many open spaces. In other words, the jar has a high porosity. Now, imagine pouring sand over the pebbles in the jar. The sand can fill the spaces between the pebbles, leaving little open space in the jar. The jar has a low porosity.

Like the jar, the sizes of the particles in a rock affect the rock's porosity. Rocks made of the same-sized particles tend to have high porosity, like the jar of pebbles. Rocks made of different-sized particles tend to have low porosity, like the jar with sand and pebbles.

If the open spaces in the rock layer are connected, water can move through the rock. A rock's ability to let water pass through is called **permeability**. Rock that water can not flow through is called *impermeable*.

The size of rock particles also affects permeability. Rock made of large particles tends to have a high permeability. This is because the large particles produce less friction on the water moving through them. *Friction* is a force that slows down moving objects. Rock particles produce friction on water when the water touches the rock particles.

READING CHECK

3. Define What is porosity?

_____

_____

_____

Critical Thinking

4. Apply Concepts Shale has a very high porosity. However, shale does not form many aquifers, because water cannot move through it easily. Explain why this might be the case.
(Hint: What is permeability?)

_____

_____

_____

_____

_____

_____

_____

TAKE A LOOK

5. Explain Why do small particles produce more friction on the water than large particles?

_____

_____

The large particles touch the water in only a few places. They produce little friction on the water, so they have a high permeability.

The small particles touch the water in many places. They produce a lot of friction on the water, so they have a low permeability.

RECHARGE ZONES

Like rivers, aquifers depend on precipitation to keep their water level constant. Precipitation that falls onto land can flow through the ground and into an aquifer. The ground surface where water enters an aquifer is called the aquifer's **recharge zone**. Recharge zones are found where the soil and rock above an aquifer are permeable. ☑

Some aquifers are small, but many cover large underground areas. Many cities and farms depend on aquifers for fresh water.

People's actions can affect the amount and quality of water in an aquifer. If people build roads or buildings in a recharge zone, less water can enter the aquifer. If people dump chemicals in a recharge zone, the chemicals can enter the aquifer and pollute the water in it.

What Is a Spring?

Like all water, groundwater tends to move downhill. Remember that the water table can be at different depths in different places. Groundwater tends to move to follow the slope of the water table. When the water table meets the Earth's surface, water flows out and forms a *spring*. Springs are important sources of drinking water. ☑

In some places, an aquifer is found between two layers of impermeable rock. This is called an *artesian formation*. The top layer of impermeable rock is called the *cap rock*. If the water in the aquifer flows through a crack in the cap rock, it forms an **artesian spring**.

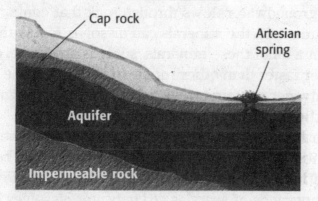

The water from most springs is cool. However, rock far below Earth's surface can be very hot. Therefore, water that flows through deep aquifers may be very hot. When this water reaches the surface, it can form a *hot spring*.

✓ READING CHECK

6. Identify Where does the water in aquifers come from?

_____

_____

✓ READING CHECK

7. Describe How does a spring form?

_____

_____

_____

TAKE A LOOK

8. Explain What causes an artesian spring to form?

_____

_____

_____

How Do People Get Water Out of the Ground?

When people need a supply of water, they often dig a well. A *well* is a hole dug by people that is deeper than the water table. If the well does not reach below the water table, the well will not produce water. In addition, too many wells may remove water from an aquifer more quickly than the aquifer can refill. Then, the water table can drop and all the wells can go dry.

TAKE A LOOK

9. Identify Two people drilled wells to try to get water out of the ground. The white bars in the figure show where the two people drilled their wells. Which of the wells will probably produce water? Explain your answer.

_____

_____

_____

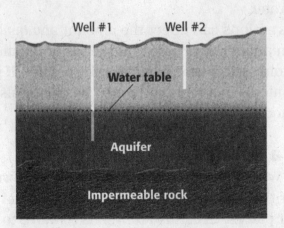

How Can Groundwater Cause Weathering?

Remember that streams can cause physical weathering when particles in the stream hit other particles and cause them to break. Groundwater can also cause weathering. However, instead of physical weathering, groundwater causes chemical weathering by dissolving rock.

Many minerals, such as calcite, can dissolve in water. When groundwater flows through rock that contains these minerals, the minerals can dissolve. Rocks that contain a lot of these minerals, such as limestone, can weather faster than other rocks. In addition, some groundwater contains weak acids. These acids can dissolve the rock more quickly than pure water can.

Weathering by groundwater can form large caves. In fact, most of the caves in the world have formed because of weathering by groundwater. The caves form slowly over thousands of years as groundwater dissolves limestone. The figure on the next page shows how large some of these caves can become.

 **Say It**

Investigate Learn more about an area that has large underground caves. What kind of rock did the caves form in? How did the caves form? Share your findings with a small group.

This cave in New Mexico formed when groundwater dissolved huge volumes of limestone.

TAKE A LOOK
10. Identify Name one kind of rock that can be dissolved by groundwater.

_____

What Structures Can Groundwater Form?

Many caves show signs of deposition as well as weathering. Water flowing through caves can have many minerals dissolved in it. If the water drips from a crack in the cave's ceiling, it can deposit these minerals. These deposits form icicle-shaped structures that hang from the ceilings. They are called *stalactites*.

Water that falls to the cave floor can build cone-shaped structures called *stalagmites*. Sometimes, a stalactite and a stalagmite join together to form a *dripstone column*.

Sometimes, the roof of a cave can collapse. This forms a circular *depression*, or pit, on the Earth's surface called a *sinkhole*. Sinkholes can damage buildings, roads, and other structures on the surface. Streams can "disappear" into sinkholes and flow through the cave underground. In areas where the water table is high, lakes can form inside sinkholes.

Critical Thinking
11. Compare How is a stalactite different from a stalagmite?

_____

_____

_____

This sinkhole formed in Florida when a cave collapsed.

TAKE A LOOK
12. Identify Relationships How are caves and sinkholes related?

_____

_____

Section 3 Review

SECTION VOCABULARY

aquifer a body of rock or sediment that stores groundwater and allows the flow of groundwater	**porosity** the percentage of the total volume of a rock or sediment that consists of open spaces
artesian spring a spring whose water flows from a crack in the cap rock over an aquifer	**recharge zone** an area in which water travels downward to become part of an aquifer
permeability the ability of a rock or sediment to let fluids pass through its open spaces, or pores	**water table** the upper surface of underground water; the upper boundary of the zone of saturation

1. Explain Why is it important to know where the water table is located?

_____

_____

_____

2. Describe How does particle size affect the porosity of an aquifer?

_____

_____

_____

3. Infer How could building new roads affect the recharge zone of an aquifer?

_____

_____

_____

4. Compare What is the difference between a spring and a well?

_____

_____

_____

_____

5. List What are two features that are formed by underground weathering?

_____

_____

_____

6. Describe How does a dripstone column form?

_____

_____

_____

Name _____ Class _____ Date _____

CHAPTER 3 | The Flow of Fresh Water

SECTION 4 | **Using Water Wisely**

BEFORE YOU READ

After you read this section, you should be able to answer these questions:

- What are two forms of water pollution?
- How is wastewater cleaned?
- How is water conserved?

Is Water A Limited Resource?

Although the Earth is covered with oceans, lakes, and rivers, only 3% of the Earth's water is fresh water. Most of that 3% is frozen in the polar icecaps. Therefore, people must take care of and protect their water resources.

Cities, factories, and farms can pollute water. Water can become so polluted that it is no longer safe to use. Two types of water pollution are point-source pollution and nonpoint-source pollution.

STUDY TIP

Identify As you learn about different types of pollution, think about where you live. Make a list of some sources of pollution in your community.

Where Does Water Pollution Come From?

Pollution that comes from one specific site is called **point-source pollution**. For example, a leak from a sewer pipe is point-source pollution. Because point-source pollution comes from a single place, it is easier to control than nonpoint-source pollution. ☑

Pollution that comes from many sources is called **nonpoint-source pollution**. Most nonpoint-source pollution gets into water by runoff. *Runoff* is water that flows over the ground into rivers, streams, or oceans. As runoff flows over the ground, it can pick up chemicals and other pollutants. These pollutants are carried to clean bodies of water.

READING CHECK

1. Define What is point-source pollution?

_____

_____

People use chemicals, such as fertilizers, on the land. Runoff can carry these chemicals to clean bodies of water.

TAKE A LOOK
2. Explain How can chemicals that are spilled on land end up in oceans or other water bodies?

_____

_____

SECTION 4 **Using Water Wisely** *continued*

Why Does Water Have To Be Clean?

Water is important to many organisms. If the water is not clean, the organisms using it will not be healthy. Three important properties of water that affect water quality are dissolved oxygen, nitrates, and pH.

DISSOLVED OXYGEN

Fish and other organisms that live in water need oxygen to survive. The oxygen that is dissolved in water is called *dissolved oxygen*, or *DO*. If the DO in water is too low, many organisms can become sick or die. ☑

Pollution can cause the DO level in water to decrease. An increase in water temperature can also cause the DO level to decrease. Many energy facilities, such as nuclear power plants, release hot water into the environment. This can increase the temperature of water in natural water bodies. This *thermal pollution* can decrease DO levels.

NITRATES

Nitrates are naturally formed compounds of nitrogen and oxygen. All water contains some nitrates. However, too much nitrate in the water can be harmful to organisms. An increase in nitrate levels can also cause the DO level to decrease. Some kinds of pollution, such as animal wastes, increase the level of nitrates in water.

Animal wastes contain a lot of nitrates. Runoff can carry these nitrates to water bodies, causing water pollution.

pH

The *pH* of water is a measure of how acidic the water is. Most organisms cannot live in very acidic water. Acid rain and some kinds of wastes can make water bodies more acidic. Water with a high *alkalinity*, or ability to react with acids, can protect organisms from acid rain and other pollution.

☑ **READING CHECK**

3. Identify What is dissolved oxygen?

_____

_____

TAKE A LOOK

4. Identify What is one source of nitrates in water?

_____

How Can Dirty Water Be Cleaned?

What happens to water that you flush down the toilet or wash down the drain? If you live in a city or a large town, the water probably flows through sewer pipes to a sewage treatment plant. **Sewage treatment plants** are facilities that clean waste materials out of water. After water has passed through a sewage treatment plant, it can safely be released into the environment. ☑

A sewage treatment plant cleans water in two ways. The first steps are called the *primary treatment*. First, the dirty water is passed through a large screen. This screen catches solid objects, such as paper, rags, and bottle caps.

The water is then placed in a large tank. As the water sits, small pieces of material sink to the bottom of the tank. These small pieces, such as food or soil, are filtered out. Any material that floats on the surface is also removed.

After going through primary treatment, the water is ready for *secondary treatment*. During secondary treatment, the water is placed in an *aeration tank*. There, the water mixes with oxygen and bacteria. The bacteria use the oxygen to consume the wastes dissolved in the water.

The water is then placed in a settling tank. Any dirt in the water sinks to the bottom of the tank and is removed. The water is then mixed with chlorine to disinfect it. Finally, it is sent to a river, lake, or ocean.

✓ READING CHECK

5. Identify What do sewage treatment plants do?

_____

_____

Critical Thinking

6. Infer Why is secondary treatment necessary?

_____

_____

_____

_____

_____

Primary treatment — Secondary treatment

Raw sewage   Settling tank   Aeration tank   Settling tank   Chlorinator

Sludge   Air pump

TAKE A LOOK
7. Identify What is the purpose of the chlorinator?

_____

_____

SEPTIC TANKS

If you live in an area without a sewage treatment plant, your waste probably goes into a septic tank. A **septic tank** is a large underground tank that cleans the wastewater from one household. The wastewater flows into the tank, where the solids sink to the bottom. Bacteria break down these solids. The water then flows into pipes buried underground. The pipes take the water to nearby ground called a *drain field*.

TAKE A LOOK
8. Identify Why do people use septic tanks?

_____

_____

_____

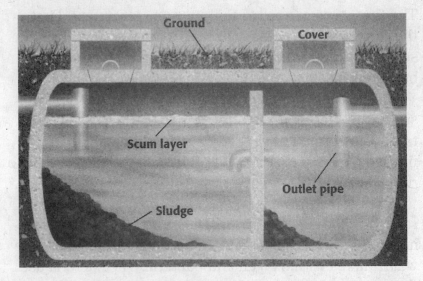

How Do People Use Water?

The average household in the United States uses 100 gallons of water each day. The graph below shows how an average household uses water.

Math Focus
9. Read a Graph What does an average United States household use most of its water for?

_____

_____

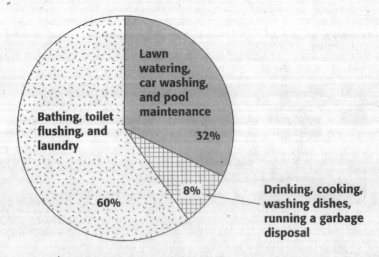

The average U.S. household uses 100 gallons of water per day.

SECTION 4 Using Water Wisely *continued*

INDUSTRY AND AGRICULTURE

About 19% of the water used in the world is used by industries. Water is used in manufacturing, mining minerals, and electricity generation. Most industries recycle and reuse water. Recycling helps keep more water in the environment.

Many farmers get their water from aquifers. When farmers use too much water from an aquifer, less water may be available for farming. For example, the Ogallala aquifer provides water for about one-fifth of the cropland in the United States. Recently, the water level in the aquifer has dropped a great deal. Scientists estimate that it would take at least 1,000 years for the water level to get back to normal if no more water is taken from the aquifer.

When farmers water their crops, a large amount of water is lost through evaporation and runoff. Drip irrigation, in which water is placed directly on the plant's roots, wastes less water.

How Can You Conserve Water at Home?

Individual families can conserve water also. Toilets and shower heads that use less water are good choices for conservation. If you have to water your lawn, water it at night and use a drip watering system.

Each person can do his or her part to conserve water. Simple choices, such as taking shorter showers and turning off the water when you brush your teeth, are helpful. If everyone tries to conserve water, we can make a big difference.

Things You Can Do to Conserve Water

• Use water-saving toilets and showerheads.
• Water the lawn at night or don't water it at all.
•
•
•
•

Math Focus
10. Calculate What fraction of the water used in the world is NOT used by industry?

Say It
Investigate Find out more about ways that farmers can help conserve water. Share what you learn with a small group.

TAKE A LOOK
11. Brainstorm With a partner, come up with other ways that you can help conserve water. Write your ideas in the table.

Section 4 Review

SECTION VOCABULARY

nonpoint-source pollution pollution that comes from many sources rather than from a single, specific site	**septic tank** a tank that separates solid waste from liquids and that has bacteria that break down solid waste
point-source pollution pollution that comes from a specific site	**sewage treatment plant** a facility that cleans the waste materials found in water that comes from sewers or drains

1. Explain How can pollution affect the level of oxygen in water? Why is this important?

_____

_____

_____

2. Compare Complete the Venn Diagram to compare how a sewage treatment plant and a septic tank clean wastewater.

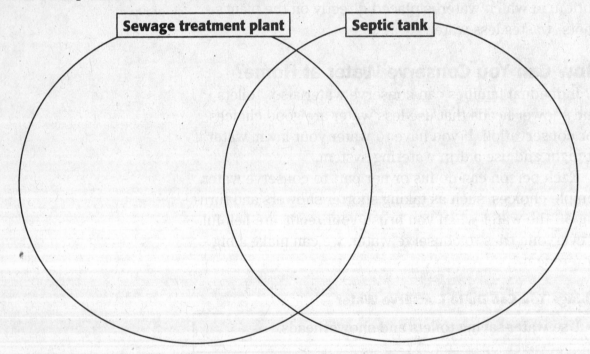

Sewage treatment plant   Septic tank

3. Apply Concepts Many farms are found along the banks of the Mississippi River. Describe what kind of pollution you might find in this river. Is the pollution point-source pollution or nonpoint-source pollution? Explain your answer.

_____

_____

_____

CHAPTER 4 Agents of Erosion and Deposition

SECTION 1 Shoreline Erosion and Deposition

National Science
Education Standards
ES 1c

BEFORE YOU READ

After you read this section, you should be able to answer these questions:

- What is a shoreline?
- How do waves shape shorelines?

How Do Waves Form?

Waves form when wind blows over the surface of the ocean. Strong winds produce large waves. The waves move toward land. When waves crash into the land over a long time, they can break rock down into smaller pieces. These pieces are called *sand*.

A **shoreline** is a place where the land and the water meet. Most shorelines contain sand. The motion of waves helps to shape shorelines. During *erosion*, waves remove sand from shorelines. During *deposition*, waves add sand to shorelines. ☑

WAVE TRAINS

Waves move in groups called *wave trains*. The waves in a wave train are separated by a period of time called the *wave period*. You can measure the wave period by counting the seconds between waves breaking on the shore. Most wave periods are 10 to 20 s long.

When a wave reaches shallow water, the bottom of the wave drags against the sea floor. As the water gets shallower, the wave gets taller. Soon, it can't support itself. The bottom slows down. The top of the wave begins to curl, fall over, and break. Breaking waves are called *surf*.

Waves travel in groups called wave trains. The time between one wave and the next is the wave period.

> **STUDY TIP**
>
> **Summarize** Read this section quietly to yourself. Talk about what you learned with a partner. Together, try to figure out the answers to any questions that you have.

> ✓ **READING CHECK**
>
> **1. Compare** How is wave erosion different from wave deposition?
>
> _____
> _____
> _____

Math Focus

2. Calculate A certain wave train contains 6 waves. The time between the first wave and the last wave is 72 seconds. What is the wave period?

_____

POUNDING SURF

The energy in waves is constantly breaking rock into smaller and smaller pieces. Crashing waves can break solid rock and throw the pieces back toward the shore. Breaking waves can enter cracks in the rock and break off large boulders. Waves also pick up fine grains of sand. The loose sand wears down other rocks on the shore through abrasion. ☑

What Are the Effects of Wave Erosion?

Wave erosion can produce many features along a shoreline. For example, *sea cliffs* form when waves erode rock to form steep slopes. As waves strike the bottom of the cliffs, the waves wear away soil and rock and make the cliffs steeper.

How fast sea cliffs erode depends on how hard the rock is and how strong the waves are. Cliffs made of hard rock, such as granite, erode slowly. Cliffs made of soft rock, such as shale, erode more quickly.

During storms, large, high-energy waves can erode the shore very quickly. These waves can break off large chunks of rock. Many of the features of shorelines are shaped by storm waves. The figures below and on the next page show some features that form because of wave erosion.

READING CHECK

3. Identify Give two ways that waves can break rock into smaller pieces.

_____

_____

_____

_____

Critical Thinking

4. Identify Relationships When may a storm not produce high-energy waves?

_____

_____

_____

TAKE A LOOK

5. Compare How is a sea stack different from a sea arch?

_____

_____

_____

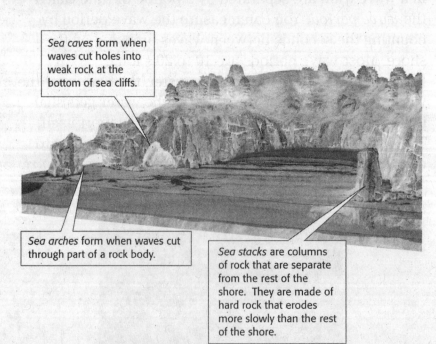

Sea caves form when waves cut holes into weak rock at the bottom of sea cliffs.

Sea arches form when waves cut through part of a rock body.

Sea stacks are columns of rock that are separate from the rest of the shore. They are made of hard rock that erodes more slowly than the rest of the shore.

SECTION 1 Shoreline Erosion and Deposition *continued*

Headlands are finger-shaped bodies of rock that stick out into the sea. They are made of harder rock than the rest of the shore.

Wave-cut terraces form when sea cliffs are worn back from the shore. This produces a nearly flat platform beneath the water at the base of the cliff.

STANDARDS CHECK

ES 1c Land forms are the result of a combination of constructive and destructive forces. Constructive forces include crustal deformation, volcanic eruption, and deposition of sediment, while destructive forces include weathering and erosion.

6. Define What is a headland?

_____

_____

What Are the Effects of Wave Deposition?

Waves carry many materials, such as sand, shells, and small rocks. When the waves deposit these materials on the shoreline, a beach forms. A **beach** is any area of shoreline that is made of material deposited by waves. Some beach material is deposited by rivers and moves down the shoreline by the action of waves. ☑

Many people think that all beaches are made of sand. However, beaches may be made of many materials, not just sand. The size and shape of beach material depend on how far the material traveled before it was deposited. They also depend on how the material is eroded. For example, beaches in stormy areas may be made of large rocks because smaller particles are removed by the waves.

The color of a beach can vary, too. A beach's color depends on what particles make up the beach. Light-colored sand is the most common beach material. Most light-colored sand is made of the mineral quartz. Many Florida beaches are made of quartz sand. On many tropical beaches, the sand is white. It is made of finely ground white coral. ☑

Beaches can also be black or dark-colored. Black-sand beaches are found in Hawaii. Their sands are made of eroded lava from volcanoes. This lava is rich in dark-colored minerals, so the sand is also dark-colored. The figures on the next page show some examples of beaches.

READING CHECK

7. Define Write your own definition for *beach*.

_____

_____

_____

_____

READING CHECK

8. Identify What mineral is most light-colored sand made of?

_____

_____

This beach in New England is made of large rocks. Smaller sand particles are washed away during storms.

This beach in Florida is made of light-colored quartz sand.

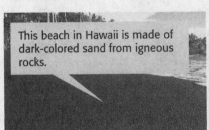

This beach in Hawaii is made of dark-colored sand from igneous rocks.

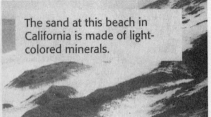

The sand at this beach in California is made of light-colored minerals.

TAKE A LOOK
9. Explain Why are some beaches made mostly of larger rock pieces, instead of sand?

_____

_____

WAVE ANGLE AND SAND MOVEMENT

Waves can move sand along a beach. The movement of the sand depends on the angle at which the waves hit the shore. *Longshore currents* form when waves hit the shore at an angle. The waves wash sand onto the shore at the same angle that the waves are moving. However, when the waves wash back into the ocean, they move sand directly down the slope of the beach. This causes the sand to move in a zigzag pattern, as shown in the figure below.

TAKE A LOOK
10. Infer Why don't longshore currents form in places where waves hit the shore head-on?

_____

_____

_____

_____

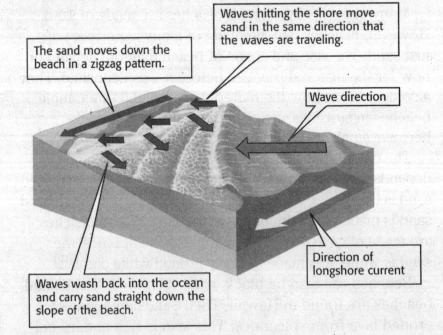

Waves hitting the shore move sand in the same direction that the waves are traveling.

The sand moves down the beach in a zigzag pattern.

Wave direction

Direction of longshore current

Waves wash back into the ocean and carry sand straight down the slope of the beach.

OFFSHORE DEPOSITS

Longshore currents can carry beach material offshore. This process can produce landforms in open water. These landforms include sandbars, barrier spits, and barrier islands.

A *sandbar* is a ridge of sand, gravel, or broken shells that is found in open water. Sandbars may be completely under water or they may stick up above the water. ☑

A *barrier spit* is a sandbar that sticks up above the water and is connected to the shoreline. Cape Cod, Massachusetts, is an example of a barrier spit. It is shown in the figure below.

READING CHECK

11. Define What is a sandbar?

_____

_____

_____

Cape Cod, Massachusetts, is an example of a barrier spit. Barrier spits form when sandbars are connected to the shoreline.

TAKE A LOOK
12. Identify What is a barrier spit?

_____

_____

A *barrier island* is a long, narrow island that forms parallel to the shoreline. Most barrier islands are made of sand.

Santa Rosa Island in Florida is an example of a barrier island.

TAKE A LOOK
13. Compare What is the difference between a barrier island and a barrier spit?

_____

_____

_____

_____

Name ______________________ Class ______________ Date ____________

Section 1 Review

NSES ES 1c

SECTION VOCABULARY

beach an area of the shoreline that is made up of deposited sediment	**shoreline** the boundary between land and a body of water

1. Compare How is a shoreline different from a beach?

____

____

2. Explain Where does the energy to change the shoreline come from? Explain your answer.

____

____

____

____

3. Identify Give two examples of different-colored beach sand and explain why each kind is a certain color.

____

____

____

____

4. Explain How do longshore currents move sand?

____

____

____

5. List Give five landforms that are produced by wave erosion.

____

____

____

____

____

CHAPTER 4 | Agents of Erosion and Deposition

SECTION 2

Wind Erosion and Deposition

National Science
Education Standards
ES 1c, 2a

BEFORE YOU READ

After you read this section, you should be able to answer
these questions:

• How can wind erosion shape the landscape?

• How can wind deposition shape the landscape?

How Can Wind Erosion Affect Rocks?

Wind can move soil, sand, and small pieces of rock.
Therefore, wind can cause erosion. However, some areas
are more likely to have wind erosion than other areas.
For example, plant roots help to hold soil and rock in
place. Therefore, areas with few plants, such as deserts
and coastlines, are more likely to be eroded by wind.
These areas also may be made of small, loose rock particles. Wind can move these particles easily. ☑

Wind can shape rock pieces in three ways: saltation,
deflation, and abrasion.

SALTATION

Wind moves large grains of soil, sand, and rock by saltation. **Saltation** happens when sand-sized particles skip
and bounce along in the direction that the wind is moving. When moving sand grains hit one another, some of
the grains bounce up into the air. These grains fall back
to the ground and bump other grains. These other grains
can then move forward.

STUDY TIP

Learn New Words As you
read this section, underline
words you don't understand.
When you figure out what
they mean, write the words
and their definitions in your
notebook.

READING CHECK

1. Explain How do plant
roots help to prevent wind
erosion?

_____

_____

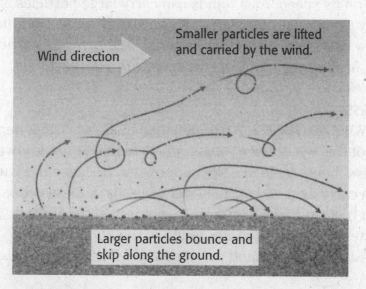

Wind direction

Smaller particles are lifted
and carried by the wind.

Larger particles bounce and
skip along the ground.

TAKE A LOOK

2. Apply Concepts Why
can't the wind lift and carry
large particles?

_____

_____

_____

DEFLATION

Wind can blow tiny particles away from larger rock pieces during deflation. **Deflation** happens when wind removes the top layers of fine sediment or soil and leaves behind larger rock pieces. ☑

Deflation can form certain land features. It can produce *desert pavement*, which is a surface made of pebbles and small, broken rocks. In some places, the wind can scoop out small, bowl-shaped areas in sediment on the ground. These areas are called *deflation hollows*.

ABRASION

Wind can grind and wear down rocks by abrasion. **Abrasion** happens when rock or sand wears down larger pieces of rock. Abrasion happens in areas where there are strong winds, loose sand, and soft rocks. The wind blows the loose sand against the rocks. The sand acts like sandpaper to erode, smooth, and polish the rocks.

Process	Description
	Large particles bounce and skip along the ground.
Deflation	
Abrasion	

What Landforms Are Produced by Wind Deposition?

Wind can carry material over long distances. The wind can carry different amounts and sizes of particles depending on its speed. Fast winds can carry large particles and may move a lot of material. However, all winds eventually slow down and drop their material. The heaviest particles fall first, while light material travels the farthest.

LOESS

Wind can deposit extremely fine material. Thick deposits of this windblown, fine-grained sediment are known as **loess**. Loess feels like talcum powder. Because the wind can carry light-weight material so easily, a loess deposit can be found far away from its source. In the United States, loess deposits are found in the Midwest, the Mississippi Valley, and in Oregon and Washington states.

READING CHECK

3. Define What is deflation?

_____

_____

_____

TAKE A LOOK

4. Complete Fill in the blank spaces in the table.

Critical Thinking

5. Infer What do you think is the reason that fast winds can carry larger particles than slower winds?

_____

_____

_____

DUNES

Barriers, such as plants and rocks, can cause wind to slow down. As it slows, the wind deposits particles on top of the barrier. As the dropped material builds up, the barrier gets larger. The barrier causes the wind to slow down even more. More and more material builds up on the barrier until a mound forms.

A mound of wind-deposited sand is called a **dune**. Dunes are common in sandy deserts and along sandy shores of lakes and oceans.

THE MOVEMENT OF DUNES

Wind conditions affect a dune's shape and size. As the wind blows sand through a desert, it is removed from some places and deposited in others. This can cause dunes to seem to move across the desert.

In general, dunes move in the same direction the wind is blowing. A dune has one gently sloped side and one steep side. The gently sloped side faces the wind. It is called the *windward slope*. The wind constantly moves sand up this side. As sand moves over the top of the dune, the sand slides down the steep side. The steep side is called the *slip face*. ☑

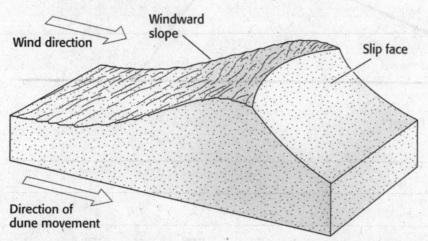

The wind blows sand up the windward slope of the dune. The sand moves over the top of the dune and falls down the steep slip face. In this way, dunes move across the land in the direction that the wind blows.

STANDARDS CHECK

ES 1c Land forms are the result of a combination of constructive and destructive forces. Constructive forces include crustal deformation, volcanic eruption, and deposition of sediment, while destructive forces include weathering and erosion.

6. Define What is a dune?

_____

_____

✓ **READING CHECK**

7. Identify In what direction do dunes generally move?

_____

_____

TAKE A LOOK

8. Compare How is the windward slope of a dune different from the slip face?

_____

_____

_____

Section 2 Review

SECTION VOCABULARY

abrasion the grinding and wearing away of rock surfaces through the mechanical action of other rock or sand particles	**loess** fine-grained sediments of quartz, feldspar, hornblende, mica, and clay deposited by the wind
deflation a form of wind erosion in which fine, dry soil particles are blown away	**saltation** the movement of sand or other sediments by short jumps and bounces that is caused by wind or water
dune a mound of wind-deposited sand that moves as a result of the action of wind	

1. Identify Give two land features that can form because of deflation.

_____

_____

2. Describe What areas are most likely to be affected by wind erosion? Give two examples.

_____

_____

3. Identify The figure shows a drawing of a sand dune. Label the windward slope and the slip face. Draw an arrow to show the direction of the wind.

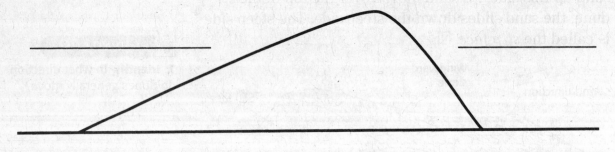

4. Explain How do dunes form?

_____

_____

_____

5. Apply Concepts Wind can transport particles of many different sizes. What sized particles are probably carried the farthest by the wind? Explain your answer.

_____

_____

_____

CHAPTER 4 Agents of Erosion and Deposition

SECTION 3 Erosion and Deposition by Ice

After you read this section, you should be able to answer these questions:

• What are glaciers?

• How do glaciers affect the landscape?

National Science Education Standards
ES 1c, 2a

What Are Glaciers?

A **glacier** is a huge piece of moving ice. The ice in glaciers contains most of the fresh water on Earth. Glaciers are found on every continent except Australia.

There are two kinds of glaciers: continental and alpine. *Continental glaciers* are ice sheets that can spread across entire continents. *Alpine glaciers* are found on the tops of mountains. Both continental and alpine glaciers can greatly affect the landscape. ☑

Glaciers form in areas that are so cold that snow stays on the ground all year round. For example, glaciers are common in polar areas and on top of high mountains. In these areas, layers of snow build up year after year. Over time, the weight of the top layers pushes down on the lower layers. The lower layers change from snow to ice.

STUDY TIP

Compare As you read, make a table comparing the landforms that glaciers can produce.

READING CHECK

1. Identify What are the two kinds of glaciers?

_____

_____

HOW GLACIERS MOVE

Glaciers can move in two ways: by sliding and by flowing. As more ice builds up on a slope, the glacier becomes heavier. The glacier can start to slide downhill, the way a skier slides downhill. Glaciers can also move by flowing. The solid ice in glaciers can move slowly, like soft putty or chewing gum.

Thick glaciers move faster than thin glaciers. Glaciers on steep slopes move faster than those on gentler slopes.

This is McBride Glacier in Alaska.

TAKE A LOOK
2. Define What is a glacier?

_____

_____

How Do Glaciers Affect the Landscape?

Glaciers can produce many different features as they move over Earth's surface. As a glacier moves, it can pick up and carry the rocks in its path. Glaciers can carry rocks of many different sizes, from dust all the way up to boulders. These rocks can scrape grooves into the land below the glacier as the glacier moves.

Continental glaciers tend to flatten the land that they pass over. However, alpine glaciers can produce sharp, rugged landscapes. The figure below shows some of the features that alpine glaciers can form.

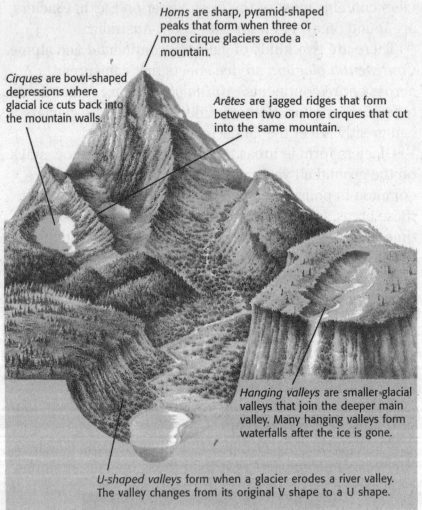

Horns are sharp, pyramid-shaped peaks that form when three or more cirque glaciers erode a mountain.

Cirques are bowl-shaped depressions where glacial ice cuts back into the mountain walls.

Arêtes are jagged ridges that form between two or more cirques that cut into the same mountain.

Hanging valleys are smaller glacial valleys that join the deeper main valley. Many hanging valleys form waterfalls after the ice is gone.

U-shaped valleys form when a glacier erodes a river valley. The valley changes from its original V shape to a U shape.

TAKE A LOOK
4. Explain How are horns, cirques, and arêtes related?

_____

_____

_____

GLACIAL DEPOSITS

As a glacier melts, it drops all of the material that it is carrying. The material that is carried and deposited by glaciers is called **glacial drift**. There are two kinds of glacial drift: till and stratified drift.

TILL DEPOSITS

Till is unsorted rock material that is deposited by melting glacial ice. It is called "unsorted" because the rocks are of all different sizes. Till contains fine sediment as well as large boulders. When the ice melts, it deposits this material onto the ground. ☑

The most common till deposits are *moraines*. Moraines form ridges along the edges of glaciers. There are many types of moraines. They are shown in the figure below.

☑ READING CHECK

5. Explain Why is till considered unsorted?

_____

_____

Lateral moraines form along each side of a glacier.

Medial moraines form when valley glaciers that have lateral moraines meet.

Ground moraines form from unsorted materials left beneath a glacier.

Terminal moraines form when sediment is dropped at the front of the glacier.

📣 Say It

Learn New Words Look up the words *lateral, medial,* and *terminal* in a dictionary. In a group, talk about why these words are used to describe different kinds of moraines.

STRATIFIED DRIFT

When a glacier melts, the water forms streams that carry rock material away from the glacier. The streams deposit the rocks in different places depending on their size. Larger rocks are deposited closer to the glacier. The rocks form a sorted deposit called **stratified drift**. The large area where the stratified drift is deposited is called an *outwash plain.* ☑

In some cases, a block of ice is left in the outwash plain as the glacier melts. As the ice melts, sediment builds up around it. The sediment forms a bowl-shaped feature called a *kettle*. Kettles can fill with water and become ponds or lakes.

☑ READING CHECK

6. Define Write your own definition for *stratified drift*.

_____

_____

_____

_____

Section 3 Review

NSES ES 1c, 2a

SECTION VOCABULARY

glacial drift the rock material carried and deposited by glaciers **glacier** a large mass of moving ice	**stratified drift** a glacial deposit that has been sorted and layered by the action of streams or meltwater **till** unsorted rock material that is deposited directly by a melting glacier

1. List Give two kinds of glacial drift.

_____

_____

2. Identify What are four kinds of moraines?

_____

_____

_____

_____

3. Compare How are continental glaciers different from alpine glaciers?

_____

_____

_____

_____

4. Explain How do glaciers form?

_____

_____

_____

_____

5. Describe How does a kettle form?

_____

_____

_____

6. Infer How can a glacier deposit both unsorted and sorted material?

_____

_____

_____

CHAPTER 4 Agents of Erosion and Deposition
SECTION 4

The Effect of Gravity on Erosion and Deposition

National Science
Education Standards
ES 1c, 2a

BEFORE YOU READ

After you read this section, you should be able to answer these questions:

- What is mass movement?
- How does mass movement shape Earth's surface?
- How can mass movement affect living things?

What Is Mass Movement?

Gravity can cause erosion and deposition. Gravity makes water and ice move. It also causes rock, soil, snow, or other material to move downhill in a process called **mass movement**.

STUDY TIP

Ask Questions As you read this section, write down any questions you have. Talk about your questions in a small group.

ANGLE OF REPOSE

Particles in a steep sand pile move downhill. They stop when the slope of the pile becomes stable. The *angle of repose* is the steepest angle, or slope, at which the loose material no longer moves downhill. If the slope of a pile of material is larger than the angle of repose, mass movement happens. ☑

READING CHECK

1. Define What is the angle of repose?

_____

_____

_____

The slope of this pile of sand is equal to the sand's angle of repose. The sand pile is stable. The sand particles are not moving.

The slope of this pile of sand is larger than the angle of repose. Therefore, particles of sand move down the slope of the pile.

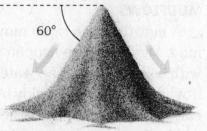

TAKE A LOOK

2. Explain Why are sand particles moving downhill in the bottom picture?

_____

_____

The angle of repose can be different in different situations. The composition, size, weight, and shape of the particles in a material affect its angle of repose. The amount of water in a material can also change the material's angle of repose.

What Are the Kinds of Mass Movement?

Mass movement can happen suddenly and quickly. Rapid mass movement can be very dangerous. It can destroy or bury everything in its path.

LANDSLIDES

A **landslide** happens when a large amount of rock and soil moves suddenly and rapidly downhill. Landslides can carry away or bury plants and animals and destroy their habitats. Several factors can make landslides more likely. ☑

- Heavy rains can make soil wet and heavy, which makes the soil more likely to move downhill.
- Tree roots help to keep land from moving. Therefore, *deforestation*, or cutting down trees, can make landslides more likely.
- Earthquakes can cause rock and soil to start moving.
- People may build houses and other buildings on unstable hillsides. The extra weight of the buildings can cause a landslide. ☑

The most common kind of landslide is a *slump*. Slumps happen when a block of material moves downhill along a curved surface.

ROCK FALLS

A **rock fall** happens when loose rocks fall down a steep slope. Many such slopes are found on the sides of roads that run through mountains. Gravity can cause the loose and broken rocks above the road to fall. The rocks in a rock fall may be many different sizes.

MUDFLOWS

A **mudflow** is a rapid movement of a large amount of mud. Mudflows can happen when a lot of water mixes with soil and rock. The water makes the slippery mud flow downhill very quickly. A mudflow can carry away cars, trees, houses, and other objects that are in its path.

Mudflows are common in mountain regions when a long dry season is followed by heavy rain. Mudflows may also happen when trees and other plants are cut down. Without plant roots to hold soil in place and help water drain away, large amounts of mud can quickly form.

☑ **READING CHECK**

3. Describe How can landslides affect wildlife habitats?

_____

_____

_____

☑ **READING CHECK**

4. Identify Give three factors that can make landslides more likely.

_____

_____

_____

_____

_____

Critical Thinking

5. Infer Does water probably increase or decrease the angle of repose of soil? Explain your answer.

_____

_____

_____

_____

SECTION 4 The Effect of Gravity on Erosion and Deposition *continued*

LAHARS

Volcanic eruptions can produce dangerous mudflows called *lahars*. A volcanic eruption on a snowy peak can suddenly melt a great amount of snow and ice. The water mixes with soil and ash to produce a hot flow that rushes downhill. Lahars can travel faster than 80 km/h.

CREEP

Not all mass movement is fast. In fact, very slow mass movement is happening on almost all slopes. **Creep** is the name given to this very slow movement of material downhill. Even though creep happens very slowly, it can move large amounts of material over a long period of time. ☑

Many factors can affect creep. Water can loosen soil and rock so that they move more easily. Plant roots can cause rocks to crack and can push soil particles apart. Burrowing animals, such as moles and gophers, can loosen rock and soil particles. All of these factors may make creep more likely.

✓ **READING CHECK**

6. Compare How is creep different from the other kinds of mass movement that are discussed in this section?

_____

_____

Type of Mass Movement	Description
Landslide	Material moves suddenly and rapidly down a slope.
Rock fall	
Mudflow	
	Water mixes with volcanic ash to produce a fast-moving, dangerous mudflow.
	Material moves downhill very slowly.

TAKE A LOOK

7. Describe Fill in the blank spaces in the table.

83

Agents of Erosion and Deposition

Section 4 Review

NSES ES 1c, 2a

SECTION VOCABULARY

creep the slow downhill movement of weathered rock material	**mudflow** the flow of a mass of mud or rock and soil mixed with a large amount of water
landslide the sudden movement of rock and soil down a slope	**rock fall** the rapid mass movement of rock down a steep slope or cliff
mass movement the movement of a large mass of sediment or a section of land down a slope	

1. List What are four kinds of mass movement?

_____

_____

_____

2. Infer Why is it important for people to think about mass movement when they decide how to use land?

_____

_____

3. Identify Relationships How is mass movement related to the angle of repose?

_____

_____

4. Identify What force causes mass movements?

_____

5. Compare How are landslides different from mudflows?

_____

_____

_____

6. List Give four things that can affect a material's angle of repose.

_____

_____

_____

CHAPTER 5 | The Atmosphere
SECTION 1 | # Characteristics of the Atmosphere

BEFORE YOU READ

After you read this section, you should be able to answer these questions:

• What is Earth's atmosphere made of?

• How do air pressure and temperature change as you move away from Earth's surface?

• What are the layers of the atmosphere?

National Science Education Standards
ES 1h

What Is Earth's Atmosphere Made Of?

An **atmosphere** is a layer of gases that surrounds a planet or moon. On Earth, the atmosphere is often called just "the air." When you take a breath of air, you are breathing in atmosphere.

The air you breathe is made of many different things. Almost 80% of it is nitrogen gas. The rest is mostly oxygen, the gas we need to live. There is also water in the atmosphere. Some of it is invisible, in the form of a gas called *water vapor*. ☑

Water is also found in the atmosphere as water droplets and ice crystals, like those that make up clouds. The atmosphere also contains tiny *particles*, or solid pieces. These particles are things like dust and dirt from continents, salt from oceans, and ash from volcanoes.

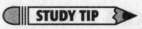

STUDY TIP

Define When you come across a word you don't know, circle it. When you figure out what it means, write the word and its definition in your notebook.

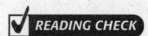

READING CHECK

1. List Which two gases make up most of Earth's atmosphere?

_____

_____

Gases in Earth's Atmosphere

Nitrogen
78%

Oxygen
21%

Other
1%

Math Focus

2. Analyze Data About what fraction of the Earth's atmosphere is NOT made of nitrogen? Give your answer as a reduced fraction.

_____

_____

Where Do the Gases in the Atmosphere Come From?

The gases in Earth's atmosphere come from many different sources. The table below shows some of those sources.

Gas	Where the gas comes from
Oxygen	Plants give off oxygen as they grow.
Nitrogen	Nitrogen is given off when dead plants and animals decay.
Water vapor	Liquid water evaporates and becomes water vapor. Plants give off water vapor as they grow. Water vapor comes out of the Earth during volcanic eruptions.
Carbon dioxide	Carbon dioxide comes out of the Earth during volcanic eruptions. When animals breathe, they give off carbon dioxide. Carbon dioxide is given off when we burn things that were once plant or animal material.

TAKE A LOOK
3. Identify Name two gases that volcanoes contribute to the atmosphere.

_____

_____

Why Does Air Pressure Change with Height?

Air pressure is how much the air above you weighs. It is a measure of how hard air molecules push on a surface. We don't normally notice air pressure, because our bodies are used to it. ☑

As you move up from the ground and out toward space, there are fewer gas molecules pressing down from above. Therefore, the air pressure drops. The higher you go, the lower the air pressure gets.

✓ **READING CHECK**
4. Define Write your own definition for air pressure.

_____

_____

TAKE A LOOK
5. Compare How is the air pressure around the tree different from the air pressure around the plane?

_____

_____

_____

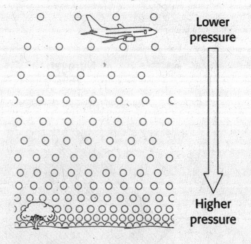

Lower pressure

Higher pressure

SECTION 1 Characteristics of the Atmosphere *continued*

Why Does Air Temperature Change with Height?

Like air pressure, air temperature changes as you move higher in the atmosphere. Air pressure always gets lower as you move higher, but air temperature can get higher or lower. The air can get hotter or colder. ☑

There are different layers of the atmosphere. Each layer is made of a different combination of gases. Air temperature depends on the gases in the atmosphere. Some gases absorb energy from the sun better than others. When a gas absorbs energy from the sun, the air temperature goes up.

What Are the Layers of the Atmosphere?

There are four main layers of the atmosphere: troposphere, stratosphere, mesosphere, and thermosphere. You cannot actually see these different layers. The divisions between the layers are based on how each layer's temperature changes with height.

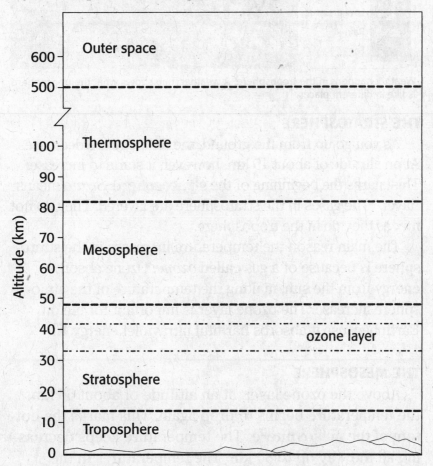

✓ READING CHECK

6. Compare How are the changes in air temperature with height different from changes in air pressure with height?

_____

_____

_____

📣 Say It

Make Up a Memory Trick In groups of two or three, make up a sentence to help you remember the order of the layers of the atmosphere. The words in the sentence should start with T, S, M, and T. For example, "Tacos Sound Mighty Tasty." A sentence like this is called a *mnemonic*.

TAKE A LOOK

7. Identify At what altitude does the mesosphere end and the thermosphere begin?

_____

Critical Thinking

8. Explain Why is the troposphere important to people?

_____

_____

_____

_____

TAKE A LOOK

9. Analyze What does the map tell you about the air temperature in the troposphere?

_____

_____

_____

READING CHECK

10. Explain Why is ozone in the stratosphere important for living things?

_____

_____

THE TROPOSPHERE

The **troposphere** is the layer of the atmosphere that we live in. It is where most of the water vapor, carbon dioxide, pollution, and living things on Earth exist. Weather conditions such as wind and rain all take place in the troposphere.

The troposphere is also the densest layer of the atmosphere. This is because the troposphere is at the bottom with all the other layers pushing down from above. Almost 90% of the gases in the atmosphere are in the troposphere. As you move higher into the troposphere (say, to the top of a mountain), both air temperature and air pressure decrease.

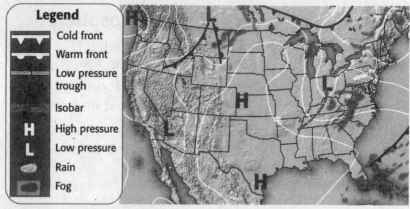

Legend
- Cold front
- Warm front
- Low pressure trough
- Isobar
- **H** High pressure
- **L** Low pressure
- Rain
- Fog

Weather happens in the troposphere. A weather map shows what the troposphere is like in different places.

THE STRATOSPHERE

As you go up from the ground, the temperature decreases. At an altitude of about 15 km, however, it starts to increase. This marks the beginning of the **stratosphere**. *Strato* means "layer." The gases in the stratosphere are layered. They do not mix as they do in the troposphere.

The main reason the temperature increases in the stratosphere is because of a gas called *ozone*. Ozone absorbs energy from the sun, making the temperature of the atmosphere increase. The ozone layer is important for life on Earth because it absorbs harmful ultraviolet energy. ☑

THE MESOSPHERE

Above the ozone layer, at an altitude of about 50 km, the temperature begins to drop again. This marks the bottom of the **mesosphere**. The temperature keeps decreasing all the way up to 80 km. The temperatures in the mesosphere can be as low as −93°C.

THE THERMOSPHERE

The **thermosphere** is the uppermost layer of the atmosphere. In the thermosphere, temperatures begin to rise again. The thermosphere gets its name from its extremely high temperatures, which can be above 1,000°C. *Therm* means "heat." The temperatures in the thermosphere are so high because it contains a lot of oxygen and nitrogen, which absorb energy from the sun. ☑

THE IONOSPHERE—ANOTHER LAYER

The troposphere, stratosphere, mesosphere, and thermosphere are the four main layers of the atmosphere. However, scientists also sometimes study a region called the ionosphere. The *ionosphere* contains the uppermost part of the mesosphere and the lower part of the thermosphere. It is made of nitrogen and oxygen *ions*, or electrically charged particles.

The ionosphere is where auroras occur. *Auroras* are curtains and ribbons of shimmering colored lights. They form when charged particles from the sun collide with the ions in the ionosphere. The ionosphere is important to us because it can reflect radio waves. An AM radio wave can travel all the way around the Earth by bouncing off the ionosphere.

✓ **READING CHECK**

11. Explain Why is the thermosphere called the thermosphere?

_____

_____

_____

Layer	How temperature and pressure change as you move higher	Important features
Troposhere	temperature decreases pressure decreases	
Stratosphere		gases are arranged in layers contains the ozone layer
		has the lowest temperatures
Thermosphere	temperature increases pressure decreases	

TAKE A LOOK

12. Identify Use the information from the text to fill in the table.

Section 1 Review

NSES ES 1h

SECTION VOCABULARY

air pressure the measure of the force with which air molecules push on a surface	**stratosphere** the layer of the atmosphere that is above the troposphere and in which temperature increases as altitude increases
atmosphere a mixture of gases that surrounds a planet or moon	**thermosphere** the uppermost layer of the atmosphere, in which temperature increases as altitude increases
mesosphere the layer of the atmosphere between the stratosphere and the thermosphere and in which temperature decreases as altitude increases	**troposphere** the lowest layer of the atmosphere, in which temperature decreases at a constant rate as altitude increases

1. Define Write your own definition for atmosphere.

_____

_____

2. Explain Why does air temperature change as you move up from the Earth's surface?

_____

_____

_____

3. Make a Graph The graph below shows how the temperature changes as you move up through the atmosphere. On the graph, draw a curve showing how the pressure changes.

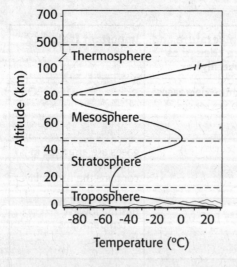

4. Identify Relationships How does the sun affect air temperatures?

_____

_____

CHAPTER 5 | The Atmosphere

SECTION
2 **Atmospheric Heating**

BEFORE YOU READ

After you read this section, you should be able to answer
these questions:

• How does energy travel from the sun to Earth?

• What are the differences between radiation,
conduction, and convection?

• Why is Earth's atmosphere so warm?

How Does Energy Travel from the Sun to Earth?

Most of the heat energy on Earth's surface comes
from the sun. Energy travels from the sun to Earth by
radiation, which means that it travels through space as
waves. As solar energy (energy from the sun) is absorbed
by air, water, and land, it turns into heat energy. This
energy causes winds, the water cycle, ocean currents,
and changes in the weather.

What Happens to Radiation from the Sun?

Not all of the radiation from the sun reaches
Earth's surface. Much of it gets absorbed by the
atmosphere. Some of it is scattered and reflected by
clouds and gases.

About **25%** is
scattered and
reflected by
clouds and air.

About **20%** is absorbed
by ozone, clouds, and
atmospheric gases.

About **50%** is absorbed
by Earth's surface.

About **5%** is reflected
by Earth's surface.

STUDY TIP

Outline In your notebook,
write an outline of this chapter. Use the questions in bold
to make your outline. As you
read, fill in information about
each question.

TAKE A LOOK

1. Identify How much of the
sunlight that gets to Earth is
absorbed by Earth's surface?

_____

2. Summarize What
happens to the sunlight
that is not absorbed by
Earth's surface?

_____

_____

_____

SECTION 2 Atmospheric Heating *continued*

How Is Heat Transferred by Contact?

Once sunlight is absorbed by Earth's surface, it is *converted*, or changed, into heat energy. Then, the heat can be transferred to other objects and moved to other places. When a warm object touches a cold object, heat moves from the warm object to the cold one. This movement of heat is called **thermal conduction**.

When you touch the sidewalk on a hot, sunny day, heat energy is conducted from the sidewalk to you. The same thing happens to air molecules in the atmosphere. When they touch the warm ground, the air molecules heat up. ☑

How Is Heat Energy Transferred by Motion?

If you have ever watched a pot of water boil, you have seen convection. During **convection**, warm material, such as air or water, carries heat from one place to another.

When you turn on the stove under a pot of water, the water closest to the pot heats up. As the water heats up, its density decreases. The warm water near the pot is not as dense as the cool water near the air. Therefore, the cool water sinks while the warm water rises.

As it rises, the warm water begins to cool. When it cools, its density increases. It becomes denser than the layer below, so it sinks back to the bottom of the pot. This forms a circular movement called a *convection current*.

Convection currents also move heat through the atmosphere. In fact, most heat energy in the atmosphere is transferred by convection. Air close to the ground is heated by conduction from the ground. It becomes less dense than the cooler air above it. The warmer air rises while the cooler air sinks. The ground warms up the cooler air by conduction, and the warm air rises again.

READING CHECK

3. List Name two ways that air gets heated.

_____

_____

Critical Thinking

4. Apply Concepts Before the water in the pot can heat up, the pot itself must heat up. Does the pot heat up by conduction, convection, or radiation? Explain your answer.

_____

_____

_____

TAKE A LOOK

5. Describe What happens to warm air as it moves through the atmosphere?

_____

Convection Current

Warm air cools down.

Warm, less dense air rises.

Cooler, denser air sinks.

Cool air warms up.

How Does the Earth Stay Warm?

A gardener who needs to keep plants warm uses a glass building called a greenhouse. Light travels through the glass into the building, and the air and plants inside absorb the energy. The energy is converted to heat, which cannot travel back through the glass as easily as light came in. Much of the heat energy stays trapped within the greenhouse, keeping the air inside warmer than the air outside.

Earth's atmosphere acts like the glass walls of a greenhouse. Sunlight travels through the atmosphere easily, but heat does not. Gases in the atmosphere, such as water vapor and carbon dioxide, absorb heat energy coming from Earth. Then, they radiate it back to Earth's surface. This is known as the **greenhouse effect**. ☑

The Greenhouse Effect

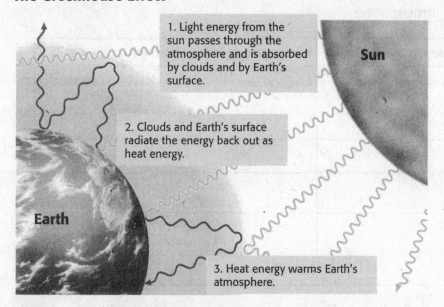

1. Light energy from the sun passes through the atmosphere and is absorbed by clouds and by Earth's surface.

2. Clouds and Earth's surface radiate the energy back out as heat energy.

Sun

Earth

3. Heat energy warms Earth's atmosphere.

What Is Global Warming?

Many scientists are worried that Earth has been getting warmer over the past hundred years. This increase in temperatures all over the world is called **global warming**.

Scientists think that human activities may be causing global warming. When we burn fossil fuels, we release greenhouse gases, such as carbon dioxide, into the atmosphere. Because greenhouse gases trap heat in the atmosphere, adding more of them can make Earth even warmer. Global warming can have a strong effect on weather and climate.

✓ READING CHECK

6. List Name two gases in Earth's atmosphere that absorb heat.

_____

_____

TAKE A LOOK

7. Identify On the drawing, label the light coming from the sun with an **L**. Label the heat energy that is trapped by Earth's atmosphere with an **H**.

 **Say It**

Predict How might global warming affect your community? What can you do to slow global warming? In groups of two or three, discuss how global warming might affect your lives.

Section 2 Review

SECTION VOCABULARY

convection the transfer of thermal energy by the circulation or movement of a liquid or gas	**radiation** the transfer of energy as electro-magnetic waves
global warming a gradual increase in average global temperature	**thermal conduction** the transfer of energy as heat through a material
greenhouse effect the warming of the surface and lower atmosphere of Earth that occurs when water vapor, carbon dioxide, and other gases absorb and reradiate thermal energy	

1. **Apply Concepts** A person is camping outside. The person toasts a marshmallow by holding it above the flames of the fire. Does the marshmallow cook because of convection, conduction, or radiation? Explain your answer.

_____

_____

2. **Compare** Fill in the table below to name and describe the three ways energy is transferred in Earth's atmosphere.

Type of energy transfer	How energy is transferred
	Energy travels as electromagnetic waves.
Conduction	

3. **Explain** How does most of the heat in Earth's atmosphere move from place to place?

_____

_____

4. **Identify Relationships** Explain how global warming and the greenhouse effect are related.

_____

_____

_____

CHAPTER 5 | The Atmosphere

SECTION
3 **Global Winds and Local Winds**

National Science
Education Standards
ES 1j

BEFORE YOU READ

After you read this section, you should be able to answer
these questions:

• What causes wind?

• What is the Coriolis effect?

• What are the major global wind systems on Earth?

What Causes Wind?

Wind moving air caused by differences in air pressure. Air moves from areas of high pressure to areas of
low pressure. The greater the pressure difference, the
faster the air moves, and the stronger the wind blows. ☑

You can see how air moves if you blow up a balloon
and then let it go. The air inside the balloon is at a higher
pressure than the air around the balloon. If you open the
end of the balloon, air will rush out.

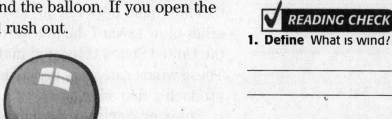

What Causes Differences in Air Pressure?

Most differences in air pressure are caused by differences in air temperature. Temperature differences happen because some parts of Earth get more energy from
the sun than others. For example, the sun shines more
directly on the equator than on the poles. As a result, the
air is warmer near the equator. ☑

The warm air near the equator is not as dense as the cool
air near the poles. Because it is less dense, the air at the
equator rises, forming areas of low pressure. The cold air
near the poles sinks, forming areas of high pressure. The air
moves in large circular patterns called *convection cells.* The
drawing on the next page shows these convection cells.

STUDY TIP
Underline Each heading in
this section is a question.
Underline the answer to each
question when you find it in
the text.

☑ **READING CHECK**
1. Define What is wind?
_____
_____

TAKE A LOOK
2. Identify On the drawing,
label the high-pressure area
with an **H** and the low-
pressure area with an **L**.

☑ **READING CHECK**
3. Explain Why isn't all
the air on Earth at the same
temperature?
_____
_____
_____

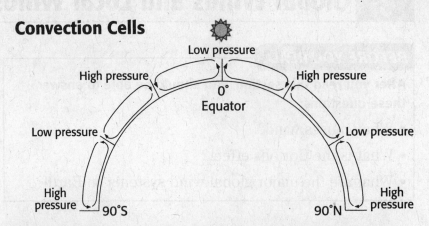

Convection Cells

TAKE A LOOK
4. Describe Is air rising or sinking in areas of high pressure?

_____

What Are the Major Global Wind Systems?

Global winds are large-scale wind systems. There are three pairs of major global wind systems, or wind belts: trade winds, westerlies, and polar easterlies. ☑

Trade winds are wind belts that blow from 30° latitude almost to the equator. They curve to the west as they blow toward the equator. **Westerlies** are wind belts that are found between 30° and 60° latitude. The westerlies blow toward the poles from west to east. Most of the United States is located in the belt of westerly winds. These winds can carry moist air over the United States, producing rain and snow.

Polar easterlies are wind belts that extend from the poles to 60° latitude. They form as cold, sinking air moves away from the poles. In the Northern Hemisphere, polar easterlies can carry cold arctic air over the United States. This can produce snow and freezing weather.

☑ **READING CHECK**
5. Identify What are the three main global wind belts?

_____

_____

Wind belt	Location (latitude)	Toward the equator or toward the poles?
Trade winds	0° to 30°	toward the equator
Westerlies		
	60° to 90°	

TAKE A LOOK
6. Describe Fill in the blanks in the table.

The figure on the next page shows the locations of these different wind belts. Notice that the winds do not move in straight lines. The paths of the wind belts are controlled by convection cells and by the Earth's rotation.

SECTION 3 Global Winds and Local Winds *continued*

The trade winds meet and rise near the equator in a region known as the doldrums. The wind in the doldrums is very weak.

The region between the trade winds and the westerlies is known as the horse latitudes. Here, cool air sinks, creating a region of high pressure. The winds here are very weak.

→ Cool air
→ Warm air
→ Wind direction

There are three pairs of major global wind belts on Earth: the polar easterlies, the westerlies, and the trade winds.

STANDARDS CHECK

ES 1j Global patterns of atmospheric movement influence local weather. Oceans have a <u>major</u> effect on climate, because water in the oceans holds a large amount of heat.

Word Help: <u>major</u>
of great importance or large scale

7. Explain Use the map to explain why surface winds are generally very weak near the equator.

_____
_____
_____
_____
_____

Why Do Global Winds Curve?

Remember that pressure differences can cause air to move and form winds. If Earth did not rotate, these winds would blow in straight lines. However, because Earth does rotate, the winds follow curved paths. This *deflection*, or curving, of moving objects from a straight path because of Earth's rotation is called the **Coriolis effect**. ☑

As Earth rotates, places near the equator travel faster than places closer to the poles. This difference in speed causes the Coriolis effect. Wind moving from the poles to the equator is deflected to the west. Wind moving from the equator to the poles is deflected east.

✓ **READING CHECK**

8. Describe How does Earth's rotation affect the paths of global winds?

_____
_____
_____

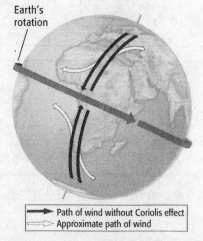

→ Path of wind without Coriolis effect
⇨ Approximate path of wind

The Coriolis effect causes wind and water to move along curved paths.

TAKE A LOOK
9. Apply Ideas If air is moving south from California, which way will it tend to curve?

_____

What Are Jet Streams?

The polar easterlies, prevailing westerlies, and trade winds are all winds that we feel on the ground. However, wind systems can also form at high altitude. **Jet streams** are narrow belts of very high-speed winds in the upper troposphere and lower stratosphere. They blow from west to east all the way around the Earth. ☑

Jet streams can reach speeds of 400 km/h. Pilots flying east over the United States or the Atlantic Ocean try to catch a jet stream. This wind pushes airplanes along, helping them fly faster and use less fuel. Pilots flying west try to avoid the jet streams.

The global wind systems are always found in about the same place every day. Unlike these global wind systems, jet streams can be in different places on different days. Because jet streams can affect the movements of storms, meteorologists try to track the jet streams. They can sometimes predict the path of a storm if they know where the jet streams are.

✓ READING CHECK

10. Identify In what two layers of the atmosphere are the jet streams found?

_____

_____

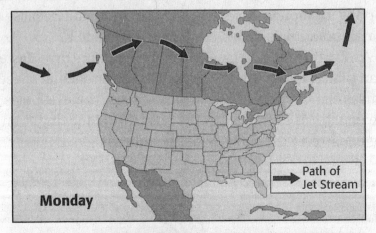

Monday

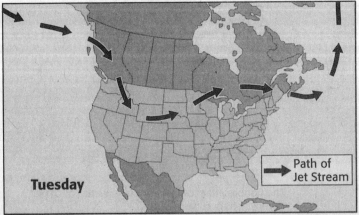

Tuesday

Jet streams form between hot and cold air masses. Unlike the other wind systems, jet streams are found in slightly different places every day.

TAKE A LOOK

11. Infer Why would a pilot flying across North America take a different route on Tuesday than on Monday?

_____

_____

_____

_____

Name _____ Class _____ Date _____

SECTION 3 Global Winds and Local Winds *continued*

What Are Local Winds?

Most of the United States is in the belt of prevailing westerly winds, which move from west to east. However, you've probably noticed that the wind in your neighborhood does not always blow from the west to the east. This is because global winds are not the only winds that blow. Local winds are also important. *Local winds* are winds that generally move over short distances and can blow from any direction.

Like the other wind systems, local winds are caused by differences in temperature. Many of these temperature differences are caused by geographic features, such as mountains and bodies of water. The figure below shows how water and mountains can affect local winds.

Critical Thinking

12. Compare Describe one difference between global winds and local winds.

_____

_____

_____

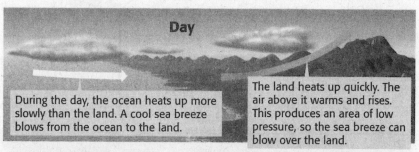

During the day, the ocean heats up more slowly than the land. A cool sea breeze blows from the ocean to the land.

The land heats up quickly. The air above it warms and rises. This produces an area of low pressure, so the sea breeze can blow over the land.

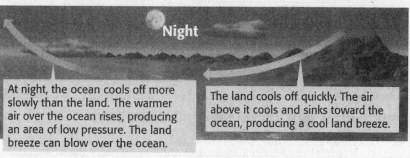

At night, the ocean cools off more slowly than the land. The warmer air over the ocean rises, producing an area of low pressure. The land breeze can blow over the ocean.

The land cools off quickly. The air above it cools and sinks toward the ocean, producing a cool land breeze.

Say It

Share Experiences Have you ever been in a very strong wind? In groups of two or three, discuss the strongest or worst wind you've ever been in.

TAKE A LOOK

13. Identify In the figures, label the high-pressure areas with an **H** and the low-pressure areas with an **L**.

MOUNTAIN BREEZES AND VALLEY BREEZES

Mountain and valley breezes are other examples of local winds caused by geography. During the day, the sun warms the air on mountain slopes. The warm air rises up the mountain slopes, producing a warm valley breeze. At night, the air on the slopes cools. The cool air moves down the slopes, producing a cool mountain breeze.

Section 3 Review

NSES   ES 1j

SECTION VOCABULARY

Coriolis effect the curving of the path of a moving object from an otherwise straight path due to the Earth's rotation	**trade winds** prevailing winds that blow from east to west from 30° latitude to the equator in both hemispheres
jet stream a narrow band of strong winds that blow in the upper troposphere	**westerlies** prevailing winds that blow from west to east between 30° and 60° latitude in both hemispheres
polar easterlies prevailing winds that blow from east to west between 60° and 90° latitude in both hemispheres	**wind** the movement of air caused by differences in air pressure

1. Identify The drawing below shows a convection cell. Put arrows on the cell to show which way the air is moving. Label high pressure areas with an **H** and low pressure areas with an **L**. Label cold air with a **C** and warm air with a **W**.

2. Identify Which global wind system blows toward the poles between 30° and 60° latitude?

_____

3. Explain Why does wind tend to blow down from mountains at night?

_____

_____

4. Apply Concepts Would there be winds if Earth's surface were the same temperature everywhere? Explain your answer.

_____

_____

_____

_____

CHAPTER 5 | The Atmosphere

SECTION 4 | **Air Pollution**

BEFORE YOU READ

After you read this section, you should be able to answer these questions:

• What is air pollution?

• What causes air pollution?

• How does air pollution affect the environment?

• How can people reduce air pollution?

What Is Air Pollution?

Air pollution is the addition of harmful substances to the atmosphere. An *air pollutant* is anything in the air that can damage the environment or make people or other organisms sick. Some air pollution comes from natural sources. Other forms of air pollution are caused by things people do.

There are two kinds of air pollutants: primary pollutants and secondary pollutants. Primary pollutants are pollutants that are put directly into the air. Dust, sea salt, volcanic ash, and pollen are primary pollutants that come from natural sources. Chemicals from paint and other materials and vehicle exhaust are primary pollutants that come from human activities.

Secondary pollutants form when primary pollutants react with each other or with other substances in the air. Ozone is an example of a secondary pollutant. It forms on sunny days when chemicals from burning gasoline react with each other and with the air. Ozone damages human lungs and can harm other living things as well. ☑

STUDY TIP

Describe As you read, make a table describing the sources of air pollution discussed in this section.

Pollutant	Primary pollutant or secondary pollutant?	Natural or caused by people?
Car exhaust	primary	human-caused
Dust		
Ozone		
Paint chemicals		
Pollen		
Sea salt		
Volcanic ash		

READING CHECK

1. **Explain** Why is ozone called a secondary pollutant?

_____

_____

_____

TAKE A LOOK

2. **Describe** Fill in the blanks in the table.

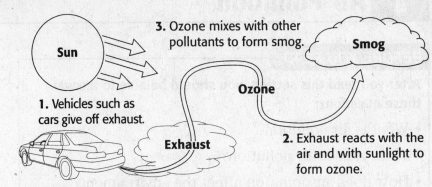

3. Ozone mixes with other pollutants to form smog.

Sun

Smog

Ozone

1. Vehicles such as cars give off exhaust.

Exhaust

2. Exhaust reacts with the air and with sunlight to form ozone.

TAKE A LOOK
3. Identify What is the primary pollutant in this figure?

_____

_____

What Is Smog?

On a hot, still, sunny day, yellowish brown air can cover a city. This is called *smog*. Smog forms when ozone mixes with other pollutants. During summer in cities such as Los Angeles, a layer of warm air can trap smog near the ground. In the winter, a storm can clear the air.

Say It
Discuss In a small group, discuss how the pollution shown in this photograph formed.

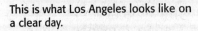

This is what Los Angeles looks like on a clear day.

This is what Los Angeles looks like when smog is trapped near the ground.

How Do Humans Cause Air Pollution?

Many of our daily activities cause air pollution. The main source of human-caused air pollution in the United States is motor vehicles. Cars, motorcycles, trucks, buses, trains, and planes all give off exhaust. *Exhaust* is a gas that contains pollutants that create ozone and smog. ☑

Factories and power plants that burn coal, oil, and gas also give off pollutants. Businesses that use chemicals, such as dry cleaners and auto body shops, can add to air pollution.

4. Identify What is the main source of human-caused air pollution in the United States?

_____

_____

What Causes Air Pollution Indoors?

Sometimes the air inside a building can be more polluted than the air outside. There is no wind to blow pollutants away and no rain to wash them out of the air indoors. Therefore, they can build up inside. It is important to air out buildings by opening the windows or using fans that bring fresh air in from outside. ☑

Sources of Indoor Air Pollution

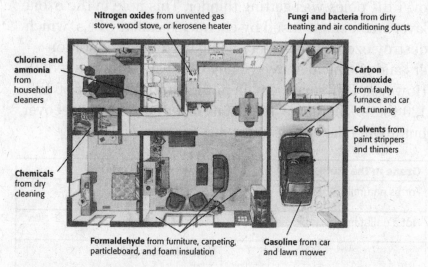

Nitrogen oxides from unvented gas stove, wood stove, or kerosene heater

Fungi and bacteria from dirty heating and air conditioning ducts

Chlorine and ammonia from household cleaners

Carbon monoxide from faulty furnace and car left running

Solvents from paint strippers and thinners

Chemicals from dry cleaning

Formaldehyde from furniture, carpeting, particleboard, and foam insulation

Gasoline from car and lawn mower

✓ **READING CHECK**

5. Explain Why can air pollution indoors be worse than air pollution outdoors?

_____

_____

_____

TAKE A LOOK

6. Identify Name two sources of indoor air pollution shown here that may be in your own home.

_____

_____

What Is Acid Precipitation?

Acid precipitation is rain, sleet, or snow that contains acids from air pollution. When we burn fossil fuels, such as coal, pollutants such as sulfur dioxide are released into the air. These pollutants combine with water in the atmosphere to form acids.

Acid precipitation can kill or damage plants, damage soil, and poison water. When acid rain flows into lakes, it can kill fish and other aquatic life.

```
┌─────────────────────────────┐
│ People burn coal for energy. │
└─────────────────────────────┘
              │
              ▼
┌─────────────────────────────┐
│                             │
│                             │
└─────────────────────────────┘
              │
              ▼
┌─────────────────────────────┐
│                             │
│                             │
└─────────────────────────────┘
              │
              ▼
┌─────────────────────────────┐
│  Acid rain falls in the lake. │
└─────────────────────────────┘
              │
              ▼
     ┌──────────────┐
     │  Fish die.   │
     └──────────────┘
```

TAKE A LOOK

7. Sequence Complete the graphic organizer to show how burning coal can cause fish to die.

What Is the Ozone Hole?

Close to the ground, ozone is a pollutant formed by human activities. However, high in the stratosphere, ozone is an important gas that forms naturally. The ozone layer absorbs harmful ultraviolet (UV) radiation from the sun. Ultraviolet radiation can harm living things. For example, it can cause skin cancer in humans. ☑

In the 1980s, scientists noticed that the ozone layer over the poles was getting thinner. This hole in the ozone layer was being caused by chemicals called CFCs, which destroy ozone. CFCs were being used in air conditioners and chemical sprays. Many CFCs are now banned. However, CFCs can remain in the atmosphere for 60 to 120 years. Therefore, the ozone layer may slowly recover, but it will take a long time.

✔ **READING CHECK**

8. Explain How is the ozone layer helpful to humans?

_____

_____

_____

_____

TAKE A LOOK
9. Compare Fill in the chart to show the differences between ozone in the atmosphere and ozone near the ground.

Ozone in the statosphere	Ozone near the ground
Forms naturally	
Not a pollutant	
	harmful to living things

How Does Air Pollution Affect Human Health?

Air pollution can cause many health problems. Some are short-term problems. They happen quickly and go away when the air pollution clears up or the person moves to a cleaner location. Others are long-term health problems. They develop over long periods of time and are not cured easily. The table below lists some of the effects of air pollution on human health. ☑

✔ **READING CHECK**

10. Compare What is the difference between short-term effects and long-term effects of air pollution?

_____

_____

_____

_____

Long-term effects	Short-term effects
Emphysema (a lung disease)	Headache
Lung cancer	Nausea and vomiting
Asthma	Eye, nose, and throat irritation
Permanent lung damage	Coughing
Heart disease	Difficulty breathing
Skin cancer	Upper respiratory infections
	Asthma attacks
	Worsening of emphysema

SECTION 4 Air Pollution *continued*

What Can We Do About Air Pollution?

Air pollution in the United States is not as bad now as it was 30 years ago. People today are much more aware of how they can cause or reduce air pollution. Air pollution can be reduced by new laws, by technology, and by people changing their lifestyles.

The United States government and the governments of other countries have passed laws to control air pollution. These laws limit the amount of pollution that sources such as cars and factories are allowed to release. For example, factories and power plants now have scrubbers on smokestacks. A *scrubber* is a tool that helps remove pollutants from smoke before it leaves the smokestack.

Many cars are more efficient now than they used to be, so they produce less pollution. Individuals can do a lot on their own to reduce air pollution, as well. For example, we can walk or bike instead of driving.

Critical Thinking

11. Analyze Processes
Electric cars don't give off any exhaust. They don't cause pollution in the cities where they are driven. However, driving them can cause pollution in other places. How? (Hint: Where does most electricity come from?)

_____

_____

_____

_____

_____

_____

In Copenhagen, Denmark, companies lend bicycles for anyone to use for free. The program helps reduce automobile traffic and air pollution.

Section 4 Review

SECTION VOCABULARY

acid precipitation rain, sleet, or snow that contains a high concentration of acids	**air pollution** the contamination of the atmosphere by the introduction of pollutants from human and natural sources

1. Identify Relationships How are fossil fuels related to air pollution and acid precipitation?

_____

_____

_____

2. Compare Complete the table below to compare different pollutants.

Pollutant	Source	Negative effects	Solutions
CFCs			banning CFCs
Ozone			
Sulfur dioxide	burning of fossil fuels		

3. Infer Name three things, other than humans, that can be harmed by air pollution.

_____

_____

_____

4. Explain Why is the hole in the ozone layer dangerous?

_____

_____

_____

CHAPTER 6 Understanding Weather
SECTION 1

Water in the Air

BEFORE YOU READ

After you read this section, you should be able to answer these questions:

• What is weather?

• How does water in the air affect the weather?

What Is Weather?

Knowing about the weather is important in our daily lives. Your plans to go outside can change if it rains. Being prepared for extreme weather conditions, such as hurricanes and tornadoes, can even save your life.

Weather is the condition of the atmosphere at a certain time and place. Weather depends a lot on the amount of water in the air. Therefore, to understand weather, you need to understand the water cycle. ☑

THE WATER CYCLE

The movement of water between the atmosphere, the land, and the oceans is called the *water cycle*. The sun is the main source of energy for the water cycle. The sun's energy heats Earth's surface. This causes liquid water to *evaporate*, or change into water vapor (a gas). When the water vapor cools, it may change back into a liquid and form clouds. This is called **condensation**. The liquid water may fall as rain, snow, sleet, or hail on the land.

STUDY TIP

Outline Before you read, make an outline of this section using the questions in bold. As you read, fill in the main ideas of the chapter in your outline.

☑ **READING CHECK**

1. Define Write your own definition for *weather*.

_____

_____

_____

_____

STANDARDS CHECK

ES 1i Clouds, formed by the condensation of water vapor, <u>affect</u> weather and climate.

Word Help: <u>affect</u> to change; to act upon

2. Identify By what process do clouds form?

_____

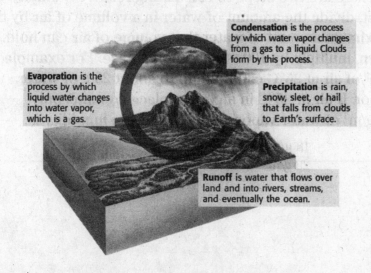

Condensation is the process by which water vapor changes from a gas to a liquid. Clouds form by this process.

Evaporation is the process by which liquid water changes into water vapor, which is a gas.

Precipitation is rain, snow, sleet, or hail that falls from clouds to Earth's surface.

Runoff is water that flows over land and into rivers, streams, and eventually the ocean.

What Is Humidity?

Water vapor makes up only a small fraction of the mass of the atmosphere. However, this small amount of water vapor has an important effect on weather and climate.

When the sun's energy heats up Earth's surface, water in oceans and water bodies evaporates. The amount of water vapor in the air is called **humidity**. Warmer air can hold more water vapor than cooler air can. ☑

READING CHECK

3. Identify How does air temperature affect how much water vapor the air can hold?

_____

_____

Amount of Water Vapor That Air Can Hold at Various Temperatures

Math Focus

4. Read a Graph How much water vapor can air at 30°C hold?

_____

RELATIVE HUMIDITY

Scientists often describe the amount of water in the air using relative humidity. **Relative humidity** is the ratio of the amount of water vapor in the air to the greatest amount the air can hold.

There are two steps to calculating relative humidity. First, divide the amount of water in a volume of air by the maximum amount of water that volume of air can hold. Then, multiply by 100 to get a percentage. For example, 1 m^3 of air at 25°C can hold up to about 23 g of water vapor. If air at 25°C in a certain place contains only 18 g/m^3 of water vapor, then the relative humidity is:

Math Focus

5. Calculate What is the relative humidity of 25°C air that contains 10 g/m^3 of water vapor? Show your work.

_____

_____

_____

$$\frac{18 \text{ g/m}^3}{23 \text{ g/m}^3} \times 100 = 78\% \text{ relative humidity}$$

SECTION 1 Water in the Air *continued*

FACTORS AFFECTING RELATIVE HUMIDITY

Temperature and humidity can affect relative humidity. As humidity increases, relative humidity increases if the temperature stays the same. Relative humidity decreases as temperature rises and increases as temperature drops if the humidity stays the same.

MEASURING RELATIVE HUMIDITY

Scientists measure relative humidity using special tools. One of these tools is called a *psychrometer*. A psychrometer contains two thermometers. The bulb of one thermometer is covered with a wet cloth. This is called a *wet-bulb thermometer*. The other thermometer bulb is dry. This thermometer is a *dry-bulb thermometer*.

You are probably most familiar with dry-bulb thermometers. Wet-bulb thermometers work differently than dry-bulb thermometers. As air passes through the cloth on a wet-bulb thermometer, some of the water in the cloth evaporates. As the water evaporates, the cloth cools. The wet-bulb thermometer shows the temperature of the cloth.

If humidity is low, the water evaporates more quickly. Therefore, the temperature reading on the wet-bulb thermometer is much lower than the reading on the dry-bulb thermometer. If the humidity is high, less water evaporates. Therefore, the temperature changes very little.

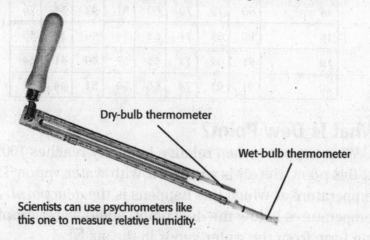

Dry-bulb thermometer

Wet-bulb thermometer

Scientists can use psychrometers like this one to measure relative humidity.

The difference in temperature readings between the dry-bulb and wet-bulb thermometers is a measure of the relative humidity. The larger the difference between the readings, the lower the relative humidity.

Critical Thinking

6. Compare How is relative humidity different from humidity?

_____

_____

_____

_____

_____

TAKE A LOOK
7. Identify What are two parts of a psychrometer?

_____

_____

USING A RELATIVE-HUMIDITY TABLE

Scientists use tables like the one below to determine relative humidity. Use the table to work through the following example.

The dry-bulb thermometer on a psychrometer reads 10°C. The wet-bulb thermometer reads 7°C. Therefore, the difference between the thermometer readings is 3°C. In the first column of the table, find the row head for 10°C, the dry-bulb reading. Then, find the column head for 3°C, the difference between the readings. Find the place where the row and column meet. The number in the table at this point is 66, so the relative humidity is 66%.

Relative Humidity (%)								
Dry-bulb reading (°C)	Difference between wet-bulb reading and dry-bulb reading (°C)							
	1	2	3	4	5	6	7	8
0	81	64	46	29	13			
2	84	68	52	37	22	7		
4	85	71	57	43	29	16		
6	86	73	60	48	35	24	11	
8	87	75	63	51	40	29	19	8
10	88	77	66	55	44	34	24	15
12	59	78	68	58	48	39	29	21
14	90	79	70	60	51	42	34	26
16	90	81	71	63	54	46	38	30
18	91	82	73	65	57	49	41	34
20	91	83	74	66	59	51	44	37

TAKE A LOOK

8. Apply Concepts The dry-bulb reading on a psychrometer is 8°C. The wet-bulb reading is 7°C. What is the relative humidity?

What Is Dew Point?

What happens when relative humidity reaches 100%? At this point, the air is *saturated* with water vapor. The temperature at which this happens is the *dew point*. At temperatures below the dew point, liquid water droplets can form from the water vapor in the air. ☑

Condensation happens when air is saturated with water vapor. Air can become saturated if water evaporates and enters the air as water vapor. Air can also become saturated when it cools below its dew point.

READING CHECK

9. Explain What happens when the temperature of air is below its dew point?

SECTION 1 Water in the Air *continued*

AN EVERYDAY EXAMPLE

You have probably seen air become saturated because of a temperature decrease. For example, when you add ice cubes to a glass of juice, the temperatures of the juice and the glass decrease. The glass absorbs heat from the air, so the temperature of the air near the glass decreases. When the air's temperature drops below its dew point, water vapor condenses on the glass. The condensed water forms droplets on the glass.

The glass absorbs heat from the air. The air cools to below its dew point. Water vapor condenses onto the side of the glass.

How Do Clouds Form?

A **cloud** is a group of millions of tiny water droplets or ice crystals. Clouds form as air rises and cools. When air cools below the dew point, water droplets or ice crystals form. Water droplets form when water condenses above 0°C. Ice crystals form when water condenses below 0°C.

DIFFERENT KINDS OF CLOUDS

Scientists classify clouds by shape and altitude. The three main cloud shapes are stratus clouds, cumulus clouds, and cirrus clouds. The three altitude groups are low clouds, middle clouds, and high clouds. The figure on the next page shows these different cloud types. ☑

Critical Thinking

10. Apply Concepts People who wear glasses may notice that their glasses fog up when they come indoors on a cold day. Why does this happen?

_____

_____

_____

_____

_____

_____

TAKE A LOOK

11. Describe Where did the liquid water on the outside of the glass come from?

_____

_____

_____

✓ **READING CHECK**

12. Explain How are clouds classified?

_____

_____

Name _____ Class _____ Date _____

SECTION 1 **Water in the Air** *continued*

 **Say It**

Observe and Describe Look at the clouds every day for a week. Each day, write down the weather and what the clouds looked like. At the end of the week, share your observations with a small group. How was the weather related to the kinds of clouds you saw each day?

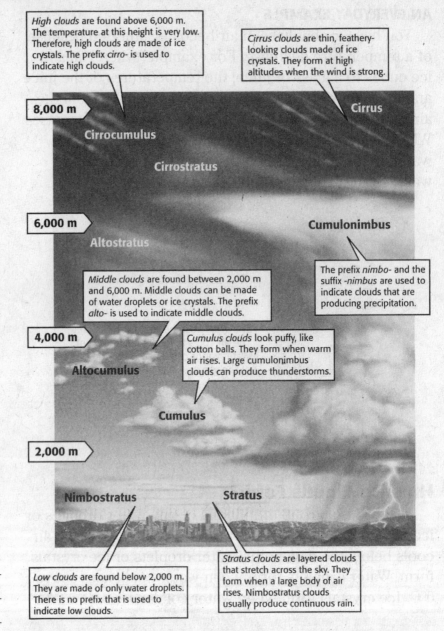

High clouds are found above 6,000 m. The temperature at this height is very low. Therefore, high clouds are made of ice crystals. The prefix *cirro-* is used to indicate high clouds.

Cirrus clouds are thin, feathery-looking clouds made of ice crystals. They form at high altitudes when the wind is strong.

8,000 m

Cirrocumulus

Cirrus

Cirrostratus

6,000 m

Altostratus

Cumulonimbus

Middle clouds are found between 2,000 m and 6,000 m. Middle clouds can be made of water droplets or ice crystals. The prefix *alto-* is used to indicate middle clouds.

The prefix *nimbo-* and the suffix *-nimbus* are used to indicate clouds that are producing precipitation.

4,000 m

Cumulus clouds look puffy, like cotton balls. They form when warm air rises. Large cumulonimbus clouds can produce thunderstorms.

Altocumulus

Cumulus

2,000 m

Nimbostratus

Stratus

Stratus clouds are layered clouds that stretch across the sky. They form when a large body of air rises. Nimbostratus clouds usually produce continuous rain.

Low clouds are found below 2,000 m. They are made of only water droplets. There is no prefix that is used to indicate low clouds.

TAKE A LOOK

13. Compare How is a nimbostratus cloud different from a stratus cloud?

_____

_____

_____

What Is Precipitation?

Water in the air can return to Earth's surface through precipitation. **Precipitation** is solid or liquid water that falls to Earth's surface from clouds. There are four main kinds of precipitation: rain, snow, sleet, and hail. Rain and snow are the most common kinds of precipitation. Sleet and hail are less common. ☑

 **READING CHECK**

14. Define What is precipitation?

_____

_____

SECTION 1 Water in the Air *continued*

RAIN

Water droplets in clouds are very tiny. Each droplet is smaller than the period at the end of this sentence. These tiny droplets can combine with each other. As the droplets combine, they become larger. When a droplet reaches a certain size, it can fall to Earth's surface as *rain*. ☑

SLEET

Sleet forms when rain falls through a layer of very cold air. If the air is cold enough, the rain freezes in the air and becomes falling ice. Sleet can make roads very slippery. When it lands on objects, sleet can coat the objects in ice.

SNOW

Snow forms when temperatures are so low that water vapor turns directly into a solid. That is, the water vapor in the cloud turns into an ice crystal without becoming a liquid first. Snow can fall as single ice crystals. In many cases, the crystals join together to form larger snowflakes. ☑

HAIL

Balls or lumps of ice that fall from clouds are called *hail*. Hail forms in cumulonimbus clouds. Hail can become very large. Hail grows larger in a cycle, as shown in the chart below.

☑ **READING CHECK**

15. Explain What happens to water droplets in clouds when they combine?

_____

_____

☑ **READING CHECK**

16. Identify What is a snowflake?

_____

_____

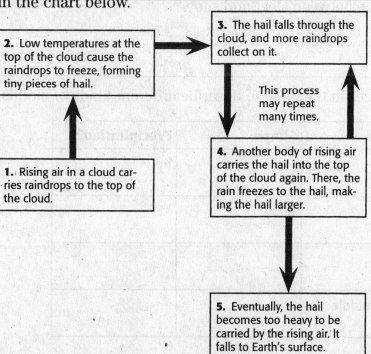

2. Low temperatures at the top of the cloud cause the raindrops to freeze, forming tiny pieces of hail.

3. The hail falls through the cloud, and more raindrops collect on it.

This process may repeat many times.

1. Rising air in a cloud carries raindrops to the top of the cloud.

4. Another body of rising air carries the hail into the top of the cloud again. There, the rain freezes to the hail, making the hail larger.

5. Eventually, the hail becomes too heavy to be carried by the rising air. It falls to Earth's surface.

TAKE A LOOK

17. Identify When does hail fall to the ground?

_____

_____

_____

Section 1 Review

SECTION VOCABULARY

cloud a collection of small water droplets or ice crystals suspended in the air, which forms when the air is cooled and condensation occurs	**precipitation** any form of water that falls to Earth's surface from the clouds
condensation the change of state from a gas to a liquid	**relative humidity** the ratio of the amount of water vapor in the air to the amount of water vapor needed to reach saturation at a given temperature
humidity the amount of water vapor in the air	**weather** the short-term state of the atmosphere, including temperature, humidity, precipitation, wind, and visibility

1. Identify Relationships How is dew point related to condensation?

_____

_____

2. Identify What is the main source of energy for the water cycle?

_____

3. Explain How do clouds form?

_____

_____

_____

4. Compare What is the difference between sleet and snow?

_____

_____

5. Apply Concepts Fill in the spaces in the table to describe different kinds of clouds.

Name	Altitude	Shape	Precipitation?
Cirrostratus	high		no
Altocumulus		puffy	
Nimbostratus			
Cumulonimbus	low to middle		

CHAPTER 6 Understanding Weather

SECTION 2 Air Masses and Fronts

National Science
Education Standards
ES 1j

BEFORE YOU READ

After you read this section, you should be able to answer these questions:

- How is an air mass different from a front?
- How do fronts affect weather?

What Are Air Masses?

Have you ever been caught outside when it suddenly started to rain? What causes such an abrupt change in the weather? Changes in weather are caused by the movement of bodies of air called air masses. An **air mass** is a very large volume of air that has a certain temperature and moisture content.

There are many types of air masses. Scientists classify air masses by the water content and temperature of the air. These features depend on where the air mass forms. The area over which an air mass forms is called a *source region*. One source region is the Gulf of Mexico. Air masses that form over this source region are wet and warm. ☑

Each type of air mass forms over a certain source region. On maps, meteorologists use two-letter symbols to represent different air masses. The first letter indicates the water content of the air mass. The second letter indicates its temperature. The figure below shows the main air masses that affect North America.

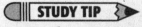

STUDY TIP
Summarize As you read, make a chart comparing the four kinds of fronts. In your chart, describe how each kind of front forms and what kind of weather it can cause.

READING CHECK
1. **Identify** How do scientists classify air masses?

_____

_____

mP   cP   mP

mT   cT   mT

maritime (m) forms over water; wet

continental (c) forms over land; dry

polar (P) forms over the polar regions; cold

tropical (T) forms over the Tropics; warm

TAKE A LOOK
2. **Apply Concepts** Describe the temperature and moisture content of a cT air mass.

_____

_____

COLD AIR MASSES

Most of the cold winter weather in the United States comes from three polar air masses. Continental polar (cP) air masses form over northern Canada. They bring extremely cold winter weather. In the summer, cP air masses can bring cool, dry weather. ☑

Maritime polar (mP) air masses form over the North Pacific Ocean. They are cool and very wet. They bring rain and snow to the Pacific Coast in winter. They bring fog in the summer.

Maritime polar air masses also form over the North Atlantic Ocean. They bring cool, cloudy weather and precipitation to New England.

WARM AIR MASSES

Four warm air masses influence the weather in the United States. Maritime tropical (mT) air masses form over warm areas in the Pacific Ocean, the Gulf of Mexico, and the Atlantic Ocean. They move across the East Coast and into the Midwest. In summer they bring heat, humidity, hurricanes, and thunderstorms to these areas.

Continental tropical air masses (cT) form over deserts and move northward. They bring clear, dry, hot weather in the summer.

Air mass	How it affects weather
cP from northern Canada	
mP from the North Pacific Ocean	
mT from the Gulf of Mexico	
cT from the deserts	

What Are Fronts?

The place where two or more air masses meet is called a **front**. When air masses meet, the less dense air mass rises over the denser air mass. Warm air is less dense than cold air. Therefore, a warm air mass will generally rise above a cold air mass. There are four main kinds of fronts: cold fronts, warm fronts, occluded fronts, and stationary fronts. ☑

READING CHECK
3. Identify What is the source region for cP air masses?

Critical Thinking
4. Infer Why don't warm air masses form over the North Atlantic or Pacific oceans?

TAKE A LOOK
5. Identify Fill in the blank spaces in the table.

READING CHECK
6. Define What is a front?

COLD FRONTS

A *cold front* forms when a cold air mass moves under a warm air mass. The cold air pushes the warm air mass up. The cold air mass replaces the warm air mass. Cold fronts can move quickly and bring heavy precipitation. When a cold front has passed, the weather is usually cooler. This is because a cold, dry air mass moves in behind the cold front.

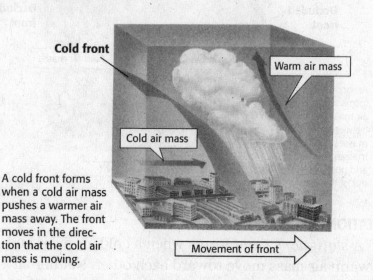

Cold front

Warm air mass

Cold air mass

A cold front forms when a cold air mass pushes a warmer air mass away. The front moves in the direction that the cold air mass is moving.

Movement of front

TAKE A LOOK
7. Describe What happens to the warm air mass at a cold front?

_____

_____

WARM FRONTS

A *warm front* forms when a warm air mass moves in over a cold air mass that is leaving an area. The warm air replaces the cold air as the cold air moves away. Warm fronts can bring light rain. They are followed by clear, warm weather. ☑

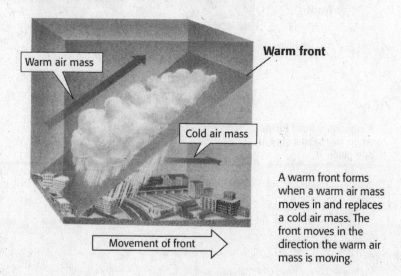

Warm air mass

Warm front

Cold air mass

A warm front forms when a warm air mass moves in and replaces a cold air mass. The front moves in the direction the warm air mass is moving.

Movement of front

READING CHECK
8. Define What is a warm front?

_____

_____

_____

OCCLUDED FRONTS

An *occluded front* forms when a warm air mass is caught between two cold air masses. Occluded fronts bring cool temperatures and large amounts of rain and snow.

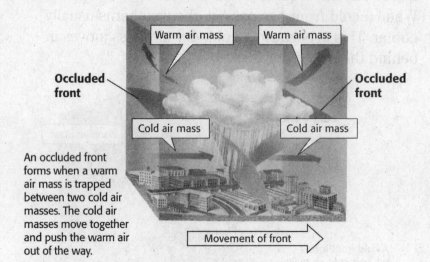

An occluded front forms when a warm air mass is trapped between two cold air masses. The cold air masses move together and push the warm air out of the way.

TAKE A LOOK
9. Describe What happens to the warm air mass in an occluded front?

_____

_____

_____

STATIONARY FRONT

A *stationary front* forms when a cold air mass and a warm air mass move toward each other. Neither air mass has enough energy to push the other out of the way. Therefore, the two air masses remain in the same place. Stationary fronts cause many days of cloudy, wet weather.

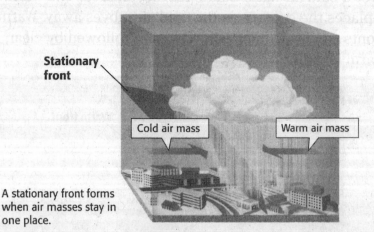

A stationary front forms when air masses stay in one place.

TAKE A LOOK
10. Infer What do you think is the reason that stationary fronts bring many days of the same weather?

_____

_____

_____

_____

How Does Air Pressure Affect Weather?

Remember that air produces pressure. However, air pressure is not always the same everywhere. Areas with different pressures can cause changes in the weather. These areas may have lower or higher air pressure than their surroundings.

A **cyclone** is an area of the atmosphere that has lower pressure than the surrounding air. The air in the cyclone rises. As the air rises, it cools. Clouds can form and may cause rainy or stormy weather.

An **anticyclone** is an area of the atmosphere that has higher pressure than the surrounding air. Air in anticyclones sinks and gets warmer. Its relative humidity decreases. This warm, sinking air can bring dry, clear weather.

Cyclones and anticyclones can affect each other. Air moving out from the center of an anticyclone moves toward areas of low pressure. This movement can form a cyclone. The figure below shows how cyclones and anticyclones can affect each other.

Critical Thinking

11. Compare Give two differences between cyclones and anticyclones.

_____

_____

_____

_____

_____

_____

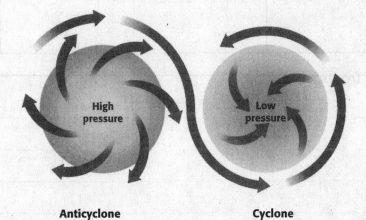

High pressure

Low pressure

Anticyclone   **Cyclone**

TAKE A LOOK
12. Identify In which direction does air move: from a cyclone to an anticyclone, or from an anticyclone to a cyclone?

_____

Section 2 Review

NSES   NSES ES 1j

SECTION VOCABULARY

air mass a large body of air throughout which temperature and moisture content are similar **anticyclone** the rotation of air around a high pressure center in the direction opposite to Earth's rotation.	**cyclone** an area in the atmosphere that has lower pressure than the surrounding areas and has winds that spiral toward the center **front** the boundary between air masses of different densities and usually different temperatures

1. Identify Relationships How are fronts and air masses related?

_____

_____

2. Compare Fill in the table to describe cyclones and anticyclones.

Name	Compared to surrounding air pressure, the pressure in the middle is...	What does the air inside it do?	What kind of weather does it cause?
cyclone	...lower than surrounding pressure.		
anticyclone		sinks and warms	

3. List What are four kinds of fronts?

_____

_____

_____

4. Identify What are the source regions for the mT air masses that affect weather in the United States?

_____

_____

_____

5. Describe What kind of air mass causes hot, clear, dry summer weather in the United States?

_____

_____

CHAPTER 6 | Understanding Weather
SECTION 3 Severe Weather

National Science
Education Standards
ES 1i, 1j

BEFORE YOU READ

After you read this section, you should be able to answer these questions:

• What are some types of severe weather?

• How can you stay safe during severe weather?

What Causes Thunderstorms?

A **thunderstorm** is an intense storm with strong winds, heavy rain, lightning, and thunder. Many thunderstorms happen along cold fronts. However, thunderstorms can also happen in other areas. Two conditions are necessary for a thunderstorm to form: warm, moist air near Earth's surface and an unstable area of the atmosphere.

The atmosphere is unstable when a body of cold air is found above a body of warm air. The warm air rises and cools as it mixes with the cool air. When the warm air reaches its dew point, the water vapor condenses and forms cumulus clouds. If the warm air keeps rising, the clouds may become dark cumulonimbus clouds.

LIGHTNING

As a cloud grows bigger, parts of it begin to develop electrical charges. The upper parts of the cloud tend to become positively charged. The lower parts tend to become negatively charged. When the charges get big enough, electricity flows from one area to the other. Electricity may also flow between the clouds and the ground. These electrical currents are **lightning**. ☑

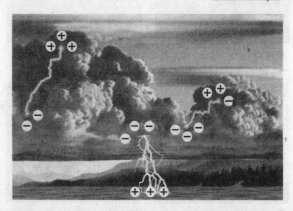

Different parts of thunderclouds and the ground can have different electrical charges. When electricity flows between these areas, lightning forms.

STUDY TIP

Describe After you read this section, make a flow chart showing how a tornado forms.

Critical Thinking

1. Infer Why does air near the surface have to be moist in order for a thunderstorm to form?

_____

_____

_____

READING CHECK

2. Describe How does lightning form?

_____

_____

_____

_____

THUNDER

You have probably seen large lightning bolts that travel between the clouds and the ground. When lightning moves through the air, the air gets very hot. The hot air expands rapidly. As it expands, it makes the air vibrate. The vibrations release energy in the form of sound waves. The result is **thunder**. ☑

✓ **READING CHECK**

3. Define What is thunder?

_____
_____
_____
_____

SEVERE THUNDERSTORMS

Severe thunderstorms can cause a lot of damage. They can produce strong winds, hail, flash floods, or tornadoes. Hail can damage crops, cars, and windows. Flash flooding from heavy rain can cause serious property damage. Flash flooding is the leading cause of weather-related deaths. Lightning can start fires and cause injuries and deaths.

How Do Tornadoes Form?

Fewer than 1% of thunderstorms produce tornadoes. A **tornado** can form when a rapidly spinning column of air, called a *funnel cloud*, touches the ground. The air in the center of a tornado has low pressure. When the area of low pressure touches the ground, material from the ground can be sucked up into the tornado. ☑

✓ **READING CHECK**

4. Explain Why can material be sucked up into a tornado?

_____
_____

A tornado begins as a funnel cloud that pokes through the bottom of a cumulonimbus cloud. The funnel cloud becomes a tornado when the funnel cloud touches the ground. The pictures below show how a tornado forms.

❶ Wind moving in opposite directions causes a layer of air in the middle of a cloud to begin to spin.

❷ Strong vertical winds cause the spinning column of air to turn into a vertical position.

❸ The spinning column of air moves to the bottom of the cloud and forms a funnel cloud.

❹ The funnel cloud becomes a tornado when it touches down on the ground.

TAKE A LOOK

5. Describe When does a funnel cloud become a tornado?

_____
_____

TORNADO FACTS

About 75% of the world's tornadoes happen in the United States. Most happen in the spring and early summer. During these times, cold, dry air from Canada meets warm, moist air from the Tropics. This causes the thunderstorms that produce tornadoes.

Most tornadoes last for only a few minutes. However, their strong, spinning winds can cause a lot of damage. An average tornado has wind speeds between 120 km/h and 180 km/h, but some can be much higher. Winds from tornadoes can tear up trees and destroy buildings. They can even be strong enough to lift cars and trailers up into the air. The area damaged by a tornado is usually about 8 km long and 10 to 60 m wide.

How Do Hurricanes Form?

A **hurricane** is a large, rotating tropical weather system. Hurricanes have wind speeds of over 120 km/h. They can be 160 km to 1,500 km in diameter and can travel for thousands of miles. They are the most powerful storms on Earth. Hurricanes are also called typhoons and cyclones.

Most hurricanes form between 5°N and 20°N latitude or between 5°S and 20°S latitude. They form over the warm, tropical oceans found at these latitudes. At higher latitudes, the water is too cold for hurricanes to form. ☑

Hurricanes can be so large that they are visible from space. This photograph of a hurricane was taken by a satellite.

Math Focus

6. Convert What is the average wind speed in a tornado in miles per hour?

1 km = 0.62 mi.

_____

✓ **READING CHECK**

7. Explain Why don't hurricanes form at high latitudes?

_____

_____

SECTION 3 Severe Weather *continued*

HOW HURRICANES FORM

A hurricane begins as a group of thunderstorms traveling over tropical ocean waters. Winds traveling in two different directions meet and cause the storm to spin. Because of the Coriolis effect, hurricanes rotate counterclockwise in the Northern Hemisphere and clockwise in the Southern Hemisphere. ☑

Hurricanes are powered by solar energy. The sun's energy causes ocean water to evaporate. As the water vapor rises in the air, it cools and condenses. A group of thunderstorms form and produce a large, spinning storm. A hurricane forms as the storm gets stronger.

✓ **READING CHECK**

8. Explain What causes hurricanes to rotate in different directions in the Northern and Southern Hemispheres?

_____

_____

TAKE A LOOK

9. Define What is the eye of a hurricane?

_____

_____

_____

_____

At the center of the hurricane is the eye. The *eye* is a core of warm, relatively calm air with low pressure and light winds. There are updrafts and downdrafts in the eye. An *updraft* is a current of rising air. A *downdraft* is a current of sinking air.

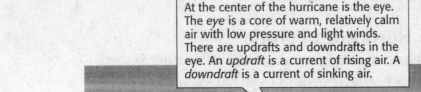

Around the eye is a group of cumulonimbus clouds called the *eye wall*. These clouds produce heavy rain and strong winds. The winds can be up to 300 km/h. The eye wall is the strongest part of the hurricane.

Outside the eye wall are spiraling bands of clouds called *rain bands*. These bands also produce heavy rain and strong wind. They circle the center of the hurricane.

The hurricane will continue to grow as long as it is over warm ocean water. When the hurricane moves over colder waters or over land, the storm loses energy. This is why hurricanes are not common in the middle of continents. The storms lose their energy quickly when they move over land.

SECTION 3 Severe Weather *continued*

DAMAGE CAUSED BY HURRICANES

Hurricanes can cause serious damage when they move near or onto land. The strong winds from hurricanes can knock down trees and telephone poles. They can damage or destroy buildings and homes.

Many people think that the winds are the most damaging part of a hurricane. However, most of the damage from hurricanes is actually caused by flooding from heavy rains and storm surges. A *storm surge* is a rise in sea level that happens during a storm. A storm surge from a hurricane can be up to 8 m high. The storm-surge flooding from Hurricane Katrina in 2005 caused more damage than the high-speed winds from the storm. ☑

How Can You Stay Safe During Severe Weather?

Severe weather can be very dangerous. During severe weather, it is important for you to listen to a local TV or radio station. Severe-weather announcements will tell you where a storm is and if it is getting worse. Weather forecasters use watches and warnings to let people know about some kinds of severe weather. A *watch* means that severe weather may happen. A *warning* means that severe weather is happening somewhere nearby.

The table below gives ways to stay safe during different kinds of severe weather. ☑

Severe weather	How to stay safe
Thunderstorms	If you are outside, stay away from tall objects that can attract lightning. If you are in an open area, crouch down. Stay away from water. If you are inside, stay away from windows.
Tornadoes	During a tornado warning, find shelter quickly in a basement or cellar. If you cannot get to a basement, go to a windowless room in the center of the building (such as a closet or bathroom). If you are outside, lie down in an open field or a deep ditch.
Floods	Find a high place to wait out the flood. Always stay out of floodwaters.
Hurricanes	Protect the windows in your home by covering them with wood. Stay inside during the storm. If you are told to leave your home, do so quickly and calmly.

READING CHECK

10. Define What is a storm surge?

_____

_____

READING CHECK

11. Explain Why should you listen to weather reports during severe weather?

_____

_____

Section 3 Review

SECTION VOCABULARY

hurricane a severe storm that develops over tropical oceans and whose strong winds of more than 120 km/h spiral in toward the intensely low-pressure storm center **lightning** an electric discharge that takes place between two oppositely charged surfaces, such as between a cloud and the ground, between two clouds, or between two parts of the same cloud	**thunder** the sound caused by the rapid expansion of air along an electrical strike **thunderstorm** a usually brief, heavy storm that consists of rain, strong winds, lightning, and thunder **tornado** a destructive, rotating column of air that has very high wind speeds and that may be visible as a funnel-shaped cloud

1. Explain Why do thunder and lightning usually happen together?

_____

_____

2. Identify How can severe thunderstorms cause damage?

_____

_____

3. Identify Where do most tornadoes happen?

_____

4. Explain Why do most tornadoes happen in the spring and early summer?

_____

_____

5. Analyze How does energy from the sun power hurricanes?

_____

_____

_____

_____

6. Describe When do hurricanes lose energy?

_____

7. Identify Give three ways to stay safe if you are caught outside in a thunderstorm.

_____

_____

_____

CHAPTER 6 | Understanding Weather

SECTION 4 **Forecasting Weather**

National Science
Education Standards
ES 1i, 1j

BEFORE YOU READ

After you read this section, you should be able to answer these questions:

• What instruments are used to forecast weather?

• How do you read a weather map?

What Is a Weather Forecast?

Weather affects how you dress and how you plan your day. Severe weather can put people in danger. Therefore, accurate weather forecasts are important. A *weather forecast* is a prediction of weather conditions over the next few days. Meteorologists make weather forecasts using information on atmospheric conditions. ☑

Meteorologists use special instruments to collect data. Some of these instruments are far above the ground. Others are tools you may be familiar with from everyday use.

WEATHER BALLOONS

Weather balloons carry electronic equipment. The equipment on a weather balloon can measure weather conditions as high as 30 km above Earth's surface. This equipment measures temperature, air pressure, and relative humidity. It transmits the information to meteorologists using radio signals. Meteorologists can track the path of the balloons to measure wind speed and direction.

STUDY TIP

Compare As you read this section, make a chart comparing the different tools that meteorologists use to collect weather data.

READING CHECK

1. Explain What do meteorologists use to forecast the weather?

_____

_____

Weather balloons carry equipment into the atmosphere. They use radio signals to transmit information on weather conditions to meteorologists on the ground.

TAKE A LOOK

2. Describe How do meteorologists obtain the information from weather balloons?

_____

_____

_____

SECTION 4 Forecasting Weather *continued*

THERMOMETERS AND BAROMETERS

Remember that air temperature and pressure can affect the weather. Therefore, meteorologists must be able to measure temperature and pressure accurately. They use **thermometers** to measure temperature, just like you do. They use tools called **barometers** to measure air pressure.

WINDSOCKS, WIND VANES, AND ANEMOMETERS

Meteorologists can use windsocks and wind vanes to measure wind direction. A *windsock* is a cone-shaped cloth bag that is open at both ends. The wind enters through the wide end and leaves through the narrow end. The wide end always points into the wind.

A *wind vane* is shaped like an arrow. It is attached to a pole. The wind pushes the tail of the arrow. The vane spins until the arrows points into the wind.

An **anemometer** measures wind speed. It has three or four cups connected to a pole with spokes. The wind pushes on the open sides of the cups. This makes them spin on the pole. The spinning of the pole produces an electric current, which is displayed on a dial. The faster the wind speed, the stronger the electric current, and the further the dial moves.

Critical Thinking

3. Infer Why is it important for meteorologists to be able to measure wind direction?

_____

_____

_____

TAKE A LOOK
4. Identify What is an anemometer?

_____

_____

Meteorologists use anemometers to measure wind speed.

RADAR AND SATELLITES

Scientists use *radar* to locate fronts and air masses. Radar can locate a weather system and show the direction it is moving. It can show how much precipitation is falling, and what kind of precipitation it is. Most television stations use radar to give information about weather systems. ☑

Weather satellites orbiting Earth produce images of weather systems. Satellites can also measure wind speeds, humidity, and temperatures from different altitudes. Meteorologists use weather satellites to track storms.

✓ **READING CHECK**
5. Describe Give two things that meteorologists can use radar to do.

_____

_____

SECTION 4 Forecasting Weather *continued*

What Are Weather Maps?

In the United States, two main groups of scientists collect weather data. One group is the National Weather Service (NWS). The other group is the National Oceanic and Atmospheric Administration (NOAA). These groups gather information from about 1,000 weather stations across the United States to produce weather maps. ☑

READING A WEATHER MAP

Some weather maps contain station models. A *station model* is a symbol that shows the weather at a certain location. Station models look like circles with numbers and symbols around them. The numbers and symbols stand for different measurements, as shown below.

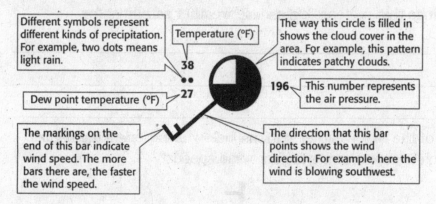

Different symbols represent different kinds of precipitation. For example, two dots means light rain.

Temperature (°F)
38

Dew point temperature (°F)
27

The way this circle is filled in shows the cloud cover in the area. For example, this pattern indicates patchy clouds.

196 — This number represents the air pressure.

The markings on the end of this bar indicate wind speed. The more bars there are, the faster the wind speed.

The direction that this bar points shows the wind direction. For example, here the wind is blowing southwest.

Some weather maps, such as those you see on television, show lines called isobars. *Isobars* are lines that connect points of equal air pressure. They are similar to contour lines on a topographic map. Isobars that form closed circles represent areas of high (H) or low (L) pressure. Weather maps also show fronts.

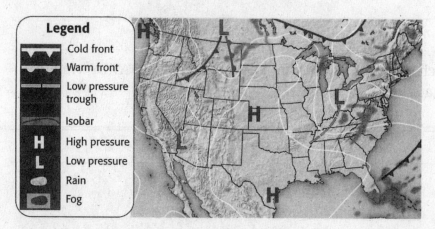

Legend

Cold front
Warm front
Low pressure trough
Isobar
H High pressure
L Low pressure
Rain
Fog

✓ **READING CHECK**

6. Identify What are two groups that collect weather data in the United States?

_____

_____

TAKE A LOOK

7. Use a Model What is the dew point temperature for the station shown in the figure?

_____

8. Infer Will condensation happen in the air at the station in the figure? Explain your answer.

_____

_____

_____

_____

TAKE A LOOK

9. Read a Map On the map, circle the areas of high pressure.

Section 4 Review

SECTION VOCABULARY

anemometer an instrument used to measure wind speed	**thermometer** an instrument that measures and indicates temperature
barometer an instrument that measures atmospheric pressure	

1. Compare How is an anemometer different from a windsock or a wind vane?

_____

2. Identify What three atmospheric conditions do weather balloons measure?

_____

_____

_____

3. Describe Give three things that meteorologists use weather satellites for.

_____

_____

_____

4. Apply Concepts Which of the two weather stations below is experiencing higher air temperatures? Which is experiencing higher wind speeds?

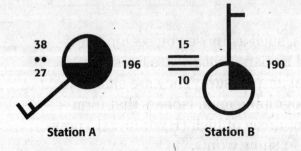

Station A   Station B

_____

_____

5. Apply Concepts In which direction is the wind blowing at station A? In which direction is it blowing at station B?

_____

_____

CHAPTER 7 Climate
SECTION 1 **What Is Climate?**

National Science Education Standards
ES 1f, 1j, 3d

BEFORE YOU READ

After you read this section, you should be able to answer these questions:
- What is climate?
- What factors affect climate?
- How do climates differ around the world?

What Is Climate?

How is weather different from climate? **Weather** is the condition of the atmosphere at a certain time. The weather can change from day to day. In contrast, **climate** describes the average weather conditions in a region over a long period of time. The climate of an area includes the area's average temperature and amount of precipitation. Different parts of the world have different climates.

What Factors Affect Climate?

Climate is mainly determined by temperature and precipitation. Many factors affect temperature and precipitation, including latitude, wind patterns, landforms, and ocean currents. ☑

SOLAR ENERGY AND LATITUDE

Remember that the **latitude** of an area is its distance north or south of the equator. In general, the temperature of an area depends on its latitude. Latitudes closer to the poles tend to have colder climates. Latitude affects temperature because latitude determines how much direct solar energy an area gets, as shown in the figure below.

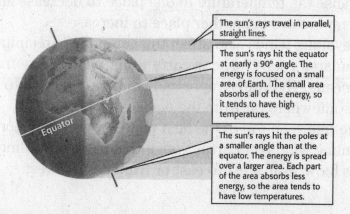

The sun's rays travel in parallel, straight lines.

The sun's rays hit the equator at nearly a 90° angle. The energy is focused on a small area of Earth. The small area absorbs all of the energy, so it tends to have high temperatures.

The sun's rays hit the poles at a smaller angle than at the equator. The energy is spread over a larger area. Each part of the area absorbs less energy, so the area tends to have low temperatures.

STUDY TIP

Ask Questions As you read this section, write down any questions that you have. When you finish reading, talk about your questions in a small group.

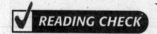

READING CHECK

1. List What are the two main things that determine climate?

_____

_____

TAKE A LOOK

2. Explain Why do areas near the equator tend to have high temperatures?

_____

_____

SECTION 1 **What Is Climate?** *continued*

STANDARDS CHECK

ES 3d The sun is the <u>major</u> source of <u>energy</u> for <u>phenomena</u> on the earth's surface, such as growth of plants, winds, ocean currents, and the water cycle. Seasons result from variations in the amount of the sun's <u>energy</u> hitting the surface, due to the tilt of the earth's rotation on its axis and the length of the day.

Word Help: <u>major</u>
of great importance or large scale

Word Help: <u>energy</u>
the ability to make things happen

Word Help: <u>phenomenon</u>
any fact or event that can be sensed or described scientifically (plural, *phenomena*)

3. Explain Why don't areas near the equator have large seasonal changes in weather?

_____

_____

_____

☑ **READING CHECK**

4. Identify What causes wind to form?

_____

_____

LATITUDE AND SEASONS

Most places in the United States have four seasons during the year. However, some places in the world do not have such large seasonal changes. For example, places near the equator have about the same temperatures and amounts of daylight all year.

Seasons happen because Earth is tilted on its axis by about 23.5°. This tilt affects how much solar energy an area gets as Earth orbits the sun. The figure below shows how Earth's tilt affects the seasons.

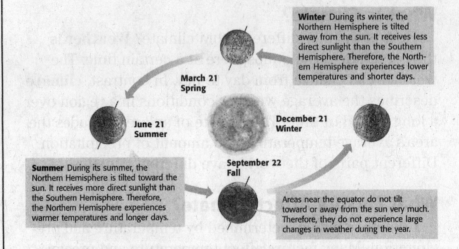

March 21 Spring

June 21 Summer

December 21 Winter

September 22 Fall

Winter During its winter, the Northern Hemisphere is tilted away from the sun. It receives less direct sunlight than the Southern Hemisphere. Therefore, the Northern Hemisphere experiences lower temperatures and shorter days.

Summer During its summer, the Northern Hemisphere is tilted toward the sun. It receives more direct sunlight than the Southern Hemisphere. Therefore, the Northern Hemisphere experiences warmer temperatures and longer days.

Areas near the equator do not tilt toward or away from the sun very much. Therefore, they do not experience large changes in weather during the year.

PREVAILING WINDS

Prevailing winds are winds that blow mainly in one direction. The wind patterns on Earth are caused by the uneven heating of Earth's surface. This uneven heating forms areas with different air pressures. *Wind* forms when air moves from areas of high pressure to areas of low pressure. ☑

Prevailing winds affect climate and weather because they move solar energy from one place to another. This can cause the temperature in one place to decrease and the temperature in another place to increase.

Prevailing winds also affect the amount of precipitation an area gets. They can carry water vapor away from the oceans. The water vapor can condense and fall to the land somewhere far from the ocean.

The figure on top of the next page shows the major prevailing winds on Earth. Notice that most prevailing winds blow from west to east or from east to west.

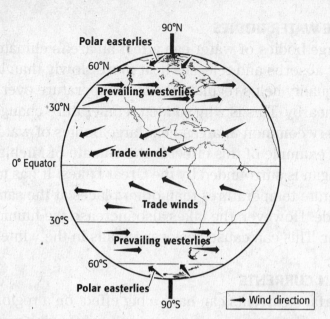

TOPOGRAPHY

The sizes and shapes of the land-surface features of a region form its *topography*. The topography of an area affects its climate because topography can affect temperature and precipitation. For example, elevation is a feature of topography that can have a large impact on temperature. **Elevation** is the height of an area above sea level. As elevation increases, temperature tends to decrease. ☑

Mountains can also affect precipitation. As air rises to move over a mountain, it cools. The cool air condenses, forming clouds. Precipitation may fall. This process causes the *rain-shadow effect*, which is illustrated in the figure below.

Air rises to flow over mountains. The air cools as it rises, and water vapor can condense to form clouds. The clouds can release the water as precipitation. Therefore, this side of the mountain tends to be wetter, with more vegetation.

The air on this side of the mountain contains much less water vapor. As the air sinks down the side of the mountain, it becomes warmer. The warm air absorbs moisture from the land. Therefore, this side of the mountain tends to be drier and more desert-like.

TAKE A LOOK
5. Read a Map In which direction do the Prevailing Westerlies blow?

_____

_____

✓ **READING CHECK**
6. Describe In general, how does elevation affect temperature?

_____

_____

_____

TAKE A LOOK
7. Explain Why do clouds form as air moves over a mountain?

_____

_____

_____

SECTION 1 What Is Climate? *continued*

LARGE WATER BODIES

Large bodies of water can affect an area's climate. Water absorbs and releases heat more slowly than land. This quality helps regulate the air temperature over the land nearby. This is why sudden temperature changes are not very common in areas near large bodies of water. ☑

An example of this effect is the climate of Michigan. Michigan is surrounded by the Great Lakes. It has more-moderate temperatures than other places at the same latitude. However, the lakes also increase the humidity of the air. This can cause heavy snowfalls in the winter.

OCEAN CURRENTS

Surface currents can have a big effect on a region's climate. **Surface currents** are paths of flowing water found near the surface of the ocean. As surface currents move, they carry warm or cool water to different places. The temperature of the water affects the temperature of the air above it. For example, warm currents can heat the surrounding air.

An example of the effects of ocean currents on climate can be seen in Iceland. Iceland is an island near the Arctic Circle. The Gulf Stream, a warm surface current, flows past Iceland. The warm water in the Gulf Stream causes Iceland's climate to be fairly mild. In contrast, the island of Greenland is at a similar latitude but is not affected by the Gulf Stream. Greenland's climate is much colder than Iceland's.

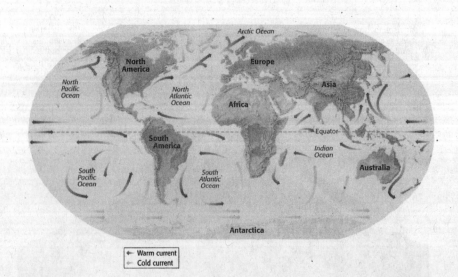

✓ **READING CHECK**

8. Explain Why aren't sudden temperature changes common near large bodies of water?

_____

_____

_____

_____

Critical Thinking

9. Describe Processes Cool surface currents can cause the air above them to become cooler. Explain how this happens.

_____

_____

_____

_____

TAKE A LOOK

10. Identify What kind of surface current is found off the East Coast of the United States?

_____

_____

SECTION 1 What Is Climate? *continued*

What Are the Different Climates Around the World?

Earth has three major climate zones: tropical, temperate, and polar. The figure below shows where these zones are found.

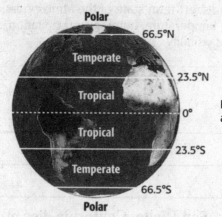

Earth's three major climate zones are determined by latitude.

TAKE A LOOK
11. Identify What determines Earth's major climate zones?

_____

Each climate zone has a certain range of temperatures. The tropical zone, near the equator, has the highest temperatures. The polar zones, located at latitudes above 66.5°, have the lowest temperatures. ☑

READING CHECK
12. Describe Which climate zone has the highest temperatures?

_____

BIOMES

Each climate zone contains several different kinds of climates. The different climates are the result of topography, winds, and ocean currents. The different climates affect the organisms that live in an area. A large area with a certain climate and types of organisms is called a **biome**. ☑

READING CHECK
13. Identify Relationships How are biomes and climate related?

_____

_____

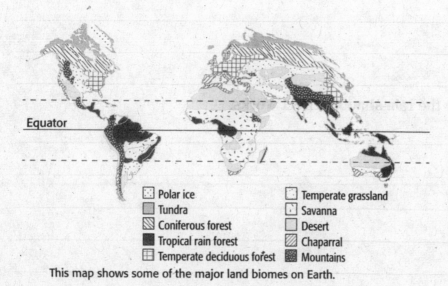

Equator

⬚ Polar ice	⬚ Temperate grassland
⬚ Tundra	⬚ Savanna
⬚ Coniferous forest	⬚ Desert
⬛ Tropical rain forest	⬚ Chaparral
⬚ Temperate deciduous forest	⬚ Mountains

This map shows some of the major land biomes on Earth.

TAKE A LOOK
14. Explain Where are most tropical rain forest biomes located?

_____

_____

Section 1 Review

NSES   ES 1f, 1j, 3d

SECTION VOCABULARY

biome a large region characterized by a specific type of climate and certain types of plant and animal communities	**prevailing winds** winds that blow mainly from one direction during a given period
climate the average weather conditions in an area over a long period of time	**surface current** a horizontal movement of ocean water that is caused by wind and that occurs at or near the ocean's surface
elevation the height of an object above sea level	**weather** the short-term state of the atmosphere, including temperature, humidity, precipitation, wind, and visibility
latitude the distance north or south from the equator; expressed in degrees	

1. Compare How is climate different from weather?

_____

_____

_____

2. Apply Concepts Nome, Alaska, lies at 64°N latitude. San Diego, California, lies at 32°N latitude. Which city receives more sunlight? Explain your answer.

_____

_____

3. Explain What causes some places on Earth to have seasons?

_____

_____

4. Identify What are four things that can affect climate?

_____

_____

_____

_____

5. Explain Describe how the rain-shadow effect works.

_____

_____

_____

_____

_____

SECTION 2 The Tropics

BEFORE YOU READ

After you read this section, you should be able to answer these questions:

• Where is the tropical zone?

• What are three biomes found in the tropical zone?

What Is the Tropical Zone?

Remember that latitudes near the equator receive more solar energy than other areas. The area between 23.5°N latitude and 23.5°S latitude receives the most solar energy. This region is called the **tropical zone**. It is also known as the *Tropics*. Because areas in the Tropics receive so much solar energy, they tend to have high temperatures.

There are three main biomes in the Tropics: tropical rain forest, tropical savanna, and tropical desert. All the tropical biomes have high temperatures. However, they receive different amounts of rain and have different types of soil. Therefore, different organisms live in each biome. The figure below shows where each of these biomes is found. ☑

✏️ **STUDY TIP**

Compare After you read this section, make a chart comparing the three kinds of tropical biomes.

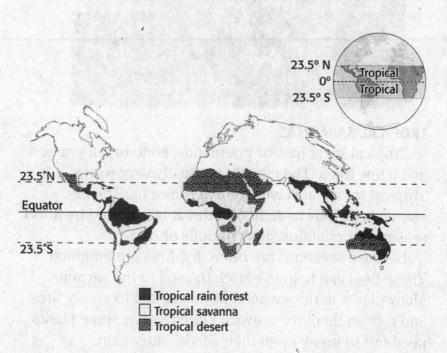
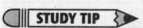

23.5° N
0°
23.5° S

Tropical
Tropical

23.5°N

Equator

23.5°S

■ Tropical rain forest
□ Tropical savanna
▨ Tropical desert

✔️ **READING CHECK**

1. Explain Why do the different tropical biomes have different organisms living in them, even though they all have high temperatures?

_____

_____

_____

TAKE A LOOK
2. Identify Where are the Tropics?

_____

_____

TROPICAL RAIN FORESTS

Tropical rain forests are warm and wet. They are located close to the equator, so they receive about the same amount of solar energy all year long. Therefore, there is little difference between the seasons. ☑

Tropical rain forests are homes to many different kinds of living things. Animals that live in tropical rain forests include monkeys, parrots, frogs, tigers, and leopards. Plants include mahogany trees, vines, ferns, and bamboo.

Many organisms live in tropical rain forests. When dead organisms decay, nutrients return to the soil. However, the nutrients are quickly used up by plants or washed away by rain. As a result, the soil is thin and poor in nutrients.

3. Explain Why is there little difference between the seasons in a tropical rain forest?

_____

_____

_____

TAKE A LOOK
4. Explain Why is the soil in tropical rainforests thin and nutrient-poor?

_____

_____

_____

Tropical Rain Forest

• **Average Temperature Range**
25°C to 28°C (77°F to 82°F)

• **Average Yearly Precipitation**
200 cm or more

• **Soil Characteristics**
thin and nutrient-poor

TROPICAL SAVANNAS

Tropical savannas, or grasslands, contain tall grasses and a few trees. The climate is usually very warm. Tropical savannas have two main seasons. The dry season lasts four to eight months. It is followed by a wet season that contains short periods of rain.

Because savannas are often dry, fires are common. These fires can help to enrich the soil in the savanna. Many plants in the savanna have adapted to yearly fires and rely on them for growth. For example, some plants need fire to break open their seeds' outer skin.

Animals that live in tropical savannas include giraffes, lions, crocodiles, and elephants. The figure on the top of the next page shows a tropical savanna.

Critical Thinking

5. Predict Consequences
What could happen to a tropical savanna if people stopped all fires from spreading? Explain your answer.

_____

_____

_____

SECTION 2 The Tropics *continued*

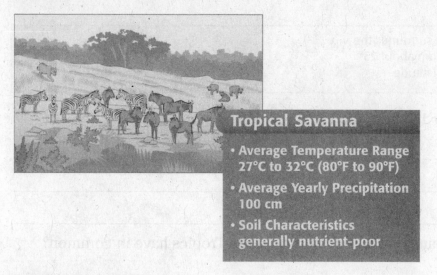

Tropical Savanna

- Average Temperature Range 27°C to 32°C (80°F to 90°F)
- Average Yearly Precipitation 100 cm
- Soil Characteristics generally nutrient-poor

Math Focus

6. Convert About how many feet of rain does a tropical savanna get in a year?

1 in. = 2.54 cm

_____

TROPICAL DESERTS

A desert is an area that receives less than 25 cm of rainfall per year. Deserts are the driest places on Earth. Tropical desert plants, such as shrubs, are adapted to living in places with little water. Animals such as camels, lizards, snakes, and scorpions also have adaptations for living in the desert.

Most tropical deserts are very hot in the daytime. They can be up to 50°C (120°F) during the day. However, the temperatures at night may be much lower. Therefore, organisms that live in deserts are also adapted to changing temperatures. ☑

✓ **READING CHECK**

7. Explain Why do tropical desert organisms have to be adapted to changing temperatures?

_____

_____

_____

Tropical Desert

- Average Temperature Range 16°C to 50°C (61°F to 120°F)
- Average Yearly Precipitation 0 cm to 25 cm
- Soil Characteristics poor in organic matter

Section 2 Review

SECTION VOCABULARY

tropical zone the region that surrounds the equator and that extends from about 23° north latitude to 23° south latitude	

1. List What are the three biomes found in the Tropics?

_____

_____

_____

2. Identify What is one thing that all the biomes in the Tropics have in common?

_____

_____

3. Compare Fill in the missing information about the features of each tropical biome.

Biome	Rainfall	Soil	Example of an animal found here	Example of a plant found here
Tropical rain forest		poor	parrot	
	100 cm per year		giraffe	
		poor		palm tree

4. Apply Concepts An area is located at 30°N latitude. It receives less than 25 cm per year of rain and has temperatures as high as 50°C during the day. Is the area a tropical desert? Explain your answer.

_____

_____

_____

5. Identify On which continent are most tropical savannas found?

_____

_____

6. Identify Which tropical biome has the largest range of temperatures? Which tropical biome has the smallest range of temperatures?

_____

_____

_____

CHAPTER 7 Climate

SECTION 3 Temperate and Polar Zones

After you read this section, you should be able to answer these questions:

• What biomes are found in the temperate zone?

• What biomes are found in the polar zone?

• What are two examples of microclimates?

What Is the Temperate Zone?

The climate zone between the tropical and the polar zones is the **temperate zone**. This zone extends from about 23.5° to about 66.5° north or south latitudes. Most of the continental United States is in the temperate zone. The temperate zone receives less solar energy than the Tropics. Therefore, temperatures in the temperate zone tend to be lower than those in the Tropics.

The four main biomes in the temperate zone are temperate forests, temperate grasslands, chaparrals, and temperate deserts. All of these biomes show seasonal changes in weather. However, some biomes have more extreme weather changes than others. For example, some areas in the United States have similar temperatures all year long. Other areas have very low temperatures in the winter and very high temperatures in the summer. ☑

STUDY TIP

Compare After you read this section, make a table comparing the four main temperate biomes.

READING CHECK

1. Identify What do the four main temperate biomes have in common?

_____

_____

_____

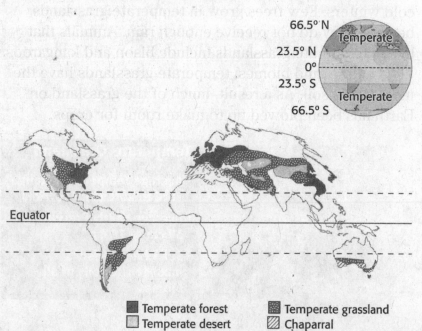

66.5° N
23.5° N
0°
23.5° S
66.5° S

Temperate

Temperate

Equator

■ Temperate forest   ▦ Temperate grassland
□ Temperate desert   ▨ Chaparral

TAKE A LOOK

2. Read a Map What kind of biome is found in northern and southern Africa?

_____

_____

TEMPERATE FORESTS

Temperate forests tend to have high amounts of rainfall and large seasonal temperature differences. The summers are warm, and the winters are cold. Animals that live in temperate forests include foxes, deer, and bears. Some trees in temperate forests lose their leaves each winter. These trees are called *deciduous* trees. Other trees, called *evergreens*, do not lose all of their leaves at once.

The soils in most temperate forests are very rich in nutrients. This is because the deciduous trees drop their leaves every winter. As the leaves decay, nutrients are added to the soil.

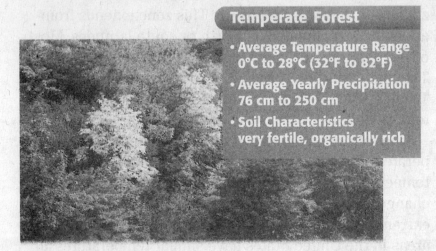

Temperate Forest

- **Average Temperature Range** 0°C to 28°C (32°F to 82°F)
- **Average Yearly Precipitation** 76 cm to 250 cm
- **Soil Characteristics** very fertile, organically rich

TEMPERATE GRASSLANDS

Temperate grasslands have warm summers and very cold winters. Few trees grow in temperate grasslands because they do not receive enough rain. Animals that live in temperate grasslands include bison and kangaroos.

Of all the land biomes, temperate grasslands have the most fertile soil. As a result, much of the grassland on Earth has been plowed up to make room for crops.

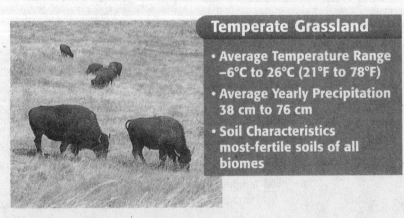

Temperate Grassland

- **Average Temperature Range** −6°C to 26°C (21°F to 78°F)
- **Average Yearly Precipitation** 38 cm to 76 cm
- **Soil Characteristics** most-fertile soils of all biomes

Critical Thinking

3. Infer A student visits a forest in Vermont in January. Most of the trees in the forest are covered with leaves. Are the trees probably deciduous trees or evergreens? Explain your answer.

_____

_____

_____

_____

_____

TAKE A LOOK

4. Identify What is the main kind of plant that grows in temperate grasslands?

_____

SECTION 3 Temperate and Polar Zones *continued*

CHAPARRALS

Chaparral regions have cool, wet winters and hot, dry summers. Animals that live in the chaparral include mountain lions, coyotes, and quail.

Fires are common during the summers in chaparrals. Some chaparral plants are adapted to these fires. Chaparral plants also have adaptations that prevent water loss during dry conditions. For example, the main kinds of plants in the chaparral are evergreen shrubs. These shrubs have thick leaves with waxy coatings. The coatings help prevent the leaves from losing water. ☑

✔ **READING CHECK**

5 Describe What adaptation do evergreen shrubs in the chaparral have to survive dry conditions?

_____

_____

Chaparral

- **Average Temperature Range** 11°C to 26°C (51°F to 78°F)
- **Average Yearly Precipitation** 48 cm to 56 cm
- **Soil Characteristics** rocky, nutrient-poor soils

TEMPERATE DESERTS

Like tropical deserts, temperate deserts are hot in the daytime and receive little rainfall. However, temperate deserts tend to have much colder nights than tropical deserts. This is because temperate deserts tend to have low humidity and cloudless skies. These conditions allow solar energy to heat the surface a lot during the day. They also allow heat to move into the atmosphere at night. ☑

Plants that live in temperate deserts include cacti, shrubs, and thorny trees. Animals include lizards, snakes, bats, and toads.

✔ **READING CHECK**

6. Compare How are temperate deserts different from tropical deserts? Give one way.

_____

_____

_____

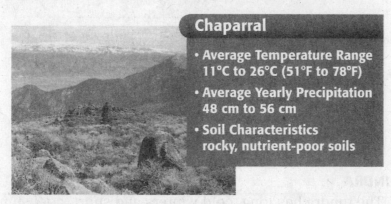

Temperate Desert

- **Average Temperature Range** 1°C to 50°C (34°F to 120°F)
- **Average Yearly Precipitation** 0 cm to 25 cm
- **Soil Characteristics** poor in organic matter

What Is the Polar Zone?

The **polar zone** is located between 66.5° and 90° north and south latitudes, near the North and South Poles. This zone has the coldest temperatures of all climate zones. There are two biomes in the polar zone: tundra and taiga.

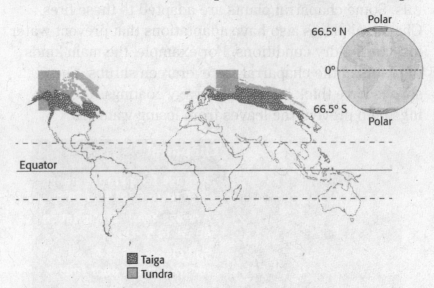

Taiga
Tundra

TAKE A LOOK

7. Identify On which continents is taiga found?

_____

_____

TUNDRA

The tundra has long, cold winters and short, cool summers. In the summer, only the top meter of soil thaws out. Below this depth is a permanently frozen layer called *permafrost*. It prevents water in the thawed soil from draining away. Therefore, the upper soil is muddy in the summer. Insects like mosquitoes thrive there. Birds migrate there in the summer to eat the insects. ☑

Other animals that live in the tundra include caribou, reindeer, and polar bears. Only small plants, such as mosses, live in the tundra.

✔ **READING CHECK**

8. Explain Why is the upper soil in the tundra muddy during the summer?

_____

_____

_____

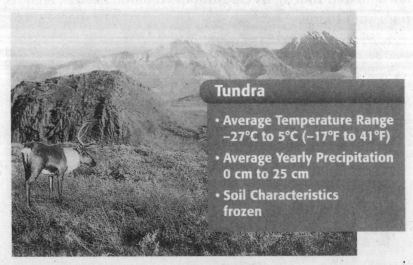

Tundra

- **Average Temperature Range** −27°C to 5°C (−17°F to 41°F)
- **Average Yearly Precipitation** 0 cm to 25 cm
- **Soil Characteristics** frozen

SECTION 3 Temperate and Polar Zones *continued*

TAIGA

Taiga biomes are found just south of tundra biomes in the Northern Hemisphere. The taiga has long, cold winters and short, warm summers. Animals that live in the taiga include moose, bears, and rabbits.

Evergreen trees called *conifers*, such as pine and spruce, are the main plants that grow in the taiga. The needle-like leaves from these trees contain acidic substances. When the needles die and decay on the ground, these substances make the soil acidic. Not very many plants can grow in acidic soils. Therefore, few plants grow on the forest floor of the taiga.

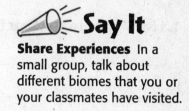

Say It

Share Experiences In a small group, talk about different biomes that you or your classmates have visited.

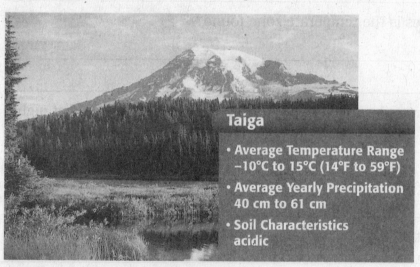

Taiga
- **Average Temperature Range** −10°C to 15°C (14°F to 59°F)
- **Average Yearly Precipitation** 40 cm to 61 cm
- **Soil Characteristics** acidic

Math Focus
9. Convert How much precipitation does the taiga get per year in inches?

1 in. = 2.54 cm

What Are Microclimates?

Remember that latitude, topography, and water help determine the climate of an area. Local conditions can also affect the climate in smaller areas. A **microclimate** is the climate of a small area. Two examples of microclimates are alpine biomes and cities. ☑

Alpine biomes are cold microclimates found near the tops of mountains. In winter, the temperatures are below freezing. In summer, they range from 10°C to 15°C. It is the high elevations of alpine biomes that cause them to be so cold. Alpine biomes are even found on mountains in the Tropics.

Cities are also microclimates. Buildings and pavement are made of dark materials. They absorb solar energy and stay warm. City temperatures can be 1°C to 2°C warmer than temperatures in other areas.

☑ READING CHECK

10. Define What is a microclimate?

_____

_____

Section 3 Review

SECTION VOCABULARY

microclimate the climate of a small area **polar zone** the North or South Pole and the surrounding region	**temperate zone** the climate zone between the Tropics and the polar zone

1. List What are the four biomes of the temperate zone?

_____

_____

_____

_____

2. Identify At what latitudes is the temperate zone found?

_____

_____

3. Explain Why are temperate deserts very hot during the day but very cold at night?

_____

_____

_____

4. Explain Why do cities often have higher temperatures than surrounding rural areas?

_____

_____

5. Explain Why are most taiga soils acidic?

_____

_____

_____

6. Compare How are temperate deserts and the tundra similar?

_____

_____

7. Explain Why do few trees grow in temperate grasslands?

_____

_____

CHAPTER 7 | Climate

SECTION 4 Changes in Climate

National Science
Education Standards
ES 1k, 2a

BEFORE YOU READ

After you read this section, you should be able to answer these questions:

• How has Earth's climate changed over time?

• What factors can cause climates to change?

How Was Earth's Climate Different in the Past?

The geologic record shows that Earth's climate in the past was different from its climate today. During some periods in the past, Earth was much warmer. During other periods, Earth was much colder. In fact, much of Earth was covered by sheets of ice during some times in the past.

An **ice age** happens when ice at high latitudes expands toward lower latitudes. Scientists have found evidence of many major ice ages in Earth's history. The most recent one began about 2 million years ago. ☑

Many people think of an ice age as a time when the temperature is always very cold. However, during an ice age, there can be periods of colder or warmer weather. A period of colder weather is called a *glacial period*. A period of warmer weather is called an *interglacial period*.

During glacial periods, large sheets of ice grow. These ice sheets form when ocean water freezes. Therefore, sea level drops during glacial periods. The figure below shows the coastlines of the continents during the last glacial period. Notice that the continental coastlines extended further into the ocean than they do today.

STUDY TIP

Learn New Words As you read, underline any words that you don't know. When you figure out what they mean, write the words and their definitions in your notebook.

READING CHECK

1. Define Write your own definition for *ice age*.

_____

_____

_____

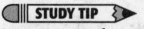

| | Extent of land mass at glacial maximum | | Extent of continental glaciation |
| | Current land mass | | Extent of sea ice |

TAKE A LOOK

2. Explain Why is more land exposed during glacial periods than at other times?

_____

_____

_____

What Can Cause Climates to Change?

Scientists have several theories to explain ice ages and other forms of climate change. Factors that can cause climate change include Earth's orbit, plate tectonics, the sun's cycles, asteroid impacts, volcanoes, and human activities.

CHANGES IN EARTH'S ORBIT

A Serbian scientist, Milutin Milankovitch, found that changes in Earth's orbit and tilt can affect Earth's climate. He modeled the way Earth moves in space and found that Earth's movements change in a regular way. These changes happen over tens of thousands of years. For example, Earth's orbit around the sun is more circular at some times than others.

These variations in Earth's orbit and tilt affect how much sunlight Earth gets. Therefore, they can also affect climate. The figure below shows how these factors can change the amount of sunlight Earth gets.

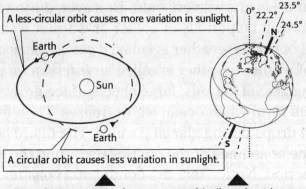

A less-circular orbit causes more variation in sunlight.

Earth

Sun

Earth

A circular orbit causes less variation in sunlight.

Earth's orbit is more circular at some times than at other times. The amount of solar energy that Earth gets from the sun varies more when Earth's orbit is less circular.

0° 22.2° 23.5° 24.5°

N

S

Earth's tilt on its axis can vary. When the tilt is greater, the poles get more solar energy.

N

S

Earth's axis wobbles slightly. This affects how much sunlight Earth's surface gets at different times of the year.

Critical Thinking

3. Infer Could changes in climate over 100 years be caused by changes in Earth's orbit and tilt? Explain your answer.

_____

_____

_____

TAKE A LOOK
4. Identify How does the shape of Earth's orbit change?

_____

_____

SECTION 4 Changes in Climate *continued*

PLATE TECTONICS

Plate tectonics and continental drift also affect Earth's climate. When a continent is closer to the equator, its climate is warmer than when it is near the poles. Also, remember that continents can deflect ocean currents and winds. When continents move, the flow of air and water around the globe changes. These changes can strongly affect Earth's climate.

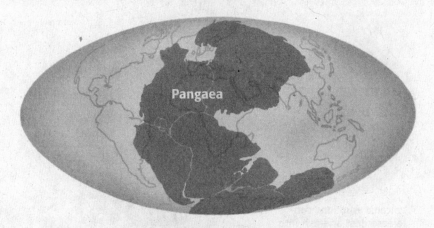

The locations of the continents can affect their climate. When India, Africa, South America, and Australia were part of Pangaea, they were covered with large ice sheets.

THE SUN

Some changes in Earth's climate are caused by changes in the sun. Many people think that the sun is always the same, but this is not true. In fact, the amount of energy that the sun gives off can change over time. The sun follows a regular cycle in how much energy it gives off. Because the sun's energy drives most cycles on Earth, these changes can affect Earth's climate. ☑

IMPACTS

Sometimes, objects from outer space, such as asteroids, crash into Earth. An *asteroid* is a small, rocky object that orbits the sun. If a large asteroid crashed into Earth, the climate of the whole planet could change.

When a large object hits Earth, particles of dust and rock fly into the atmosphere. This material can block some sunlight from reaching Earth's surface. This can cause temperatures on Earth to go down. In addition, plants may not be able to survive with less sunlight. Without plants, many animals would die off. Many scientists believe that an asteroid impact may have caused the dinosaurs to become extinct.

TAKE A LOOK
5. **Identify** How was the climate of India different when it was part of Pangaea?

_____

_____

☑ **READING CHECK**
6. **Explain** Why do changes in the sun's energy affect the climate on Earth?

_____

_____

_____

Critical Thinking
7. **Identify Relationships** Why may animals die off if there are fewer plants around?

_____

_____

VOLCANIC ERUPTIONS

Volcanic eruptions can affect Earth's climate for a short time. They send large amounts of dust and ash into the air. As with an asteroid impact, the dust and ash block sunlight from reaching Earth's surface. The figure below shows how volcanic dust can affect sunlight.

TAKE A LOOK
8. Compare How are the effects on climate of volcanic eruptions and asteroid impacts similar?

_____

_____

_____

_____

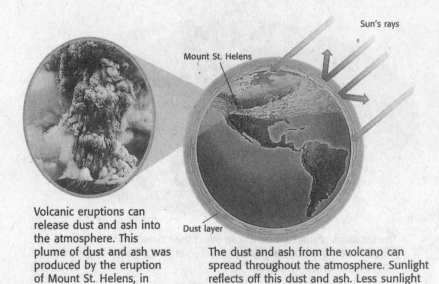

Volcanic eruptions can release dust and ash into the atmosphere. This plume of dust and ash was produced by the eruption of Mount St. Helens, in Washington, in 1980.

The dust and ash from the volcano can spread throughout the atmosphere. Sunlight reflects off this dust and ash. Less sunlight reaches Earth's surface.

What Is Global Warming?

A slow increase in global temperatures is called **global warming**. One thing that can cause global warming is an increase in the greenhouse effect. The **greenhouse effect** is Earth's natural heating process. During this process, gases in the atmosphere absorb energy in sunlight. This energy is released as heat, which helps to keep Earth warm. Without the greenhouse effect, Earth's surface would be covered in ice. ☑

One of the gases that absorbs sunlight in the atmosphere is carbon dioxide (CO_2). If there is more CO_2 in the atmosphere, the greenhouse effect can increase. This can cause global warming.

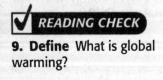

9. Define What is global warming?

_____

_____

SECTION 4 **Changes in Climate** *continued*

WHERE CO$_2$ COMES FROM

Much of the CO$_2$ in the atmosphere comes from natural processes, such as volcanic eruptions and animals breathing. However, human activities can also increase the amount of CO$_2$ in the atmosphere. ☑

When people burn fossil fuels for energy, CO$_2$ is released into the atmosphere. When people burn trees to clear land for farming, CO$_2$ is released. In addition, plants use CO$_2$ for food. Therefore, when trees are destroyed, we lose a natural way of removing CO$_2$ from the atmosphere.

PROBLEMS WITH GLOBAL WARMING

Many scientists think that if global warming continues, the ice at Earth's poles could melt. This could cause sea levels to rise. Many low-lying areas could flood. Global warming could also affect areas far from the oceans. For example, the Midwestern part of the United States could become warmer and drier. Northern areas, such as Canada, may become warmer. ☑

WHAT PEOPLE CAN DO

Many countries are working together to reduce the effects of global warming. Treaties and laws have helped to reduce pollution and CO$_2$ production. Most CO$_2$ is produced when people burn fossil fuels for energy. Therefore, reducing how much energy you use can reduce the amount of CO$_2$ produced. Here are some ways you can reduce your energy use:

- Turn off electrical devices, such as lights and computers, when you are not using them.

- Ride a bike, walk, or take public transportation instead of using a car to travel.

- Turn the heater to a lower temperature in the winter.

- Turn the air conditioner to a higher temperature in the summer.

✔ **READING CHECK**

10. Identify What are two natural sources of carbon dioxide in the atmosphere?

_____

_____

✔ **READING CHECK**

11. Explain Why may sea level rise if global warming continues?

_____

_____

_____

_____

Section 4 Review

SECTION VOCABULARY

global warming a gradual increase in average global temperature **greenhouse effect** the warming of the surface and lower atmosphere of Earth that occurs when water vapor, carbon dioxide, and other gases absorb and reradiate thermal energy	**ice age** a long period of climatic cooling during which the continents are glaciated repeatedly

1. Identify Relationships How is global warming related to the greenhouse effect?

_____

_____

_____

2. Describe What did Milutin Milankovitch's research show can affect Earth's climate?

_____

_____

3. Identify Give two ways that plate tectonics can affect an area's climate.

_____

_____

_____

4. Predict Consequences How could global warming affect cities near the oceans? Explain your answer.

_____

_____

_____

5. List Give three ways that human activities can affect the amount of CO_2 in the atmosphere.

_____

_____

_____

_____

Name _____ Class _____ Date _____

Our Solar System

National Science Education Standards
ES 1c, 3a, 3b, 3c

BEFORE YOU READ

After you read this section, you should be able to answer these questions:

• What are the parts of our solar system?

• When were the planets discovered?

• How do astronomers measure large distances?

What Is Our Solar System?

Our *solar system* includes our sun, the planets, their moons, and many other objects. At the center of our solar system is a star that we call the sun. Planets and other smaller objects move around the sun. Most planets have one or more moons that move around them. In this way, our solar system is a combination of many smaller systems.

Sun
Mercury
Earth
Mars
Venus
Jupiter
Saturn
Uranus
Neptune

The planets and the sun are some of the objects in our solar system.

How Was Our Solar System Discovered?

Until the 1600s, people thought that there were only eight bodies in our solar system. These were the sun, Earth's moon, and the planets Earth, Mercury, Venus, Mars, Jupiter, and Saturn. These are the only objects in the solar system that we can see from Earth without a telescope.

Once the telescope was invented, however, scientists were able to see many more bodies in our solar system. In the 17th century, scientists discovered some of the moons of Jupiter and Saturn. Uranus and several other moons were discovered in the 1700s. Neptune was discovered in the 1800s. Pluto was not discovered until the 1900s.

STUDY TIP

Describe As you read, make a chart showing the parts of our solar system that were discovered in the following time periods: before the 1600s; the 1700s; the 1800s; and the 1900s.

STANDARDS CHECK

ES 3a The earth is the third planet from the sun in a system that includes the moon, the sun, eight other planets and their moons, and smaller objects, such as asteroids and comets. The sun, an average star, is the central and largest body in the solar system. **Note:** In 2006, the International Astronomical Union ruled that Pluto is no longer considered to be a planet.

1. Identify List three kinds of objects that make up our solar system.

_____

_____

_____

How Do Scientists Measure Long Distances?

Remember that astronomers use light-years to measure long distances in space. To measure distances within our solar system, astronomers use two other units: the astronomical unit and the light-minute.

One **astronomical unit** (AU) is the average distance between the sun and Earth. This distance is about 150,000,000 km. Earth is 1 AU from the sun. Neptune is about 30.1 AU from the sun. Therefore, Neptune is about $30.1 \times 150,000,000$ km = 4,500,000,000 km from the sun.

Another way to measure distances in space is by using the speed of light. Light travels at about 300,000 km/s in space. In one minute, light travels about 18,000,000 km. Therefore, one *light-minute* is equal to about 18,000,000 km. Light from the sun takes 8.3 minutes to reach Earth. Therefore, Earth is 8.3 light-minutes from the sun.

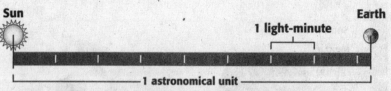

One astronomical unit equals about 8.3 light-minutes.

How Is Our Solar System Divided?

Astronomers divide our solar system into two main parts. These parts are called the *inner solar system* and the *outer solar system*.

THE INNER PLANETS

The inner solar system contains the four planets that are closest to the sun: Mercury, Venus, Earth, and Mars. The inner planets are also sometimes called the *terrestrial planets. Terrestrial* means "like Earth." Mercury, Venus, and Mars are like Earth because they have dense, rocky surfaces, as Earth does. The figure on the top of the next page shows the orbits of the inner planets.

SECTION 1 Our Solar System *continued*

The Inner Planets

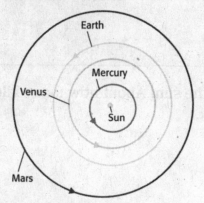

TAKE A LOOK
4. Identify Which of the inner planets is closest to the sun?

_____

5. Identify Which of the inner planets is furthest from the sun?

_____

THE OUTER PLANETS

The outer solar system contains four planets: Jupiter, Saturn, Uranus, and Neptune. The outer planets are very different from the inner planets.

T he outer planets are very large and are made mostly of gases. Therefore, Jupiter, Saturn, Uranus, and Neptune are sometimes called the *gas giant* planets, or simply the "gas giants."

The distances between the outer planets are much larger than the distances between the inner planets. For example, the distance between Jupiter and Saturn is much larger than the distance between Mars and Earth. The figure below shows the orbits of the outer planets.

Critical Thinking

6. Compare Give two differences between the inner solar system and the outer solar system.

_____

_____

_____

_____

The Outer Planets

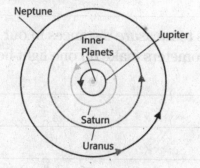

TAKE A LOOK
7. Identify What planet is farthest from the sun?

_____

Name _____ Class _____ Date _____

Section 1 Review

NSES ES 1c, 3a, 3b, 3c

SECTION VOCABULARY

astronomical unit the average distance between the Earth and the sun; approximately 150 million kilometers (symbol, AU)	

1. Calculate Mercury is about 0.39 AU from the sun. About how many kilometers from the sun is Mercury? Show your work.

2. Identify Name the four planets in the inner solar system.

_____

_____

_____

3. Identify Name the four gas giant planets.

_____

_____

_____

4. Infer Scientists sometimes use light-hours to measure distances in our solar system. What is a light-hour? About how many kilometers make up one light-hour?

_____

_____

_____

5. Explain Why do scientists use light-minutes and light-hours instead of light-years to measure distances within our solar system?

_____

_____

_____

CHAPTER 8 | A Family of Planets

SECTION 2 The Inner Planets

National Science
Education Standards
ES 1c, 3a, 3b

BEFORE YOU READ

After you read this section, you should be able to answer these questions:

• Which planets are known as the inner planets?

• What properties do the inner planets share?

Why Group the Inner Planets Together?

The inner solar system includes the only planet known to support life, Earth, and three other planets. These four inner planets are called **terrestrial planets** because they all have a chemical makeup similar to that of Earth. The terrestrial planets are much smaller, denser, and more rocky than most of the outer planets. ☑

Which Planet Is Closest to the Sun?

Mercury is the planet closest to the sun. After Earth, it is the second densest object in the solar system. This is because, like Earth, Mercury has a large iron core in its center. The surface of Mercury is covered with craters.

Mercury rotates on its axis much more slowly than Earth. Remember that the amount of time that a planet takes to rotate once is its *period of rotation*. It is the length of a day on the planet. Mercury's period of rotation is about 59 Earth days long. Therefore, a day on Mercury is about 59 Earth days long.

On Mercury, a year is not much longer than a day. Remember that the time it takes a planet to go around the sun once is the planet's *period of revolution*. It is the length of one year on the planet. A *Mercurian* year, or a year on Mercury, is equal to 88 Earth days. Therefore, each year on Mercury lasts only 1.5 Mercurian days.

STUDY TIP

Compare In your notebook, create a chart showing the similarities and differences among the inner planets.

☑ **READING CHECK**

1. Explain Why are the inner planets called terrestrial planets?

_____

_____

Critical Thinking

2. Infer Which of the facts on the table could scientists use to infer that Mercury has a core made of iron?

_____

_____

Facts About Mercury

Distance from sun	0.38 AU
Period of rotation	58 Earth days, 19 hours
Period of revolution	88 Earth days
Diameter	4,879 km
Density	5.43 g/cm³
Surface gravity	38% of Earth's

Is Venus Earth's Twin?

The second planet from the sun is Venus. In some ways, Venus is more like Earth than any of the other planets. It is about the same size as Earth. However, Venus is slightly smaller, less dense, and less massive than Earth.

If you could observe the sun from the surface of Venus, you would see it rise in the west and set in the east. That is because Venus and Earth rotate on their axes in opposite directions.

If you looked down on Earth from above the North Pole, you would see Earth spinning counterclockwise. This is called **prograde rotation**. However, if you were to look down on Venus from above its north pole, you would see it spinning clockwise. This is called **retrograde rotation**. ☑

THE ATMOSPHERE OF VENUS

Venus has the densest atmosphere of the terrestrial planets. The atmospheric pressure on Venus's surface is 90 times that on Earth. This pressure would instantly crush a human on Venus. Venus's atmosphere is mostly made of carbon dioxide and thick clouds of sulfuric acid. The thick atmosphere causes a strong greenhouse effect. As a result, surface temperatures on Venus average about 464°C. This is hot enough to melt lead and some other metals.

Facts About Venus

Distance from sun	0.72 AU
Period of rotation	243 Earth days, 16 hours
Period of revolution	224 Earth days, 17 hours
Diameter	12,104 km
Density	5.24 g/cm³
Surface gravity	91% of Earth's

MAPPING THE SURFACE OF VENUS

Because of its thick atmosphere, we cannot observe the surface of Venus from Earth with telescopes. Between 1990 and 1992, the *Magellan* spacecraft made maps of Venus using radar waves. These waves can travel through the atmosphere and bounce off the surface. Maps made from the radar data showed that Venus has craters, mountains, lava plains, and volcanoes.

✓ **READING CHECK**

3. Compare How do prograde rotation and retrograde rotation differ?

_____

_____

_____

TAKE A LOOK

4. Compare Which is longer on Venus, one day or one year?

_____

_____

Critical Thinking

5. Analyzing Methods Why did scientists use *Magellan's* radar instead of telescopes to map the surface of Venus?

_____

_____

_____

_____

h Unique?

no one knew what Earth looked
satellites and spacecraft can take
blue planet. Light reflecting off
th look blue from space. ☑

t in the solar system that can
t. This is because Earth has a
ctors that make life possible.
ndant water and just the right
sun.

fe as we know it. Earth is not
system to have water on its
he only planet that has large
its surface. Earth is close
g to the sun that all the water does not freeze. It
is far enough away that the water does not boil away. If
Earth were much closer to or farther from the sun, liquid
water—and life—could not exist here.

6. Identify What feature
of Earth causes it to appear
blue from space?

_____

_____

Facts About Earth

Distance from sun	1.0 AU
Period of rotation	23 hours, 56 minutes
Period of revolution	365 Earth days, 6 hours
Diameter	12,756 km
Density	5.52 g/cm³
Surface gravity	100% of Earth's

STUDYING EARTH FROM SPACE

NASA's Earth Science Enterprise is a program to study
Earth from space. Studying Earth from space lets scientists study the Earth as a whole system. It helps them
understand changes in Earth's atmosphere, oceans, ice,
landforms, and living things. It may also be able to help
them understand how humans affect the global environment. By studying Earth from space, scientists can learn
how different parts of the Earth interact.

Math Focus

7. Calculate Use the information on the table to
explain why every fourth year
on Earth is a leap year. Show
your work.

(Hint: Compare Earth's period
of revolution to the number of
days in a calendar year.)

_____

_____

_____

_____

What Is the Red Planet?

Besides Earth, the most studied planet in the solar system is Mars. Mars looks red, so it is sometimes known as "the red planet." Some scientists think that there could be simple life on Mars.

Scientists have learned much about Mars by observing it from Earth. However, most of our knowledge of the planet has come from unmanned spacecraft. So far, these observations have found no evidence of life.

THE ATMOSPHERE OF MARS

Because it has a thinner atmosphere than Earth and is farther from the sun, Mars is colder than Earth. In the middle of the summer, the spacecraft *Mars Pathfinder* recorded a temperature range from $-13°C$ to $-77°C$. The Martian atmosphere is made mainly of carbon dioxide. ☑

The atmospheric pressure on Mars is very low. At the surface, it is about the same as the pressure 30 km above Earth's surface. Because of the low temperatures and air pressure, liquid water cannot exist on the surface of Mars. The only water on Mars's surface is in the form of ice.

Facts About Mars

Distance from sun	1.52 AU
Period of rotation	24 hours, 37 minutes
Period of revolution	687 Earth days
Diameter	6,794 km
Density	3.93 g/cm³
Surface gravity	38% of Earth's

WATER ON MARS

Even though water cannot exist on the surface of Mars today, it may have in the past. Evidence from spacecraft suggests that some of Mars's features were formed by liquid water. For example, some of Mars's features are similar to those caused by water erosion on Earth. Other features indicate that Mars's surface contains sediments that may have been deposited by the water from a large lake. ☑

Scientists cannot prove that these features were caused by liquid water. However, they indicate that at some time in the past, Mars may have had liquid water. If this is true, it would show that Mars was once warmer and had a thicker atmosphere than it does today.

READING CHECK

8. Explain What are two reasons that the surface of Mars is colder than that of Earth?

_____

_____

_____

_____

TAKE A LOOK

9. Compare How does the length of a day on Mars compare to the length of day on Earth?

_____

_____

_____

READING CHECK

10. Identify What two features suggest that water once existed on the surface of Mars?

_____

_____

WHERE THE WATER IS NOW

Mars has two polar icecaps made of a combination of frozen water and frozen carbon dioxide. Most of the water on Mars is trapped in this ice. There is some evidence from the *Mars Global Surveyor* that water could exist just beneath the surface. If so, it may be in liquid form. If Mars does have liquid water beneath its surface, there is a possibility that some form of life may exist on Mars. ☑

VOLCANOES ON MARS

The remains of giant volcanoes exist on the surface of Mars. They show that Mars has had active volcanoes in the past. Unlike on Earth, however, the volcanoes are not spread across the whole planet. There are two large volcanic systems on Mars. The largest one is about 8,000 km long.

The largest mountain in the solar system, Olympus Mons, is a Martian volcano. It is a shield volcano, similar to Mauna Kea on the island of Hawaii. However, Olympus Mons is much larger than Mauna Kea. The base of Olympus Mons is 600 kilometers —about 370 miles—across. It is nearly 24 kilometers tall. That is three times as tall as Mount Everest! It may have grown so tall because the volcano erupted for long periods of time. ☑

MISSIONS TO MARS

Scientists sent several vehicles to Mars in the early 21st century. The figure below shows *Mars Express Orbiter*, which reached Mars in December 2003. In January 2004, the exploration rovers *Spirit* and *Opportunity* landed on Mars. These solar-powered wheeled robots have found evidence that water once existed on the Martian surface. ☑

The *Mars Express Orbiter* helps scientists map Mars and study Mars's atmosphere.

> ✓ **READING CHECK**
>
> **11. Identify** Where does water exist on Mars today?
>
> _____
>
> _____

> ✓ **READING CHECK**
>
> **12. Explain** What may have allowed Olympus Mons to grow so large?
>
> _____
>
> _____
>
> _____

> ✓ **READING CHECK**
>
> **13. Describe** What have the rovers *Spirit* and *Opportunity* found?
>
> _____
>
> _____

Section 2 Review

NSES ES 1c, 3a, 3b

SECTION VOCABULARY

prograde rotation the counterclockwise spin of a planet or moon as seen from above the planet's North Pole; rotation in the same direction as the sun's rotation	**retrograde rotation** the clockwise spin of a planet or moon as seen from above the planet's North Pole
	terrestrial planet one of the highly dense planets nearest to the sun; Mercury, Venus, Mars, and Earth

1. Identify Does Earth show prograde or retrograde rotation?

_____

2. Compare Fill in the blanks to complete the table.

Planet	Distance from sun	Period of revolution
	0.38 AU	58 Earth days, 19 hours
	0.72 AU	243 Earth days, 16 hours
	1.00 AU	365 Earth days, 6 hours
	1.52 AU	687 Earth days

3. Analyze Ideas Why do scientists think that Mars was once warmer and had a thicker atmosphere than it does today?

_____

_____

_____

4. Identify Relationships How is the period of revolution of a planet related to its distance from the sun? (Hint: examine the statistics tables.)

_____

_____

_____

5. Explain Why is the surface temperature of Venus higher than the surface temperatures of the other inner planets?

_____

_____

6. Explain Why could life probably not have developed on Earth if Earth were closer to the sun?

_____

_____

SECTION 3 # The Outer Planets

National Science Education Standards
ES 1c, 3a, 3b

BEFORE YOU READ

After you read this section, you should be able to answer these questions:

• Which planets are known as the outer planets?

• What properties do the outer planets share?

How Are the Outer Planets Different from the Inner Planets?

The outer planets are very large and are made mostly of gases. These planets are called **gas giants**. Unlike the inner planets, they have very thick atmospheres and not very much hard, rocky material on their surfaces.

Which Planet Is the Biggest?

Jupiter is the largest planet in our solar system. Its mass is twice as large as the other eight planets combined. Even though it is large, Jupiter's rotation takes less than 10 hours.

Like the sun, Jupiter is made mostly of hydrogen and helium. Jupiter's atmosphere also contains small amounts of ammonia, methane, and water. These gases form clouds in the outer part of Jupiter's atmosphere. The outer atmosphere also contains storms, such as the Great Red Spot. This huge storm is about 3 times the diameter of Earth. It has lasted for over 400 years! ☑

Deeper into Jupiter's atmosphere, the pressure is so high that hydrogen turns to liquid. Deeper still, the pressure is even higher. Because of the high pressures, the inside of Jupiter is very hot. It is so hot that Jupiter produces more heat than it gets from the sun.

The information that scientists have about Jupiter has come from five space missions: *Pioneer 1*, *Pioneer 2*, *Voyager 1*, *Voyager 2*, and *Galileo*. The Voyager probes showed that Jupiter has a thin, faint ring.

STUDY TIP

Compare In your notebook, create a chart showing the similarities and differences among the outer planets.

READING CHECK

1. List Give four gases that are found in Jupiter's atmosphere.

_____

_____

_____

Facts About Jupiter

Distance from sun	5.20 AU
Period of rotation	9 hours, 55.5 minutes
Period of revolution	11 Earth years, 313 days
Diameter	142,984 km
Density	1.33 g/cm³
Surface gravity	236% of Earth's

TAKE A LOOK

2. Identify Which of the facts in the table could you use to infer that Jupiter has a shorter day than Earth does?

_____

SECTION 3 The Outer Planets *continued*

What Are Saturn's Rings Made Of?

Saturn is the second-largest planet in the solar system. Like Jupiter, Saturn is made up mostly of hydrogen with some helium and traces of other gases and water. Saturn has about 764 times more volume than Earth and about 95 times more mass. Therefore, it is much less dense than Earth.

Critical Thinking

3. Compare About how many times does Earth revolve around the sun in the time it takes Saturn to revolve once?

_____

_____

Facts About Saturn

Distance from sun	9.54 AU
Period of rotation	10 hours, 42 minutes
Period of revolution	29 Earth years, 155 days
Diameter	120,536 km
Density	0.69 g/cm³
Surface gravity	92% of Earth's

The inside of Saturn is probably similar to the inside of Jupiter. Also, like Jupiter, Saturn gives off more heat than it gets from the sun. Scientists think that Saturn's extra energy comes from helium condensing from the atmosphere and sinking toward the core. In other words, Saturn is still forming.

Saturn is probably best known for the rings that orbit the planet above its equator. They are about 250,000 km across, but less than 1 km thick. The rings are made of trillions of particles of ice and dust. These particles range from a centimeter to several kilometers across. ☑

✔ **READING CHECK**

4. Identify What two materials make up the rings of Saturn?

_____

_____

This picture of Saturn was taken by the *Voyager 2* probe.

SECTION 3 The Outer Planets *continued*

How Is Uranus Unique?

Uranus is the third-largest planet in the solar system. It is so far from the sun that it does not reflect much sunlight. You cannot see it from Earth without using a telescope. ☑

Uranus is different from the other planets because it is "tipped" on its side. As shown in the figure below, the north and south poles of Uranus point almost directly at the sun. The north and south poles of most other planets, like Earth, are nearly at right angles to the sun.

For about half the Uranian year, one pole is constantly in sunlight, and for the other half of the year it is in darkness. Some scientists think that Uranus may have been tipped over by a collision with a massive object.

5. Identify Why can't Uranus be seen from Earth without a telescope?

_____

_____

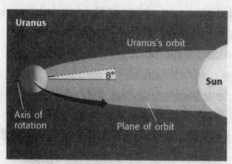

Uranus is tilted so that its poles point almost directly at the sun.

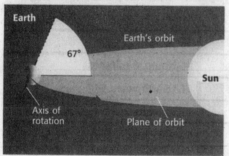

In contrast, Earth's poles, like those of most other planets, are nearly at right angles to the sun.

TAKE A LOOK
6. Explain Why do scientists say that Uranus is "tipped over"?

_____

_____

Like Jupiter and Saturn, Uranus is made mostly of hydrogen, helium, and small amounts of other gases. One of these gases, methane, filters sunlight and makes the planet look bluish-green. ☑

Facts About Uranus

Distance from sun	19.22 AU
Period of rotation	17 hours, 12 minutes
Period of revolution	83 Earth years, 273 days
Diameter	51,118 km
Density	1.27 g/cm³
Surface gravity	89% of Earth's

TAKE A LOOK
7. Compare How does the length of a year on Uranus compare to the length of a year on Earth?

_____

_____

What Is Neptune Like?

Some astronomers predicted that there was a planet beyond Uranus before the planet was observed. Uranus did not move in its orbit exactly as they expected. The force of gravity due to another large object was affecting it. Using predictions of its effect on Uranus, astronomers discovered Neptune in 1846. ☑

Neptune is the fourth-largest planet in the solar system. Like the other gas giants, Neptune is made up mostly of hydrogen, helium, and small amounts of other gases. It has a deep blue color, which is caused by methane in its atmosphere.

Clouds and weather changes are seen in the atmosphere of Neptune. The spacecraft *Voyager 2* flew past Neptune in 1989 and observed a Great Dark Spot in the southern hemisphere. This spot was a storm as large as Earth. It moved across the planet's surface at about 300 m/s. By 1994, the Great Dark Spot had disappeared. Another dark spot was then found in the northern hemisphere. *Voyager 2* images also showed that Neptune has very narrow rings.

READING CHECK

8. Explain What evidence did astronomers have that Neptune existed before they actually observed it?

_____

_____

TAKE A LOOK

9. Compare How does Neptune's average distance from the sun compare to Earth's?

_____

_____

Facts About Neptune

Distance from sun	30.06 AU
Period of rotation	16 hours, 6 minutes
Period of revolution	163 Earth years, 263 days
Diameter	49,528 km
Density	1.64 g/cm³
Surface gravity	112% of Earth's

Why Is Pluto Called a Dwarf Planet?

Pluto has been called the ninth planet since its discovery in 1930. However, in 2006, astronomers defined *planet* in a new way. Pluto does not fit the new definition of a planet. So, Pluto has been reclassified as a dwarf planet.

A *dwarf planet* is any object that orbits the sun and is round because of its own gravity, but has not cleared its orbital path. In addition to Pluto, Eris and Ceres have been classified as dwarf planets. Eris is larger than Pluto. Ceres was previously classified as an asteroid.

SECTION 3 The Outer Planets *continued*

A SMALL WORLD

Pluto is made of rock and ice and has a thin atmosphere made of methane and nitrogen. Scientists do not know if Pluto formed along with the planets.

AN UNUSUAL ORBIT

The shape of Pluto's orbit is different from the shapes of the outer planets. As shown in the figure below, sometimes Pluto is closer to the sun than Neptune. At other times, Neptune is closer to the sun.

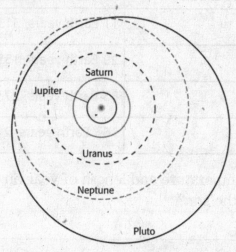

The shape of Pluto's orbit is very different from the orbits of other planets.

TAKE A LOOK
10. Compare How is Pluto's orbit different from the orbits of the other outer planets?

_____

_____

_____

_____

Facts About Pluto

Distance from sun	39.5 AU
Period of rotation	6 days, 10 hours
Period of revolution	248 Earth years, 4 days
Diameter	2,390 km
Density	1.75 g/cm³
Surface gravity	6% of Earth's

A LARGE MOON

Pluto's moon, Charon, is more than half the size of Pluto. From Earth, it is hard to separate the images of Pluto and Charon because they are so far away. Charon may be covered by frozen water.

Critical Thinking

11. Infer How do you think scientists learned that Pluto has a moon if it is difficult to separate their images?

_____

_____

_____

_____

Section 3 Review

NSES   ES 1c, 3a, 3b

SECTION VOCABULARY

gas giant a planet that has a deep, massive atmosphere, such as Jupiter, Saturn, Uranus, or Neptune

1. Identify What is the main element found in the atmosphere of a gas giant planet?

_____

2. Compare Fill in the blanks to complete the table.

Planet	Distance from sun	Period of revolution
	5.20 AU	11 Earth years, 313 days
	9.54 AU	29 Earth years, 155 days
	19.22 AU	83 Earth years, 273 days
	30.06 AU	163 Earth years, 263 days

3. Evaluate Data How are the surface temperature and length of year on a planet related to the planet's distance from the sun?

_____

_____

_____

4. Make Comparisons How do the gas giants differ from the inner planets of the solar system? In your answer, discuss composition, size, distance from the sun, length of a year, and solar energy received.

_____

_____

_____

_____

5. Identify What gives Neptune and Uranus their blue to blue-green colors?

_____

6. Describe What did the Voyager probes discover about Jupiter?

_____

_____

CHAPTER 8 A Family of Planets

SECTION 4 Moons

National Science
Education Standards
ES 1a, 3a, 3b, 3c

BEFORE YOU READ

After you read this section, you should be able to answer
these questions:

• How did Earth's moon probably form?
• How does the moon's appearance change with time?
• What moons revolve around other planets?

What Are Moons?

Satellites are natural or artificial bodies that revolve
around larger bodies in space, such as planets. Except for
Mercury and Venus, all of the planets have natural satellites called *moons*. Moons come in a wide variety of sizes,
shapes, and compositions.

What Do We Know About Earth's Moon?

Scientists have learned a lot about Earth's moon,
which is also called *Luna*. Much of what we know comes
from observations from Earth, but other discoveries have
come from visiting the moon. Some lunar rocks brought
back by Apollo astronauts were found to be almost
4.6 billion years old. These rocks have not changed much
since they were formed. This tells scientists that the solar
system itself is at least 4.6 billion years old. ☑

THE MOON'S SURFACE

The moon is almost as old as Earth. It is covered with
craters, many of which can be seen from Earth on a clear
night. Because the moon has no atmosphere and no erosion, its surface shows where objects have collided with
it. Scientists think that many of these collisions happened
about 3.8 billion years ago. They were caused by matter
left over from the formation of the solar system.

Facts About Luna

Period of rotation	27 Earth days, 9 hours
Period of revolution	27 Earth days, 7 hours
Diameter	3,475 km
Density	3.34 g/cm3
Surface gravity	16% of Earth's

STUDY TIP

Describe In your notebook,
create a Concept Map about
Earth's moon, including
information about its origin,
why it shines, phases, and
eclipses.

READING CHECK

1. Explain How do scientists
know what moon's crust is
made of?

_____

_____

_____

Math Focus

2. Identify What fraction of
Earth's gravity is the moon's
gravity?

_____

THE ORIGIN OF THE MOON

When scientists studied the rock samples brought back from the moon by astronauts, they found some surprises. The composition of the moon is similar to that of Earth's mantle. This evidence led to a theory about the moon's formation. ☑

Scientists now think that the moon formed when a large object collided with the early Earth. The object was probably about the size of Mars. The collision was so violent that a large mass of material was thrown into orbit around Earth. Gravity pulled this material into a sphere. The sphere continued to revolve around the planet. Eventually, it became the moon.

3. Identify What discovery caused scientists to revise their theory about the origin of the moon?

_____

_____

Formation of the Moon

❶ About 4.6 billion years ago, a large body collided with Earth. At this time, Earth was still mostly molten. The collision blasted part of Earth's mantle into space.

❷ Within a few hours of the collision, the debris began to orbit the Earth. The debris was made of mantle material from Earth and some iron core material from the colliding body.

❸ In time, the material began to clump together. Eventually, the moon formed. As it cooled, collisions with smaller objects produced cracks in the moon's crust. Lava flowed onto the moon's surface. This formed the dark patches, or *maria*, that we can see on the moon today.

TAKE A LOOK

4. Identify According to this theory, material was thrown from Earth in clumps. What caused the material to come together as a sphere?

_____

_____

_____

PHASES OF THE MOON

The moon revolves around the Earth once each month. It rotates on its axis in almost the same period. Therefore, we always see the same side of the moon. However, the moon does not always look the same. This is because we cannot always see all of the part that is reflecting light.

As the moon's position changes compared to the sun and Earth, it looks different to people on Earth. During a month, the face of the moon that we can see changes from a fully lit circle to a thin crescent and then back to a circle. The figure below shows how the moon's appearance changes as it moves around Earth.

5. Explain The moon does not produce its own light. How can the moon be seen from Earth?

_____

_____

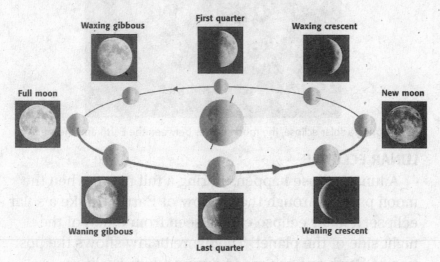

TAKE A LOOK

6. Explain Why does the moon look different on different nights?

_____

_____

The different appearances of the moon are called **phases**. When the moon is *waxing*, the part of the sunlit side that we can see increases every day. The moon appears to get bigger. When the moon is *waning*, the part of the sunlit side that we can see decreases every day. The moon appears to get smaller.

What Is an Eclipse?

An **eclipse** happens when the shadow of one body in space falls on another. A *solar eclipse* happens when the moon comes between the sun and Earth. Then, the shadow of the moon falls on part of Earth's surface. A *lunar eclipse* happens when Earth comes between the sun and the moon. Then, the shadow of Earth falls on the moon. ☑

✓ READING CHECK

7. Explain What happens during a solar eclipse?

_____

_____

_____

_____

_____

SECTION 4 Moons *continued*

SOLAR ECLIPSES

Because the moon's orbit is elliptical (oval-shaped) instead of circular, the distance between Earth and the moon changes. When the moon is close to Earth, the moon appears to be the same size as the sun. If the moon passes between the sun and Earth during that time, there is a *total solar eclipse*. If the moon is farther from earth, the eclipse is an annular eclipse. During an *annular eclipse*, a thin ring of the sun can be seen around the moon.

TAKE A LOOK
8. Explain Why can't a solar eclipse be seen from every point on Earth?

_____

_____

_____

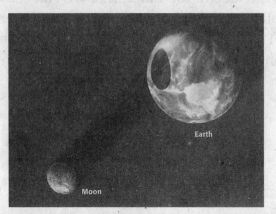

During a solar eclipse, the moon passes between the Earth and the sun.

LUNAR ECLIPSES

A lunar eclipse happens during a full moon when the moon passes through the shadow of Earth. Unlike a solar eclipse, a lunar eclipse can be seen from much of the night side of the planet. The figure below shows the position of Earth and the moon during a lunar eclipse.

TAKE A LOOK
9. Describe What happens during a lunar eclipse?

_____

_____

_____

During a lunar eclipse, the Earth passes between the sun and the moon.

Lunar eclipses are interesting to watch. At the beginning and end of a lunar eclipse, the moon is in the outer part of the shadow. In this part of the shadow, Earth's atmosphere filters out some of the blue light. As a result, the light that is reflected from the moon is red.

THE MOON'S TILTED ORBIT

The moon rotates around Earth each month, so you might expect that there would be an eclipse each month. However, eclipses happen only about once a year.

Eclipses don't happen every month because the moon's orbit is slightly tilted compared to Earth's orbit. This tilt is enough to place the moon out of Earth's shadow during most full moons. It also causes the Earth to be out of the moon's shadow during most new moons. ☑

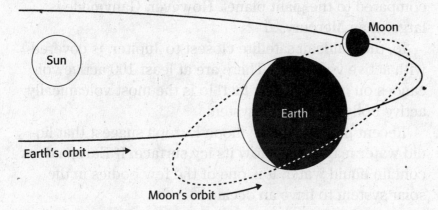

The moon's orbit is tilted compared to the Earth's. Therefore, eclipses do not happen every month.

Are Other Moons Like Earth's Moon?

All of the planets, except Mercury and Venus, have moons. Mars has two moons. All of the gas giants have many moons. Many of these moons were discovered fairly recently using spacecraft cameras or the Hubble Space Telescope. Some moons may not have been discovered yet. ☑

The solar system's moons vary widely. They range in size from very small bits of rock to objects as large as a terrestrial planet. Their orbits range from nearly circular to very elliptical. Most moons orbit in the same direction as the planets orbit the sun. However, some orbit in the opposite direction.

THE MOONS OF MARS

Mars has two moons, Phobos and Deimos. They are small, oddly shaped satellites. Both moons have dark surfaces and resemble *asteroids*, or rocky bodies in space. Phobos is about 22 km across at its largest dimension. Deimos is about 15 km across. Both moons may be asteroids that were captured by Mars's gravity. ☑

✔ **READING CHECK**

10. Explain Why don't solar eclipses occur each month?

_____

_____

_____

_____

📣 **Say It**

Discuss In a group, discuss why you can't look at the sun during a solar eclipse but you can look at the moon during a lunar eclipse.

✔ **READING CHECK**

11. Compare Which types of planets have the most moons—terrestrial planets or gas giants?

_____

_____

✔ **READING CHECK**

12. Identify What are the names of Mars's moons?

_____

_____

THE MOONS OF JUPITER

Jupiter has more than 60 moons. The four largest were discovered in 1610 by Galileo. When he observed Jupiter through a telescope, Galileo saw what looked like four dim stars that moved with Jupiter. He observed that they changed position compared to Jupiter and each other from night to night.

These moons—Ganymede, Callisto, Io, and Europa— are known as the *Galilean satellites*. They appear small compared to the giant planet. However, Ganymede is larger than Mercury. ☑

Io, the Galilean satellite closest to Jupiter, is covered with active volcanoes. There are at least 100 active volcanoes on its surface. In fact, Io is the most volcanically active body in the solar system.

Recent pictures of the moon Europa suggest that liquid water may exist below its icy surface. If Europa does contain liquid water, it is one of the few bodies in the solar system to have an ocean. ☑

✔ READING CHECK

13. Identify What are the names of the Galilean satellites?

_____

_____

_____

_____

✔ READING CHECK

14. Identify What may lie below the icy surface of Europa?

_____

_____

Critical Thinking

15. Make Inferences Would humans be able to live unprotected on the surface of Titan? Explain your answer.

_____

_____

_____

_____

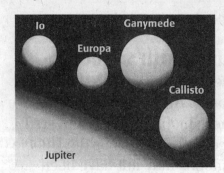

This figure shows the sizes of the four Galilean satellites compared to Jupiter.

THE MOONS OF SATURN

Saturn has more than 50 moons. Saturn's largest moon, Titan, is slightly smaller than Ganymede. Unlike most moons, Titan has an atmosphere. Its atmosphere is composed mostly of nitrogen, with small amounts of other gases, such as methane. Scientists think that Titan's atmosphere is similar to Earth's early atmosphere.

None of Saturn's other moons are as large as the Galilean moons of Jupiter. Most of them are from several kilometers to several hundred kilometers across. They are made mostly of frozen water and rocks.

SECTION 4 Moons *continued*

THE MOONS OF URANUS

Uranus has at least 27 moons. Most of them are small. They were discovered by space probes or orbiting observatories, such as the Hubble Space Telescope. Like the moons of Saturn, Uranus's largest moons are made of ice and rock. ☑

THE MOONS OF NEPTUNE

Neptune has 13 known moons. The largest, Triton, revolves in a *retrograde*, or "backward," orbit. Triton's unusual orbit suggests that it was captured by Neptune's gravity after forming somewhere else in the solar system. Triton has a thin nitrogen atmosphere. Its surface is mostly frozen nitrogen and methane. It has active "ice volcanoes" that send gas high into its atmosphere. Neptune's other moons are small and are made of ice and rock.

THE MOONS OF PLUTO

Although Pluto is not considered a planet, it does have at least three moons. The diameter of Charon, the largest moon, is about half that of Pluto. Charon revolves around Pluto in 6.4 days, the same period as Pluto's rotation. That means that Charon is always located at the same place in Pluto's sky. Two additional moons of Pluto, discovered by the Hubble telescope in 2005, are much smaller than Charon. These moons are called Hydra and Nix.

READING CHECK

16. Describe What are Uranus's largest moons made of?

_____

_____

Some of the Moons of the Solar System

Planet	Moon	Diameter	Period of revolution
Earth	Luna	3475 km	27.3 Earth days
Mars	Phobos	26 km	0.3 Earth days
Mars	Deimos	15 km	1.3 Earth days
Jupiter	Io	3636 km	1.8 Earth days
Jupiter	Europa	3120 km	3.6 Earth days
Jupiter	Ganymede	5270 km	7.1 Earth days
Jupiter	Callisto	4820 km	16.7 Earth days
Saturn	Titan	5150 km	15.9 Earth days
Uranus	Titania	1580 km	8.7 Earth days
Neptune	Triton	2700 km	5.9 Earth days
Pluto	Charon	1180 km	6.4 Earth days

Critical Thinking

17. Identify Relationships Some of the moons of the gas giants are larger than Mercury. Why are they not considered to be planets?

_____

_____

_____

Section 4 Review

NSES   ES 1a, 3a, 3b, 3c

SECTION VOCABULARY

eclipse an event in which the shadow of one celestial body falls on another	**satellite** a natural or artificial body that revolves around a planet
phase the change in the sunlit area of one celestial body as seen from another celestial body	

1. Compare How is a solar eclipse different from a lunar eclipse?

_____

_____

2. Identify Fill in the blanks to complete the chart.

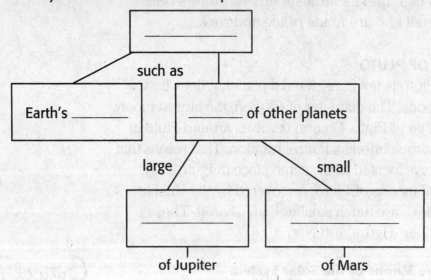

An object that revolves around a planet is called a

_____

such as

Earth's _____   _____ of other planets

large   small

_____   of Jupiter   _____   of Mars

3. Analyze Methods How can astronomers use rocks from the moon to estimate the age of the solar system?

_____

_____

_____

4. Explain Why don't eclipses happen every month?

_____

_____

CHAPTER 8 A Family of Planets

SECTION 5 # Small Bodies in the Solar System

After you read this section, you should be able to answer these questions:

• What are comets?
• What are asteroids?
• What are meteoroids?

What Are Comets?

The sun, the planets, and their moons are not the only objects in our solar system. There are also a large number of smaller bodies, including comets, asteroids, and meteoroids. Scientists study these objects to learn about the formation and composition of the solar system.

A **comet** is a small, loosely packed body of ice, rock, and dust. The *nucleus*, or core, of a comet is made of rock, metal, and ice. A comet's nucleus can range from 1 km to 100 km in diameter. A spherical cloud of gas and dust, called a *coma*, surrounds the nucleus. The coma may extend as far as 1 million kilometers from the nucleus. ☑

COMET TAILS

A comet's tail is its most spectacular feature. Sunlight changes some of the comet's ice to gas, which streams away from the nucleus. Part of the tail is made of *ions*, or charged particles. The *ion tail*, pushed by the solar wind, always points away from the sun, no matter which way the comet is moving. A second tail, the *dust tail*, follows the comet in its orbit. Some comet tails are more than 80 million kilometers long, glowing brightly with reflected sunlight.

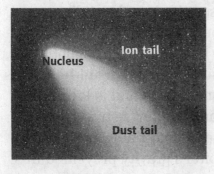

This image shows the physical features of a comet when the comet comes close to the sun. The nucleus of the comet is hidden by the brightly lit gases and dust of the coma.

 **Say It**

Compare In your notebook, create a table that compares comets, asteroids, and meteoroids.

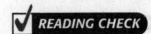 **READING CHECK**

1. Describe What are comets made of?

_____

_____

TAKE A LOOK
2. Identify Draw an arrow from the nucleus label showing the direction the comet is moving.

COMET ORBITS AND ORIGINS

Remember that the planets move in *elliptical*, or oval-shaped, orbits. Comets also move in elliptical orbits. However, the orbits of comets are much more stretched out than the orbits of planets.

Scientists think that many comets come from the Oort cloud. The *Oort cloud* is a spherical cloud of dust and ice. It surrounds the solar system, far beyond the orbit of Pluto. Pieces of the Oort cloud may fall into orbits around our sun and become comets. Some comets may also come from the *Kuiper belt*, a flat ring of objects just beyond Neptune's orbit. ☑

READING CHECK

3. Identify Where is the Oort cloud located?

_____

_____

TAKE A LOOK

4. Explain Why does the ion tail extend in different directions during most of the comet's orbit?

_____

_____

_____

Nucleus

Dust tail

Sun

Ion tail

Comets have very long orbits that take them close to the sun and well beyond Pluto.

What Are Asteroids?

Asteroids are small, rocky bodies that revolve around the sun. They range from a few meters to almost 1,000 km in diameter. More than 50,000 asteroids have been discovered. None of them can be seen from Earth without a telescope. In fact, scientists didn't know that asteroids exist until 1801.

Most asteroids orbit the sun in the **asteroid belt**. This is a region that is 300 million km wide and is located between the orbits of Mars and Jupiter. Astronomers think that asteroids are made of material from the early solar system. The pull of Jupiter's gravity prevented this material from coming together to form a planet. ☑

READING CHECK

5. Identify Where is the asteroid belt?

_____

_____

COMPOSITION OF ASTEROIDS

It is hard to determine what asteroids are made of. This is because they are small and usually far away from Earth. Mostly, they are composed of either rock or metal. Some asteroids may contain carbon and carbon compounds.

In general, asteroids do not have a spherical shape because of their small size. Gravity must be very strong to pull matter together into a spherical shape. Only the largest asteroids are spherical.

Critical Thinking

6. Make Inferences How do you think scientists know what asteroids are made of?

_____

_____

_____

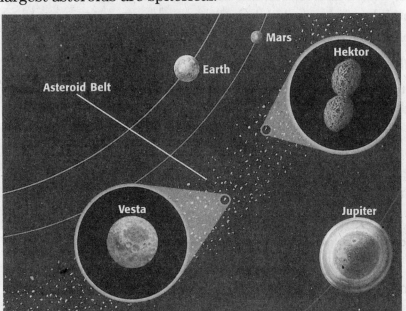

What Are Meteoroids?

Pieces of dust and debris from asteroids and comets, called **meteoroids**, are scattered throughout the solar system. Most meteoroids are about the size of a grain of sand. When a meteoroid enters Earth's atmosphere, it can reach a speed of up to 250,000 km/h.

Friction with the atmosphere heats meteoroids and the air around them, causing them to glow brightly. The glowing trails that form when meteoroids burn up in the atmosphere are called **meteors**. A meteor can be a few hundred meters in diameter and tens of kilometers long before it fades.

Sometimes, a larger meteoroid enters the atmosphere. Some of these meteoroids pass through the atmosphere without burning up completely. When they reach Earth's surface, they are called **meteorites**. ☑

TAKE A LOOK
7. Compare How do asteroid sizes compare to planet sizes?

_____

_____

✓ **READING CHECK**
8. Compare What is the difference between a meteoroid and a meteorite?

_____

_____

_____

_____

_____

TYPES OF METEORITES

Scientists classify meteorites based on composition. There are three main types of meteorites: stony, metallic, and stony-iron. *Stony meteorites* are similar to rocks on Earth. Some of them contain carbon compounds similar to those found in living organisms. Stony meteorites probably come from carbon-rich asteroids. ☑

Metallic meteorites have a distinctive metallic appearance and do not look like terrestrial rocks. They are made mainly of iron and nickel. *Stony-iron meteorites* are made of a combination of rocky material, iron, and nickel.

Three Major Types of Meteorites

Stony Meteorite:
rocky material

Metallic Meteorite:
iron and nickel

Stony-iron Meteorite:
rocky material, iron, and nickel

Scientists study meteorites to learn about the early solar system.

Remember that asteroids and comets are probably made of debris from the formation of our solar system. Meteorites are easier for scientists to study than asteroids and comets. Because meteorites are pieces of asteroids and comets, scientists study meteorites to learn about the early solar system.

METEOR SHOWERS

Meteors can be seen on most clear nights. When many small meteoroids enter the atmosphere in a short period, it is called a *meteor shower*. During some meteor showers, several meteors are visible every minute. Meteor showers happen at the same time each year. These showers happen when Earth passes through orbits of comets that have left a dust trail.

IMPACTS IN OUR SOLAR SYSTEM

Impacts are common in our solar system. An *impact* happens when an object in space collides with another object. In many cases, impacts produce impact craters. Many of the planets and moons in our solar system, including Earth, have visible impact craters.

Planets and moons with atmospheres have fewer impact craters than those without atmospheres. For example, there are only a few visible impact craters on Earth. However, the surface of our moon is covered with impact craters. Earth has fewer craters because the atmosphere acts as a shield. Most objects that enter Earth's atmosphere burn up before they reach the surface. ☑

Another reason that there are few visible impact craters on Earth is that Earth has a very active surface. Plate tectonics, weathering, erosion, and deposition act to smooth out and change Earth's surface. These processes are less common on other planets and moons.

Most objects that enter Earth's atmosphere are small and burn up completely before reaching the surface. However, scientists think that impacts powerful enough to cause a natural disaster happen every few thousand years. An impact large enough to cause a global catastrophe may happen once every 50 to 100 million years.

> ☑ **READING CHECK**
>
> **12. Identify** Why do fewer meteorites hit Earth's surface than the surface of the moon?
>
> _____
> _____
> _____
> _____

THE TORINO SCALE

Scientists can track objects that are close to Earth to learn whether they might hit Earth. Scientists use the *Torino scale* to rate the chance than an object will hit the Earth. The Torino scale ranges from 0 to 10. Zero indicates that an object has a very small chance of hitting the Earth. Ten indicates that the object will definitely hit the Earth. The Torino scale is also color coded, as shown in the table below.

Color	Number	Hazard level
White	0	very low; almost certainly will not hit the Earth
Green	1	low
Yellow	2, 3, or 4	moderate
Orange	5, 6, or 7	high
Red	8, 9, or 10	very high; almost certainly will hit the Earth

TAKE A LOOK
13. Identify Which color on the Torino scale is used to describe an object that will probably hit the Earth?

_____

Section 5 Review

SECTION VOCABULARY

asteroid a small, rocky object that orbits the sun; most asteroids are located in a band between the orbits of Mars and Jupiter	**meteor** a bright streak of light that results when a meteoroid burns up in Earth's atmosphere
asteroid belt the region of the solar system that is between the orbits of Mars and Jupiter and in which most asteroids orbit	**meteorite** a meteoroid that reaches the Earth's surface without burning up completely
comet a small body of ice, rock, and cosmic dust that follows an elliptical orbit around the sun and that gives off gas and dust in the form of a tail as it passes close to the sun	**meteoroid** a relatively small, rocky body that travels through space

1. Describe How can a comet become the source of meteoroids and meteors?

_____

_____

2. Classify Fill in the blanks to complete the table.

Object	Composition	Main Location
	Large chunk of rock or metal— much smaller than planets	
		Oort cloud and Kuiper belt
	small chunk of rock or metal	throughout the solar system

3. Identify Connections Why is information about comets, asteroids, and meteoroids important for understanding the development of the solar system?

_____

_____

_____

4. Apply Concepts Why would scientists want to know if an asteroid is on a course to collide with Earth in 20 years?

_____

_____

SECTION 1 | Characteristics of Living Things

BEFORE YOU READ

After you read this section, you should be able to answer these questions:

• What are all living things made of?

• What do all living things have in common?

National Science Education Standards

LS 1b, 1c, 1d, 2a, 2b, 2c, 3a, 3b, 3c

What Are All Living Things Made Of?

If you saw a bright yellow, slimy blob in the grass, would you think it was alive? How could you tell? All living things, or *organisms*, share several characteristics. What does a dog have in common with a bacterium? What do *you* have in common with a bright yellow slime mold?

All living things are made of one or more cells. A **cell** is the smallest unit that can carry out all the activities of life. Every cell is surrounded by a cell membrane. The *cell membrane* separates the cell from the outside environment.

Some organisms are made of trillions of cells. In these organisms, different kinds of cells do different jobs. For example, muscle cells are used for movement. Other organisms are made of only one cell. In these organisms, different parts of the cell have different jobs.

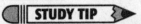
STUDY TIP

Organize As you read this section, make a list of the six characteristics of living things.

STANDARDS CHECK

LS 1b All organisms are composed of cells—the fundamental unit of life. Most organisms are single cells; other organisms, including humans, are multicellular.

1. Identify What are all living things made of?

_____

_____

Some organisms, such as the monkeys on the left, are made up of trillions of cells. The protists on the right are made up of one or a few cells. They are so small they can only be seen with a microscope.

How Do Living Things Respond to Change?

All organisms can sense changes in their environments. Each organism reacts differently to these changes. A change that affects how an organism acts is called a **stimulus** (plural, *stimuli*). Stimuli can be chemicals, light, sounds, hunger, or anything that causes an organism to react. ☑

READING CHECK

2. List Give three examples of stimuli.

_____

_____

_____

The touch of an insect is a stimulus for a Venus' flytrap. The stimulus causes the plant to close its leaves quickly.

TAKE A LOOK
3. Complete For a Venus' flytrap, the touch of an insect is a _____.

Even when things outside the body change, an organism must keep the conditions inside its body the same. The act of keeping a constant environment inside an organism is called **homeostasis**. When an organism maintains homeostasis, all the chemical reactions inside its body can work correctly.

RESPONDING TO EXTERNAL CHANGES

If it is hot outside, your body starts to sweat to cool down. If it is cold outside, your body starts to shiver to warm up. In each situation, your body reacts to the changes in the environment. It tries to return itself to normal.

Different kinds of organisms react to changes in the environment in different ways. For example, crocodiles lie in the sun to get warm. When they get too warm, they open their mouths wide to release heat.

Critical Thinking
4. Predict What would happen if your body couldn't maintain homeostasis?

_____
_____
_____
_____
_____
_____

How Do Organisms Have Offspring?

Every type of organism has *offspring*. The two ways organisms can produce offspring are by sexual or asexual reproduction. Generally, in **sexual reproduction**, two parents make offspring. The offspring get traits from both parents. In **asexual reproduction**, one parent makes offspring. The offspring are identical to the parent. ☑

Most plants and animals make offspring by sexual reproduction. However, most single-celled organisms and some multicellular organisms make offspring by asexual reproduction. For example, hydra make offspring by forming buds that break off and grow into new hydra.

 **READING CHECK**
5. Identify How many parents are generally needed to produce offspring by sexual reproduction?

_____

SECTION 1 Characteristics of Living Things *continued*

Like most animals, bears produce offspring by sexual reproduction. However, some animals, such as hydra, can reproduce asexually.

Why Do Offspring Look Like Their Parents?

All organisms are made of cells. Inside each cell, there is information about all of the organism's traits. This information is found in DNA (**d**eoxyribo**n**ucleic **a**cid). *DNA* carries instructions for the organism's traits. Offspring look like their parents because they get copies of parts of their parent's DNA. Passing traits from parent to offspring is called **heredity**. ☑

Why Do Organisms Need Energy?

All organisms need energy to live. Most organisms get their energy from the food they eat. Organisms use this energy to carry out all the activities that happen inside their bodies. For example, organisms need energy to break down food, to move materials in and out of cells, and to build cells.

An organism uses energy to keep up its metabolism. An organism's **metabolism** is all of the chemical reactions that take place in its body. Breaking down food for energy is one of these chemical reactions.

How Do Organisms Grow?

All organisms grow during some part of their lives. In a single-celled organism, the cell gets bigger and divides. This makes new organisms. An organism made of many cells gets bigger by making more cells. As these organisms grow, they get new traits. These traits often change how the organism looks. For example, as a tadpole grows into a frog, it develops legs and loses its tail. ☑

TAKE A LOOK
6. Identify How do most animals reproduce?

_____

_____

✓ READING CHECK

7. Define What is the function of DNA?

_____

_____

_____

_____

✓ READING CHECK

8. Compare How does growth differ in single-celled organisms and those made of many cells?

_____

_____

_____

_____

_____

_____

Section 1 Review

NSES   LS 1b 1c, 1d, 2a, 2b, 2c, 3a, 3b, 3c

SECTION VOCABULARY

asexual reproduction reproduction that does not involve the union of sex cells and in which one parent produces offspring that are genetically identical to the parent.	**homeostasis** the maintenance of a constant internal state in a changing environment
cell in biology, the smallest unit that can perform all life processes; cells are covered by a membrane and contain DNA and cytoplasm	**metabolism** the sum of all chemical processes that occur in an organism
heredity the passing of genetic traits from parent to offspring	**sexual reproduction** reproduction in which the sex cells from two parents unite to produce offspring that share traits from both parents
	stimulus anything that causes a reaction or change in an organism or any part of an organism

1. Summarize Complete the Spider Map to show the six characteristics of living things. Add lines to give details on each characteristic.

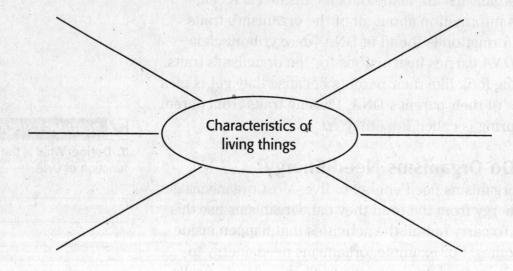

Characteristics of living things

2. Compare How does sexual reproduction differ from asexual reproduction?

_____

_____

_____

3. Explain How do the buds of an organism such as hydra compare to the parent?

_____

_____

4. Identify Relationships How is a bear's fur related to homeostasis?

_____

_____

CHAPTER 9 It's Alive!! Or Is It?

SECTION 2 # The Necessities of Life

After you read this section, you should be able to answer these questions:

• What things do almost all organisms need?

• Why do living things need food?

What Do Living Things Need?

Would it surprise you to learn that you have the same basic needs as a tree, a frog, and a fly? Almost every organism has the same basic needs: water, air, a place to live, and food.

WATER

Your body is made mostly of water. The cells that make up your body are about 70% to 85% water. Cells need water to keep their inside environments stable. Most of the chemical reactions that happen in cells need water. Your body loses water as you breathe, sweat, or get rid of wastes, such as urine. Because of this, you must replace the water that you lose.

Organisms get water from the fluids they drink and the foods they eat. However, organisms need different amounts of water. You could survive only three days without water. A kangaroo rat never drinks. It lives in the desert and gets all the water it needs from its food.

AIR

Oxygen, nitrogen, and carbon dioxide are some of the gases in air. Most organisms use oxygen to help them break down food for energy. Other organisms, such as green plants, use carbon dioxide to make food.

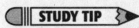

National Science Education Standards

LS 1a, 1c, 2c, 3a, 3d, 4b, 4c, 4d

STUDY TIP

Organize As you read, make a table of the basic needs of most organisms. Fill in examples of how different organisms meet those needs.

STANDARDS CHECK

LS 1c Cells carry out the many <u>functions</u> needed to sustain life. They grow and divide, thereby producing more cells. This requires that they take in nutrients, which they use to provide energy for the work that cells do and to make the materials that a cell or organism needs.

Word Help: function
use or purpose

1. Explain Why do cells need water?

_____

_____

_____

_____

_____

TAKE A LOOK

2. Infer Why do you think this diving spider surrounds itself with a bubble in the water?

_____

_____

_____

_____

A PLACE TO LIVE

Just as you do, all living things need a place to live. Organisms look for an area that has everything they need to survive. Often, many organisms live in the same area. They all must use the same resources, such as food and water. Many times, an organism will try to keep others out of its area. For example, some birds keep other birds away by singing.

FOOD

All organisms need food. Food gives organisms energy and nutrients to live and grow. However, not all organisms get food in the same way. There are three ways in which organisms can get food. ☑

Some organisms, such as plants, are producers. **Producers** make their own food using energy from their environment. For example, plants, and some bacteria and protists, use the sun's energy to make food from carbon dioxide and water. This process is called *photosynthesis*.

Many organisms are consumers. **Consumers** eat other organisms to get food. For example, a frog is a consumer because it eats insects. All animals are consumers.

A mushroom is a decomposer. Decomposers are a special kind of consumer. **Decomposers** break down dead organisms and animal wastes to get food. Although they are a kind of consumer, decomposers play a different role in an ecosystem than most other consumers. Without decomposers, dead organisms and wastes would pile up all over the Earth!

✔ **READING CHECK**

3. Explain Why do living things need food?

_____

_____

_____

Critical Thinking

4. Identify Are you a producer, consumer, or decomposer? Explain your answer.

_____

_____

_____

_____

TAKE A LOOK

5. Label On the picture, label the producer, consumer, and decomposer.

SECTION 2 The Necessities of Life *continued*

What Do Organisms Get from Food?

As you just read, organisms can get their food in three different ways. However, all organisms must break down their food to use the nutrients.

Nutrients are molecules. *Molecules* are made of two or more atoms joined together. Most molecules in living things are combinations of carbon, nitrogen, oxygen, phosphorus, and sulfur. Proteins, nucleic acids, lipids, carbohydrates, and ATP are some of the molecules needed by living things.

PROTEINS

Proteins are used in many processes inside a cell. **Proteins** are large molecules made up of smaller molecules called *amino acids*. Living things break down the proteins in food and use the amino acids to make new proteins. ☑

An organism uses proteins in many different ways. Some proteins are used to build or fix parts of an organism's body. Some proteins stay on the outside of a cell, to protect it. Proteins called *enzymes* help to start or speed up reactions inside a cell.

Some proteins help cells do their jobs. For example, a protein called *hemoglobin* is found in our red blood cells. It picks up oxygen and delivers it through the body.

Spider webs, horns, and feathers are made from proteins.

Say It

Discuss With a partner, name 10 organisms and describe what foods they eat. Discuss whether these organisms are producers, consumers, or decomposers.

READING CHECK

6. Complete Proteins are

made up of _____

_____

Math Focus

7. Calculate Each red blood cell carries about 250 million molecules of hemoglobin. If every hemoglobin molecule is attached to four oxygen molecules, how many oxygen molecules could one red blood cell carry?

_____

_____

NUCLEIC ACIDS

When you bake a cake, you follow instructions to make sure the cake is made correctly. When cells make new molecules, such as proteins, they also follow a set of instructions. The instructions for making any part of an organism are stored in *DNA*.

DNA is a nucleic acid. **Nucleic acids** are molecules made of smaller molecules called *nucleotides*. The instructions carried by DNA tell a cell how to make proteins. The order of nucleotides in DNA tells cells which amino acids to use and which order to put them in.

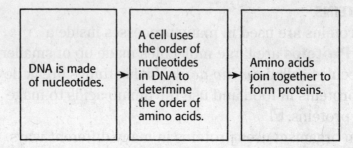

DNA is made of nucleotides. → A cell uses the order of nucleotides in DNA to determine the order of amino acids. → Amino acids join together to form proteins.

LIPIDS

Lipids are molecules that cannot mix with water. They are a form of stored energy. When lipids are stored in an animal, they are usually solid. These are called *fats*. When lipids are stored in a plant, they are usually liquid. These are called *oils*. When an organism has used up other sources of energy, it can break down fats and oils for more energy.

Lipids also form cell membranes. Cell membranes surround and protect cells. They are made of special lipids called **phospholipids**. When phospholipids are in water, the tail ends of the molecules come together and the head ends face out. This is shown in the figure below.

Phospholipid Membranes

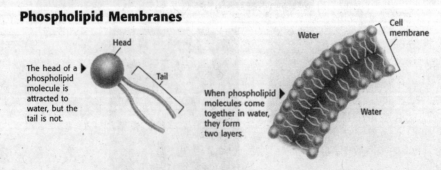

The head of a phospholipid molecule is attracted to water, but the tail is not.

When phospholipid molecules come together in water, they form two layers.

Head
Tail
Water
Cell membrane
Water

Critical Thinking

8. Identify Relationships
What is the relationship between amino acids and nucleotides?

_____
_____
_____
_____
_____

TAKE A LOOK
9. Describe Describe the structure of a phospholipid, and how it behaves in water.

_____
_____
_____
_____
_____

CARBOHYDRATES

Carbohydrates are molecules made of sugars. They provide and store energy for cells. An organism's cells break down carbohydrates to free energy. There are two types of carbohydrates: simple and complex. ☑

Simple carbohydrates are made of one or a few sugar molecules. Both table sugar and sugar in fruits are simple carbohydrates. The simple carbohydrate *glucose* is the most common source of energy for cells. The body breaks down simple carbohydrates more quickly than complex carbohydrates.

Complex carbohydrates are made of hundreds of sugar molecules linked together. When organisms such as plants have more sugar than they need, they can store the extra sugar as complex carbohydrates. For example, potatoes store extra sugar as starch. You can also find complex carbohydrates in foods such as whole-wheat bread, pasta, oatmeal, and brown rice.

READING CHECK

10. Identify What are two types of carbohydrates?

_____

_____

Type of carbohydrate	Structure	Example
	made of one or a few sugar molecules	
Complex		

TAKE A LOOK

11. Complete Complete the table to explain the two types of carbohydrates.

ATP

After carbohydrates and fats have been broken down, how does their energy get to where it is needed? The cells use **a**denosine **t**riphosphate, or ATP. **ATP** is a molecule that carries energy in cells. The energy released from carbohydrates and fats is passed to ATP molecules. ATP then carries the energy to where it is needed in the cell. ☑

READING CHECK

12. Identify What molecule carries energy in cells?

_____

It's Alive!! Or Is It?

Section 2 Review

NSES   LS 1a, 1c, 2c, 3a, 3d, 4b, 4c, 4d

SECTION VOCABULARY

ATP adenosine triphosphate, a molecule that acts as the main energy source for cell processes	**lipid** a type of biochemical that does not dissolve in water; fats and steroids are lipids
carbohydrate a class of energy-giving molecules that includes sugars, starches, and fiber; contains carbon, hydrogen, and oxygen	**nucleic acid** a molecule made up of subunits called nucleotides
consumer an organism that eats other organisms or organic matter	**phospholipid** a lipid that contains phosphorus and that is a structural component in cell membranes
decomposer an organism that gets energy by breaking down the remains of dead organisms or animal wastes and consuming or absorbing the nutrients	**producer** an organism that can make its own food by using energy from its surroundings
	protein a molecule that is made up of amino acids and that is needed to build and repair body structures and to regulate processes in the body

1. List Name four things that organisms need to survive.

_____

2. Explain Why are decomposers also consumers?

_____

_____

3. Identify What two nutrients store energy?

_____

4. Describe Describe the structure of a cell membrane.

_____

_____

_____

_____

5. Compare Name two ways that simple carbohydrates differ from complex carbohydrates.

_____

_____

_____

6. Explain Why is ATP important to cells?

_____

_____

CHAPTER 10 Cells: The Basic Units of Life

SECTION 1 The Diversity of Cells

National Science Education Standards
LS 1a, 1b, 1c, 2c, 3b, 5a

BEFORE YOU READ

After you read this section, you should be able to answer these questions:

• What is a cell?

• What do all cells have in common?

• What are the two kinds of cells?

What Is a Cell?

Most cells are so small that they cannot be seen by the naked eye. So how did scientists find cells? By accident! The first person to see cells wasn't even looking for them.

A _____ is the smallest unit that can perform all the functions necessary for life. All living things are made of cells. Some living things are made of only one cell. Others are made of millions of cells.

Robert Hooke was the first person to describe cells. In 1665, he built a microscope to look at tiny objects. One day he looked at a piece of cork. Cork is found in the bark of cork trees. Hooke thought the cork looked like it was made of little boxes. He named these boxes *cells*, which means "little rooms" in Latin.

Organize As you read this section, make lists of things that are found in prokaryotic cells, things that are found in eukaryotic cells, and things that are found in both kinds of cells.

The first cells that Hooke saw were from cork. These cells were easy to see because plant cells have cell walls. At first, Hooke didn't think animals had cells because he couldn't see them. Today we know that all living things are made of cells.

In the late 1600s, a Dutch merchant named Anton van Leeuwenhoek studied many different kinds of cells. He made his own microscopes. With them, he looked at tiny pond organisms called protists. He also looked at blood cells, yeasts, and bacteria.

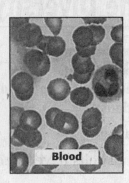
Euglena

Blood

Yeast

Bacteria

Leeuwenhoek looked at many different kinds of cells with his microscope. He was the first person to see bacteria. Bacterial cells are usually much smaller than most other types of cells.

TAKE A LOOK
2. Identify Which of these cells is probably the smallest? Explain your answer.

_____

_____

_____

What Is the Cell Theory?

Since Hooke first saw cork cells, many discoveries have been made about cells. Cells from different organisms can be very different from one another. Even cells from different parts of the same organism can be very different. However, all cells have several important things in common. These observations are known as the *cell theory*. The cell theory has three parts:

1. All organisms are made of one or more cells.
2. The cell is the basic unit of all living things.
3. All cells come from existing cells.

What Are the Parts of a Cell?

Cells come in many shapes and sizes and can have different functions. However, all cells have three parts in common: a cell membrane, genetic material, and organelles. ☑

CELL MEMBRANE

All cells are surrounded by a cell membrane. The **cell membrane** is a layer that covers and protects the cell. The membrane separates the cell from its surroundings. The cell membrane also controls all material going in and out of the cell. Inside the cell is a fluid called *cytoplasm*.

READING CHECK

3. List What three parts do all cells have in common?

_____

_____

_____

GENETIC MATERIAL

All cells contain DNA (deoxyribonucleic acid) at some point in their lives. *DNA* is the genetic material that carries information needed to make proteins, new cells, and new organisms. DNA is passed from parent cells to new cells and it controls the activities of the cell.

The DNA in some cells is found inside a structure called the **nucleus**. Most of your cells have a nucleus.

ORGANELLES

Cells have structures called **organelles** that do different jobs for the cell. Most organelles have a membrane covering them. Different types of cells can have different organelles.

Parts of a Cell

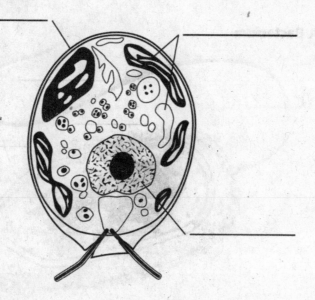

STANDARDS CHECK

LS 2c Every organism requires a set of instructions for specifying its traits. Heredity is the passage of these instructions from one generation to another.

Word Help: specify
to describe or define in detail

4. Explain What is the function of DNA?

_____

_____

_____

_____

_____

TAKE A LOOK
5. Identify Use the following words to fill in the blank labels on the figure: DNA, cell membrane, organelles.

What Are the Two Kinds of Cells?

There are two basic kinds of cells—cells with a nucleus and cells without a nucleus. Those without a nucleus are called *prokaryotic cells*. Those with a nucleus are called *eukaryotic cells*. ☑

What Are Prokaryotes?

A **prokaryote** is an organism made of one cell that does not have a nucleus or other organelles covered by a membrane. Prokaryotes are made of prokaryotic cells. There are two types of prokaryotes: bacteria and archaea.

READING CHECK

6. Compare What is one way prokaryotic and eukaryotic cells differ?

_____

_____

_____

BACTERIA

The most common prokaryotes are bacteria (singular, *bacterium*). Bacteria are the smallest known cells. These tiny organisms live almost everywhere. Some bacteria live in the soil and water. Others live on or inside other organisms. You have bacteria living on your skin and teeth and in your digestive system. The following are some characteristics of bacteria:

- no nucleus
- circular DNA shaped like a twisted rubber band
- no membrane-covered (or *membrane-bound*) organelles
- a cell wall outside the cell membrane
- a *flagellum* (plural, *flagella*), a tail-like structure that some bacteria use to help them move

Critical Thinking

7. Make Inferences Why do you think bacteria can live in your digestive system without making you sick?

_____

_____

_____

A Bacterium

TAKE A LOOK
8. Identify Label the parts of the bacterium using the following terms: DNA, flagellum, cell membrane, cell wall.

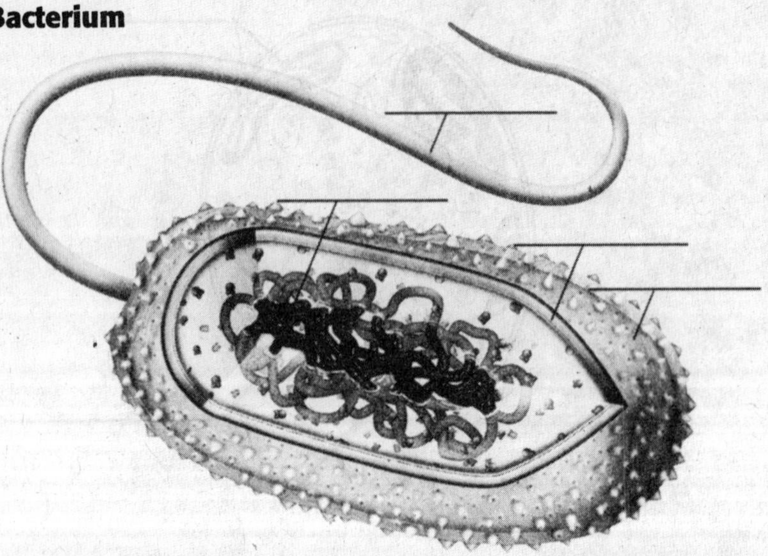

ARCHAEA

Archaea (singular, *archaeon*) and bacteria share the following characteristics:

- no nucleus
- no membrane-bound organelles
- circular DNA
- a cell wall

SECTION 1 The Diversity of Cells *continued*

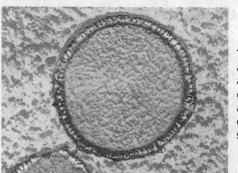

This photograph was taken with an electron microscope. This archaeon lives in volcanic vents deep in the ocean. Temperatures at these vents are very high. Most other living things could not survive there.

Archaea have some other features that no other cells have. For example, the cell wall and cell membrane of archaea are made of different substances from those of bacteria. Some archaea live in places where no other organisms could live. For example, some can live in the boiling water of hot springs. Others can live in toxic places such as volcanic vents filled with sulfur. Still others can live in very salty water in places such as the Dead Sea. ☑

What Are Eukaryotes?

Eukaryotic cells are the largest cells. They are about 10 times larger than bacteria cells. However, you still need a microscope to see most eukaryotic cells.

Eukaryotes are organisms made of eukaryotic cells. These organisms can have one cell or many cells. Yeast, which makes bread rise, is an example of a eukaryote with one cell. Multicellular organisms, or those made of many cells, include plants and animals.

Unlike prokaryotic cells, eukaryotic cells have a nucleus that holds their DNA. Eukaryotic cells also have membrane-bound organelles. ☑

☑ **READING CHECK**

9. Compare Name two ways that archaea differ from bacteria.

_____

_____

_____

_____

_____

☑ **READING CHECK**

10. Identify Name two things eukaryotic cells have that prokaryotic cells do not.

_____

_____

Eukaryotic Cell

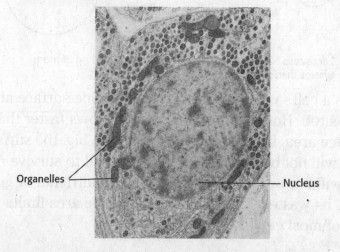

Organelles —

— Nucleus

SECTION 1 The Diversity of Cells *continued*

Organelles in a Typical Eukaryotic Cell

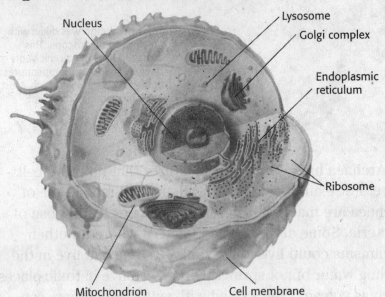

TAKE A LOOK
11. Identify Where is the genetic material found in this cell?

_____

_____

Critical Thinking
12. Apply Concepts The yolk of a chicken egg is a very large cell. Unlike most cells, egg yolks do not have to take in any nutrients. Why does this allow the cell to be so big?

_____

_____

_____

_____

_____

_____

Why Are Cells So Small?

Your body is made of trillions of cells. Most cells are so small you need a microscope to see them. More than 50 human cells can fit on the dot of this letter *i*. However, some cells are big. For example, the yolk of a chicken egg is one big cell! Why, then, are most cells small?

Cells take in food and get rid of waste through their outer surfaces. As a cell gets larger, it needs more food to survive. It also produces more waste. This means that more materials have to pass through the surface of a large cell than a small cell.

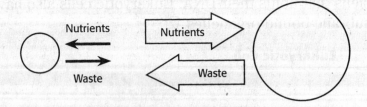

Large cells have to take in more nutrients and get rid of more wastes than small cells.

As a cell's volume increases, its outside surface area grows too. However, volume always grows faster than surface area. If the cell volume gets too big, the surface area will not be large enough for the cell to survive. The cell will not be able to take in enough nutrients or get rid of all its wastes. This means that surface area limits the size of most cells.

SECTION 1 The Diversity of Cells *continued*

SURFACE AREA AND VOLUME OF CELLS

To understand how surface area limits the size of a cell, study the figures below. Imagine that the cubes are cells. You can calculate the surface areas and volumes of the cells using these equations:

$$volume\ of\ cube = side \times side \times side$$

$$surface\ area\ of\ cube = number\ of\ sides \times area\ of\ side$$

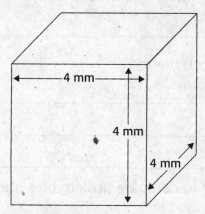

The volume of this cell is 64 mm³. Its surface area is 96 mm².

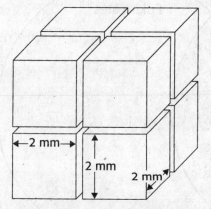

When the large cell is broken up into 8 smaller cells, the total volume stays the same. However, all of the small cells together have more surface area than the large cell. The total surface area of the small cells is 192 mm².

Math Focus
13. Calculate Ratios
Scientists say that most cells are small because of the surface area-to-volume ratio. What is this ratio for the large cell?

_____

_____

TAKE A LOOK
14. Compare Which cell has a greater surface area compared to its volume—the large cell or one of the smaller cells?

_____

_____

The large cell takes in and gets rid of the same amount of material as all of the smaller cells. However, the large cell does not have as much surface area as the smaller cells. Therefore, it cannot take in nutrients or get rid of wastes as easily as each of the smaller cells.

Name _____ Class _____ Date _____

Section 1 Review

NSES LS 1a, 1b, 1c, 2c, 3b, 5a

SECTION VOCABULARY

cell in biology, the smallest unit that can perform all life processes; cells are covered by a membrane and have DNA and cytoplasm

cell membrane a phospholipid layer that covers a cell's surface; acts as a barrier between the inside of a cell and the cell's environment

eukaryote an organism made up of cells that have a nucleus enclosed by a membrane; eukaryotes include animals, plants, and fungi, but not archaea or bacteria

nucleus in a eukaryotic cell, a membrane-bound organelle that contains the cell's DNA and that has a role in processes such as growth, metabolism, and reproduction

organelle one of the small bodies in a cell's cytoplasm that are specialized to perform a specific function

prokaryote an organism that consists of a single cell that does not have a nucleus

1. Identify What are the three parts of the cell theory?

_____

_____

_____

2. Compare Fill in the Venn Diagram below to compare prokaryotes and eukaryotes. Be sure to label the circles.

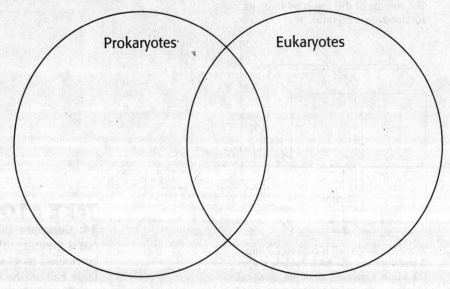

3. Apply Concepts You have just discovered a new organism. It has only one cell and was found on the ocean floor, at a vent of boiling hot water. The organism has a cell wall but no nucleus. Explain how you would classify this organism.

_____

_____

_____

CHAPTER 10 Cells: The Basic Units of Life

SECTION 2 Eukaryotic Cells

After you read this section, you should be able to answer these questions:

• What are the parts of a eukaryotic cell?

• What is the function of each part of a eukaryotic cell?

National Science Education Standards
LS 1a, 1b, 1c, 3a, 5a

What Are the Parts of a Eukaryotic Cell?

Plant cells and animal cells are two types of eukaryotic cells. A eukaryotic cell has many parts that help the cell stay alive.

CELL WALL

All plant cells have a cell wall. The **cell wall** is a stiff structure that supports the cell and surrounds the cell membrane. The cell wall of a plant cell is made of a type of sugar called cellulose.

Fungi (singular *fungus*), such as yeasts and mushrooms, also have cell walls. The cell walls of fungi are made of a sugar called *chitin*. Prokaryotic cells such as bacteria and archaea also have cell walls. ☑

STUDY TIP

Organize As you read this section, make a chart comparing plant cells and animal cells.

READING CHECK

1. Identify Name two kinds of eukaryotes that have a cell wall.

_____

_____

Plant Cell

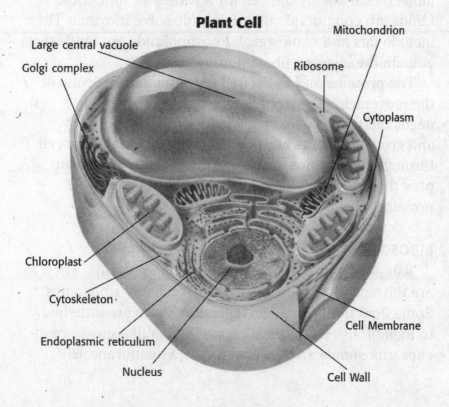

Large central vacuole
Golgi complex
Mitochondrion
Ribosome
Cytoplasm
Chloroplast
Cytoskeleton
Endoplasmic reticulum
Nucleus
Cell Membrane
Cell Wall

TAKE A LOOK

2. Identify Describe where the cell wall is located.

_____

_____

Animal Cell

Nucleus

Lysosome

Golgi complex

Endoplasmic reticulum

Cytoskeleton

Ribosome

Cytoplasm

Mitochondrion

Cell membrane

TAKE A LOOK
3. Compare Compare the pictures of an animal cell and a plant cell. Name three structures found in both.

_____

_____

_____

4. Explain What is the main function of the cell membrane?

_____

_____

_____

_____

_____

_____

5. Compare How are ribosomes different from other organelles?

_____

_____

_____

CELL MEMBRANE

All cells have a cell membrane. The cell membrane is a protective barrier that surrounds the cell. It separates the cell from the outside environment. In cells that have a cell wall, the cell membrane is found just inside the cell wall.

The cell membrane is made of different materials. It contains proteins, lipids, and phospholipids. Proteins are molecules made by the cell for a variety of functions. Lipids are compounds that do not dissolve in water. They include fats and cholesterol. Phospholipids are lipids that contain the element phosphorous.

The proteins and lipids in the cell membrane control the movement of materials into and out of the cell. A cell needs materials such as nutrients and water to survive and grow. Nutrients and wastes go in and out of the cell through the proteins in the cell membrane. Water can pass through the cell membrane without the help of proteins.

RIBOSOMES

Ribosomes are organelles that make proteins. They are the smallest organelles. A cell has many ribosomes. Some float freely in the cytoplasm. Others are attached to membranes or to other organelles. Unlike most organelles, ribosomes are not covered by a membrane. ✓

Ribosome
This organelle is where amino acids are hooked together to make proteins.

NUCLEUS

The nucleus is a large organelle in a eukaryotic cell. It contains the cell's genetic material, or DNA. DNA has the instructions that tell a cell how to make proteins.

The nucleus is covered by two membranes. Materials pass through pores in the double membrane. The nucleus of many cells has a dark area called the *nucleolus*.

The Nucleus

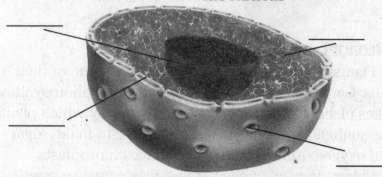

TAKE A LOOK
6. Identify Label the diagram of a nucleus using these terms: pore, DNA, nucleolus, double membrane.

ENDOPLASMIC RETICULUM

Many chemical reactions take place in the cell. Many of these reactions happen on or inside the endoplasmic reticulum. The **endoplasmic reticulum** (ER) is a system of membranes with many folds in which proteins, lipids, and other materials are made.

The ER is also part of the cell's delivery system. Its folds have many tubes and passageways. Materials move through the ER to other parts of the cell.

There are two types of ER: rough and smooth. Smooth ER makes lipids and helps break down materials that could damage the cell. Rough ER has ribosomes attached to it. The ribosomes make proteins. The proteins are then delivered to other parts of the cell by the ER. ☑

✓ READING CHECK
7. Compare What is the difference between smooth ER and rough ER?

_____

_____

_____

_____

Endoplasmic reticulum
This organelle makes lipids, breaks down drugs and other substances, and packages proteins for the Golgi complex.

MITOCHONDRIA

A **mitochondrion** (plural, *mitochondria*) is the organelle in which sugar is broken down to make energy. It is the main power source for a cell.

A mitochondrion is covered by two membranes. Most of a cell's energy is made in the inside membrane. Energy released by mitochondria is stored in a molecule called ATP. The cell uses ATP to do work.

Mitochondria are about the same size as some bacteria. Like bacteria, mitochondria have their own DNA. The DNA in mitochondria is different from the cell's DNA. ☑

Mitochondrion
This organelle breaks down food molecules to make ATP.

CHLOROPLASTS

Plants and algae have chloroplasts in some of their cells. *Chloroplasts* are organelles in which photosynthesis takes place. *Photosynthesis* is a process by which plants use sunlight, carbon dioxide, and water to make sugar and oxygen. Animal cells do not have chloroplasts.

Chloroplasts are green because they contain a green molecule called *chlorophyll*. Chlorophyll traps the energy of sunlight. Mitochondria then use the sugar made in photosynthesis to make ATP.

Critical Thinking

9. Infer Why don't animal cells need chloroplasts?

_____

_____

_____

_____

Chloroplast
This organelle uses the energy of sunlight to make food.

CYTOSKELETON

The cytoskeleton is a web of proteins inside the cell. It acts as both a skeleton and a muscle. The cytoskeleton helps the cell keep its shape. It also helps some cells, such as bacteria, to move.

VESICLES

A **vesicle** is a small sac that surrounds material to be moved. The vesicle moves material to other areas of the cell or into or out of the cell. All eukaryotic cells have vesicles.

GOLGI COMPLEX

The **Golgi complex** is the organelle that packages and distributes proteins. It is the "post office" of the cell. The Golgi complex looks like the smooth ER.

The ER delivers lipids and proteins to the Golgi complex. The Golgi complex can change the lipids and proteins to do different jobs. The final products are then enclosed in a piece of the Golgi complex's membrane. This membrane pinches off to form a vesicle. The vesicle transports the materials to other parts of the cell or out of the cell. ☑

Golgi complex
This organelle processes and transports proteins and other materials out of cell.

READING CHECK

10. Define What is the function of the Golgi complex?

_____

_____

_____

LYSOSOMES

Lysosomes are organelles that contain digestive enzymes. The enzymes destroy worn-out or damaged organelles, wastes, and invading particles.

Lysosomes are found mainly in animal cells. The cell wraps itself around a particle and encloses it in a vesicle. Lysosomes bump into the vesicle and pour enzymes into it. The enzymes break down the particles inside the vesicle. Without lysosomes, old or dangerous materials could build up and damage or kill the cell.

Lysosome
This organelle digests food particles, wastes, cell parts, and foreign invaders.

VACUOLES

A vacuole is a vesicle. In plant and fungal cells, some vacuoles act like lysosomes. They contain enzymes that help a cell digest particles. The large central vacuole in plant cells stores water and other liquids. Large vacuoles full of water help support the cell. Some plants wilt when their vacuoles lose water. ☑

Large central vacuole
This organelle stores water and other materials.

READING CHECK

11. Identify Vacuoles are found in what types of eukaryotic cells?

_____

_____

Section 2 Review

SECTION VOCABULARY

cell wall a rigid structure that surrounds the cell membrane and provides support to the cell	**lysosome** a cell organelle that contains digestive enzymes
endoplasmic reticulum a system of membranes that is found in a cell's cytoplasm and that assists in the production, processing, and transport of proteins and in the production of lipids	**mitochondrion** in eukaryotic cells, the cell organelle that is surrounded by two membranes and that is the site of cellular respiration
Golgi complex cell organelle that helps make and package materials to be transported out of the cell	**ribosome** cell organelle composed of RNA and protein; the site of protein synthesis
	vesicle a small cavity or sac that contains materials in a eukaryotic cell

1. Compare Name three parts of a plant cell that are not found in an animal cell.

_____

_____

2. Explain How does a cell get water and nutrients?

_____

_____

_____

3. Explain What would happen to an animal cell if it had no lysosomes?

_____

_____

_____

4. Apply Concepts Which kind of cell in the human body do you think would have more mitochondria—a muscle cell or a skin cell? Explain.

_____

_____

_____

5. List What are two functions of the cytoskeleton?

_____

_____

CHAPTER 10 Cells: The Basic Units of Life

SECTION 3 The Organization of Living Things

BEFORE YOU READ

After you read this section, you should be able to answer these questions:

• What are the advantages of being multicellular?

• What are the four levels of organization in living things?

• How are structure and function related in an organism?

What Is an Organism?

Anything that can perform life processes by itself is an **organism**. An organism made of a single cell is called a *unicellular organism*. An organism made of many cells is a multicellular organism. The cells in a multicellular organism depend on each other for the organism to survive. ☑

What Are the Benefits of Having Many Cells?

Some organisms exist as one cell. Others can be made of trillions of cells. A *multicellular organism* is an organism made of many cells.

There are three benefits of being multicellular: larger size, longer life, and specialization of cells.

LARGER SIZE

Most multicellular organisms are bigger than one-celled organisms. In general, a larger organism, such as an elephant, has few predators. ☑

LONGER LIFE

A multicellular organism usually lives longer than a one-celled organism. A one-celled organism is limited to the life span of its one cell. The life span of a multicellular organism, however, is not limited to the life span of any one of its cells.

SPECIALIZATION

In a multicellular organism, each type of cell has a particular job. Each cell does not have to do everything the organism needs. Specialization makes the organism more efficient.

STUDY TIP

Outline As you read, make an outline of this section. Use the heading questions from the section in your outline.

☑ **READING CHECK**

1. Define What is an organism?

_____

_____

_____

 **READING CHECK**

2. Identify Name one way that being large can benefit an organism.

_____

_____

_____

LS 1d Specialized cells perform specialized functions in multicellular organisms. Groups of specialized cells cooperate to form a tissue, such as a muscle. Different tissues are in turn grouped together and form larger functional units, called organs. Each type of cell, tissue, and organ has a distinct structure and set of functions that serve the organism as a whole.

3. List What are the four levels or organization for an organism?

_____
_____
_____
_____

What Are the Four Levels of Organization of Living Things?

Multicellular organisms have four levels of organization:

Cell

Tissue

Cells form tissues.

Organ

Tissues form organs.

Organ system

Organs form organ systems.

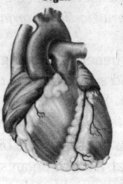

Organ systems form organisms such as you.

TAKE A LOOK

4. Explain Are the cells that make up heart tissue prokaryotic or eukaryotic? How do you know?

_____
_____
_____
_____

SECTION 3 The Organization of Living Things *continued*

CELLS WORK TOGETHER AS TISSUES

A **tissue** is a group of cells that work together to perform a specific job. Heart muscle tissue, for example, is made of many heart muscle cells.

TISSUES WORK TOGETHER AS ORGANS

A structure made of two or more tissues that work together to do a certain job is called an **organ**. Your heart, for example, is an organ made of different tissues. The heart has muscle tissues and nerve tissues that work together.

ORGANS WORK TOGETHER AS ORGAN SYSTEMS

A group of organs working together to do a job is called an **organ system**. An example of an organ system is your digestive system. Organ systems depend on each other to help the organism function. For example, the digestive system depends on the cardiovascular and respiratory systems for oxygen.

HOW DOES STRUCTURE RELATE TO FUNCTION?

In an organism, the structure and function of part are related. **Function** is the job the part does. **Structure** is the arrangement of parts in an organism. It includes the shape of a part or the material the part is made of.

Critical Thinking

5. Apply Concepts Do prokaryotes have tissues? Explain.

_____

_____

_____

_____

 **Say It**

Name With a partner, name as many of the organs in the human body as you can.

The function of the lungs is to bring oxygen to the body and get rid of carbon dioxide. The structure of the lungs helps them to perform their function.

Oxygen-poor blood

Oxygen-rich blood

The lungs contain tiny, spongy sacs that blood can flow through. Carbon dioxide moves out of the blood and into the sacs. Oxygen flows from the sacs into the blood. If the lungs didn't have this structure, it would be hard for them to perform their function.

Blood vessels

Section 3 Review

SECTION VOCABULARY

function the special, normal, or proper activity of an organ or part	**organism** a living thing; anything that can carry out life processes independently
organ a collection of tissues that carry out a specialized function of the body	**structure** the arrangement of parts in an organism
organ system a group of organisms that work together to perform body functions	**tissue** a group of similar cells that perform a common function

1. List What are three benefits of being multicellular?

_____

_____

_____

2. Apply Concepts Could an organism have organs but no tissues? Explain.

_____

_____

3. Compare How are structure and function different?

_____

_____

_____

4. Explain What does "specialization of cells" mean?

_____

_____

_____

5. Apply Concepts Why couldn't your heart have only cardiac tissue?

_____

_____

_____

6. Explain Why do multicellular organisms generally live longer than unicellular organisms?

_____

_____

CHAPTER 11 | Classification

SECTION 1

Sorting It All Out

National Science Education Standards
LS 5a

BEFORE YOU READ

After you read this section, you should be able to answer these questions:

• What is classification?

• How do scientists classify organisms?

• How do scientists name groups of organisms?

Why Do We Classify Things?

Imagine that you lived in a tropical rain forest and had to get your own food, shelter, and clothing from the forest. What would you need to know to survive? You would need to know which plants were safe to eat and which were not. You would need to know which animals you could eat and which ones could eat you. In other words, you would need to study the organisms around you and put them into useful groups. You would *classify* them.

Biologists use a *classification* system to group the millions of different organisms on Earth. **Classification** is putting things into groups based on characteristics the things share. Classification helps scientists answer several important questions:

• What are the defining characteristics of each species?

• When did the characteristics of a species evolve?

• What are the relationships between different species?

How Do Scientists Classify Organisms?

What are some ways we can classify organisms? Perhaps we could group them by where they live or how they are useful to humans. Throughout history, people have classified organisms in many different ways.

In the 1700s, a Swedish scientist named Carolus Linnaeus created his own system. His system was based on the structure or characteristics of organisms. With his new system, Linnaeus founded modern taxonomy. **Taxonomy** is the science of describing, classifying, and naming organisms. ☑

STUDY TIP

Organize As you read, make a diagram to show the eight-level system of organization.

Say It

Discuss With a partner, describe some items at home that you have put into groups. Explain why you grouped them and what characteristics you used.

READING CHECK

1. Explain How did Linnaeus classify organisms?

_____

_____

_____

SECTION 1 Sorting It All Out *continued*

CLASSIFICATION TODAY

Taxonomists use an eight-level system to classify living things based on shared characteristics. Scientists also use shared characteristics to describe how closely related living things are.

The more characteristics organisms share, the more closely related they may be. For example, the platypus, brown bear, lion, and house cat are thought to be related because they share many characteristics. These animals all have hair and mammary glands, so they are grouped together as mammals. However, they can also be classified into more specific groups.

BRANCHING DIAGRAMS

Shared characteristics can be shown in a *branching diagram*. Each characteristic on the branching diagram is shared by only the animals above it. The characteristics found higher on the diagram evolved more recently than the characteristics below them.

In the diagram below, all of the animals have hair and mammary glands. However, only the brown bear, lion, and house cat give birth to live young. More recent organisms are at the ends of branches high on the diagram. For example, according to the diagram, the house cat evolved more recently than the platypus. ☑

| Platypus | Brown bear | Lion | House cat |

Ability to purr

Retractable claws

Giving birth to live young

Hair, mammary glands

This branching diagram shows the similarities and differences between four kinds of mammals. The bottom of the diagram begins in the past, and the tips of the branches end in the present.

Critical Thinking

2. Infer What is the main difference between organisms that share many characteristics and organisms that do not?

_____

_____

_____

_____

_____

✔ **READING CHECK**

3. Identify On a branching diagram, where would you see the characteristics that evolved most recently?

_____

TAKE A LOOK

4. Identify According to the diagram, which organisms evolved before the lion? Circle these organisms.

What Are the Levels of Classification?

Scientists use shared characteristics to group organisms into eight levels of classification. At each level of classification, there are fewer organisms than in the level above. A domain is the largest, most general level of classification. Every living thing is classified into one of three domains.

Species is the smallest level of classification. A species is a group of organisms that can mate and produce fertile offspring. For example, dogs are all one species. They can mate with one another and have fertile offspring. The figure on the next page shows each of the eight levels of classification.

TWO-PART NAMES

We usually call organisms by common names. For example, "cat," "dog," and "human" are all common names. However, people who speak a language other than English have different names for a cat and dog. Sometimes, organisms are even called by different names in English. For example, cougar, mountain lion, and puma are three names for the same animal! ☑

Scientists need to be sure they are all talking about the same organism. They give organisms *scientific names*. Scientific names are the same in all languages. An organism has only one scientific name.

Scientific names are based on the system created by Linnaeus. He gave each kind of organism a two-part name. The first part of the name is the *genus*, and the second part is the *species*. All genus names begin with a capital letter. All species names begin with a lowercase letter. Both words in a scientific name are underlined or italicized. For example, the scientific name for the Asian elephant is *Elephas maximus*. ☑

☑ **READING CHECK**

5. List What are two problems with common names?

_____

_____

_____

_____

_____

☑ **READING CHECK**

6. Identify What are the two parts of a scientific name?

_____

_____

```
        ┌─────────────┐
        │  Two-part   │
        │    name     │
        └─────────────┘
          ↙         ↘
┌──────────────┐   ┌──────────────┐
│ Genus:       │   │ species:     │
│ Elephas      │   │ maximus      │
└──────────────┘   └──────────────┘
```

Levels of Classification of the House Cat

Kingdom Animalia: All animals are in the kingdom Animalia.

Phylum Chordata: All animals in the phylum Chordata have a hollow nerve cord. Most have a backbone.

Class Mammalia: Animals in the class Mammalia have a backbone. They also nurse their young.

Order Carnivora: Animals in the order Carnivora have a backbone and nurse their young. They also have special teeth for tearing meat.

Family Felidae: Animals in the family Felidae are cats. They have a backbone, nurse their young, have special teeth for tearing meat, and have retractable claws.

TAKE A LOOK

7. Identify Which level contains organisms that are more closely related: a phylum or a class?

_____

_____

8. Describe How does the number of organisms change from the level of kingdom to the level of species?

_____

_____

_____

Genus *Felis*: Animals in the genus *Felis* share traits with other animals in the same family. However, these cats cannot roar; they can only purr.

Species *Felis catus*: The species *Felis catus* is the common house cat. The house cat shares traits with all of the organisms in the levels above the species level, but it also has unique traits.

What Is a Dichotomous Key?

What could you do if you found an organism that you did not recognize? You could use a special guide called a dichotomous key. A **dichotomous key** is set of paired statements that give descriptions of organisms. These statements let you rule out out certain species based on characteristics of your specimen. There are many dichotomous keys for many different kinds of organisms. You could even make your own!

In a dichotomous key, there are only two choices at each step. To use the key, you start with the first pair of statements. You choose the statement from the pair that describes the organism. At each step, the key may identify the organism or it may direct you to another pair of statements. By working through the statements in order, you can identify the organism.

Critical Thinking

9. Infer Why couldn't one single dichotomous key be used for all of the organisms on Earth?

_____

_____

_____

_____

Dichotomous Key to 10 Common Mammals in the Eastern United States

1.	a. This mammal flies. Its "hand" forms a wing.	little brown bat
	b. This mammal does not fly. It's "hand" does not form a wing.	Go to step 2.
2.	a. This mammal has no hair on its tail.	Go to step 3.
	b. This mammal has hair on its tail.	Go to step 4.
3.	a. This mammal has a short, naked tail.	eastern mole
	b. This mammal has a long, naked tail.	Go to step 5.
4.	a. This mammal has a black mask across its face.	raccoon
	b. This mammal does not have a black mask across its face.	Go to step 6.
5.	a. This mammal has a tail that is flat and paddle shaped.	beaver
	b. This mammal has a tail that is not flat or paddle shaped.	opossum
6.	a. This mammal is brown and has a white underbelly.	Go to step 7.
	b. This mammal is not brown and does not have a white underbelly.	Go to step 8.
7.	a. This mammal has a long, furry tail that is black on the tip.	longtail weasel
	b. This mammal has a long tail that has little fur.	white-footed mouse
8.	a. This mammal is black and has a narrow white stripe on its forehead and broad white stripes on its back.	striped skunk
	b. This mammal is not black and does not have white stripes.	Go to step 9.
9.	a. This mammal has long ears and a short, cottony tail.	eastern cottontail
	b. This mammal has short ears and a medium-length tail.	woodchuck

TAKE A LOOK
10. Identify Use this dichotomous key to identify the two animals shown.

_____

_____

Section 1 Review

SECTION VOCABULARY

classification the division of organisms into groups, or classes, based on specific characteristics	**taxonomy** the science of describing, naming, and classifying organisms
dichotomous key an aid that is used to identify organisms and that consists of the answers to a series of questions	

1. List Give the eight levels of classification from the largest to the smallest.

_____

_____

2. Identify According to the branching diagram below, which characteristic do ferns have that mosses do not?

_____

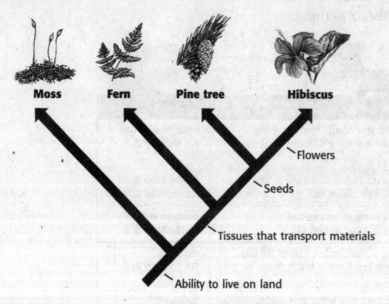

Moss   Fern   Pine tree   Hibiscus

Flowers

Seeds

Tissues that transport materials

Ability to live on land

3. Analyze Which species in the diagram above is most similar to the hibiscus? Which is the least similar?

_____

_____

4. Identify What are the two parts of a scientific name?

_____

5. Infer Could you use the dichotomous key in this section to identify a species of lizard? Explain your answer.

_____

_____

CHAPTER 11 (Classification)

SECTION
2 Domains and Kingdoms

National Science
Education Standards
LS 1f, 2a, 4b, 4c, 5b

BEFORE YOU READ

After you read this section, you should be able to answer these questions:

- How are prokaryotes classified?
- How are eukaryotes classified?

How Do Scientists Classify Organisms?

For hundreds of years, all organisms were classified as either plants or animals. However, as more organisms were discovered, scientists found some organisms that did not fit well into these two kingdoms. Some animals, for example, had characteristics of both plants and animals.

What would you call an organism that is green and makes its own food? Is it a plant? What if the organism moved and could also eat other organisms? Plants generally do neither of these things. Is the organism a plant or an animal?

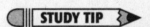

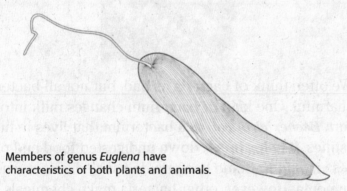

Members of genus *Euglena* have
characteristics of both plants and animals.

The organism above belongs to genus *Euglena*. Its members show all of the characteristics just described. As scientists discovered organisms, such as *Euglena*, that didn't fit easily into existing groups, they created new ones. As they added kingdoms, scientists found that members of some kingdoms were closely related to members of other kingdoms. Today, scientists group kingdoms into *domains*.

All organisms on Earth are grouped into three domains. Two domains, Bacteria and Archaea, are made up of prokaryotes. The third domain, Eukarya, is made up of all the eukaryotes. Scientists are still working to describe the kingdoms in each of the three domains.

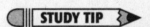

STUDY TIP

List As you read this section, make a list of the domains and kingdoms scientists use to classify organisms.

Critical Thinking

1. Apply Concepts In which domain would multicellular organisms be classified? Explain your answer.

_____

_____

_____

_____

How Are Prokaryotes Classified?

A prokaryote is a single-celled organism that does not have a nucleus. Prokaryotes are the oldest group of organisms on Earth. They make up two domains: Archaea and Bacteria.

DOMAIN ARCHAEA

Domain **Archaea** is made up of prokaryotes. The cell walls and cell membranes of archaea are made of different substances than those of other prokaryotes. Many archaea can live in extreme environments where other organisms could not survive. Some archaea can also be found in more moderate environments, such as the ocean. ☑

DOMAIN BACTERIA

All bacteria belong to domain **Bacteria**. Bacteria can be found in the air, in soil, in water, and even on and inside the human body!

We often think of bacteria as bad, but not all bacteria are harmful. One kind of bacterium changes milk into yogurt. *Escherichia coli* is a bacterium that lives in human intestines. It helps break down undigested food and produces vitamin K. Some bacteria do cause diseases, such as pneumonia. However, other bacteria make chemicals that can help us fight bacteria that cause disease. ☑

The Grand Prismatic Spring in Yellowstone National Park contains water that is about 90°C (190°F). Most organisms would die in such a hot environment.

✔ READING CHECK

2. Compare How are members of Archaea different from other prokaryotes?

_____

_____

_____

_____

_____

✔ READING CHECK

3. Explain Are all bacteria harmful? Explain your answer.

_____

_____

_____

_____

TAKE A LOOK

4. Apply Concepts What kind of prokaryotes do you think could live in this spring? Explain your answer.

_____

_____

_____

How Are Eukaryotes Classified?

Organisms that have cells with membrane-bound organelles and a nucleus are called *eukaryotes*. All eukaryotes belong to domain **Eukarya**. Domain Eukarya includes the following kingdoms: Protista, Fungi, Plantae, and Animalia.

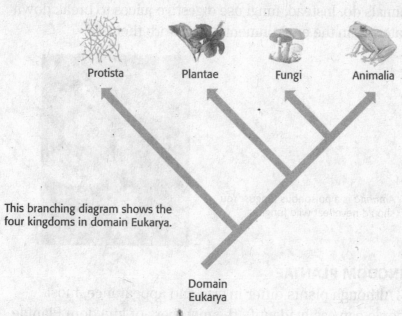

This branching diagram shows the four kingdoms in domain Eukarya.

STANDARDS CHECK

LS 5a Biological evolution accounts for the diversity of species developed through gradual processes over many generations. Species acquire many of their unique characteristics through biological adaptation, which involves the selection of naturally occurring variations in populations. Biological adaptations include changes in structures, behaviors, or physiology that enhance survival and reproductive success in a particular environment.

Word Help: selection
the process of choosing

Word Help: variation
a difference in the form or function

5. Identify Based on the branching diagram, which two kingdoms in Eukarya evolved most recently? How do you know?

_____

_____

_____

_____

_____

_____

KINGDOM PROTISTA

Members of kingdom **Protista** are either single-celled or simple multicellular organisms. They are commonly called *protists*. Scientists think that the first protists evolved from ancient bacteria about 2 billion years ago. Much later, plants, fungi, and animals evolved from ancient protists.

Kingdom Protista contains many different kinds of organisms. Some, such as *Paramecium*, resemble animals. They are called *protozoa*. Plantlike protists are called *algae*. Some algae, such as phytoplankton, are single cells. Others, such as kelp, are multicellular. Multicellular slime molds also belong to kingdom Protista.

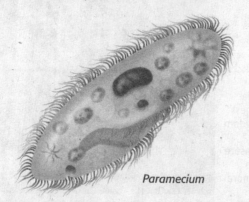

Paramecium

KINGDOM FUNGI

Molds and mushrooms are members of kingdom **Fungi**. Some fungi (singular, *fungus*) are unicellular. That is, they are single-celled organisms. Yeast is unicellular. Most other fungi are multicellular. Unlike plants, fungi do not perform photosynthesis. However, they also do not eat food, as animals do. Instead, fungi use digestive juices to break down materials in the environment and absorb them. ☑

6. Describe How do fungi get food?

_____

_____

_____

_____

Amanita is a poisonous fungus. You should never eat wild fungi.

KINGDOM PLANTAE

Although plants differ in size and appearance, most people can easily identify the members of kingdom Plantae. Kingdom **Plantae** contains organisms that are eukaryotic, have cell walls, and make food by photosynthesis. Most plants need sunlight to carry out photosynthesis. Therefore, plants must live in places where light can reach.

The food that plants make is important for the plants and also for other organisms. Many animals, fungi, protists, and bacteria get nutrients from plants. When they digest the plant material, they get the energy stored by the plant. Plants also provide homes for other organisms.

Math Focus

7. Calculate The average student's arms extend about 1.3 m. How many students would have to join hands to form a human chain around a giant sequoia?

_____

The giant sequoia is one of the largest members of kingdom Plantae. A giant sequoia can measure 30 m around the base and grow more than 91 m tall!

SECTION 2 Domains and Kingdoms *continued*

KINGDOM ANIMALIA

Kingdom **Animalia** contains complex, multicellular organisms. Organisms in kingdom Animalia are commonly called *animals*. The following are some characteristics of animals:

• Their cells do not have cell walls.
• They are able to move from place to place.
• They have sense organs that help them react quickly to their environment.

TAKE A LOOK
8. Identify Which animal characteristic from above can be seen in this bald eagle?

_____

_____

STRANGE ORGANISMS

Some organisms are not easy to classify. For example, some plants can eat other organisms to get nutrition as animals do. Some protists use photosynthesis as plants do but also move around as animals do.

Red Cup Sponge

Critical Thinking

9. Apply Concepts To get nutrients, a Venus' flytrap uses photosynthesis and traps and digests insects. Its cells have cell walls. Into which kingdom would you place this organism? Explain your answer.

_____

_____

_____

_____

_____

What kind of organism is this red cup sponge? It does not have sense organs and cannot move for most of its life. Because of this, scientists once classified sponges as plants. However, sponges cannot make their own food as plants do. They must eat other organisms to get nutrients. Today, scientists classify sponges as animals. Sponges are usually considered the simplest animals.

Section 2 Review

NSES   LS 1f, 2a, 4b, 4c, 5b

SECTION VOCABULARY

Animalia a kingdom made up of complex, multicellular organisms that lack cell walls, can usually move around, and quickly respond to their environment

Archaea in a modern taxonomic system, a domain made up of prokaryotes (most of which are known to live in extreme environments) that are distinguished from other prokaryotes by differences in their genetics and in the makeup of their cell wall; this domain aligns with the traditional kingdom Archaebacteria

Bacteria in a modern taxonomic system, a domain made up of prokaryotes that usually have a cell wall and that usually reproduce by cell division. This domain aligns with the traditional kingdom Eubacteria

Eukarya in a modern taxonomic system, a domain made up of all eukaryotes; this domain aligns with the traditional kingdoms Protista, Fungi, Plantae, and Animalia

Fungi a kingdom made up of nongreen, eukaryotic organisms that have no means of movement, reproduce by using spores, and get food by breaking down substances in their surroundings and absorbing the nutrients

Plantae a kingdom made up of complex, multicellular organisms that are usually green, have cell walls made of cellulose, cannot move around, and use the sun's energy to make sugar by photosynthesis

Protista a kingdom of mostly one-celled eukaryotic organisms that are different from plants, animals, bacteria, and fungi

1. Compare What is one major difference between domain Eukarya and domains Bacteria and Archaea?

_____

_____

2. Explain Why do scientists continue to add new kingdoms to their system of classification?

_____

_____

3. Analyze Methods Why do you think Linnaeus did not include classifications for archaea and bacteria?

_____

_____

_____

4. Apply Concepts Based on its characteristics described at the beginning of this section, in which kingdom would you classify *Euglena*?

_____

_____

CHAPTER 12 Bacteria and Viruses

SECTION 1

Bacteria and Archaea

National Science
Education Standards
LS 1b, 1c, 2a, 3a, 3b, 4b

BEFORE YOU READ

After you read this section, you should be able to answer these questions:

• What are bacteria and archaea?

• What are the characteristics of bacteria?

• How do archaea and bacteria differ?

What Are Bacteria and Archaea?

Organisms are grouped by traits they have in common. All living things can be grouped into one of three domains: Bacteria, Archaea, or Eukarya.

All organisms in domain Eukarya are eukaryotes. Each cell of a *eukaryote* has a nucleus and membrane-bound organelles. All organisms in domains Bacteria and Archaea are prokaryotes. A **prokaryote** is an organism that is single-celled and has no nucleus.

Although many prokaryotes live in groups, they are single organisms that can move, get food, and make copies of themselves. Most prokaryotes are very small and cannot be seen without a microscope. However, you can see some very large bacteria with your naked eye. ☑

STUDY TIP

Underline Use colored pencils to underline the characteristics of bacteria in red, characteristics of archaea in blue, and characteristics shared by both in green.

READING CHECK

1. Identify What are two characteristics of prokaryotes?

_____

_____

_____

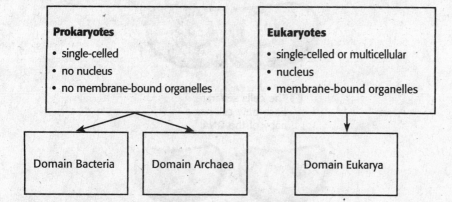

What Characteristics Do Archaea and Bacteria Share?

NO NUCLEUS

Prokaryotes do not store their DNA in a nucleus as eukaryotes do. Their DNA is stored as a circular loop inside the cell. ☑

READING CHECK

2. Describe What does the DNA of a prokaryote look like?

_____

_____

REPRODUCTION

Prokaryotes copy themselves, or reproduce, by a process called binary fission. **Binary fission** is reproduction in which a single-celled organism splits into two single-celled organisms. Before a prokaryote can reproduce, it must make a copy of its loop of DNA. After the cell splits, the two new cells are identical to the original cell. ☑

3. Identify How do prokaryotes reproduce?

_____

_____

Binary Fission

❶ The cell grows.

❷ The cell makes a copy of its DNA. Both copies attach to the cell membrane.

❸ The DNA and its copy separate as the cell grows larger.

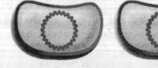
❹ The cells separate. Each new cell has a copy of the DNA.

TAKE A LOOK

4. Describe After binary fission, how do the two cells compare to the original cell?

_____

_____

What Are Some Characteristics of Bacteria?

Most of the prokaryotes that scientists have found are bacteria. Domain Bacteria has more individual members than domains Archaea and Eukarya combined have. Bacteria can be found almost everywhere.

SECTION 1 Bacteria and Archaea *continued*

SHAPE

Most bacteria are one of three shapes: bacilli, cocci, and spirilla. *Bacilli* are rod-shaped. *Cocci* are spherical. *Spirilla* are long and spiral-shaped. Different shapes help bacteria survive. Most bacteria have a stiff cell wall that gives them their shape. ☑

Some bacteria have hairlike parts called *flagella* (singular, *flagellum*). A flagellum works like a tail to push a bacterium through fluids.

☑ **READING CHECK**

5. List What are three common shapes of bacteria?

_____

_____

_____

The Most Common Shapes of Bacteria

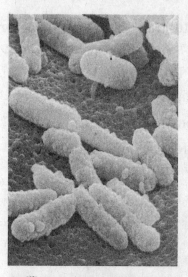

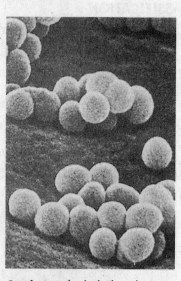

Bacilli are rod shaped. They have a large surface area, which helps them take in nutrients. However, a large surface area causes them to dry out quickly.

Cocci are spherical. They do not dry out as quickly as rod-shaped bacteria.

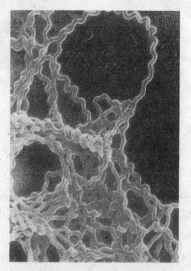

Spirilla are long and spiral-shaped. They have flagella at both ends. These tail-like structures help the bacteria move through fluids.

TAKE A LOOK

6. Compare What advantage do cocci have over bacilli?

_____

_____

_____

7. Compare What advantage do bacilli have over cocci?

_____

_____

_____

ENDOSPORES

Most bacteria do well in warm, moist places. Some species of bacteria die in dry and cold surroundings. However, some bacteria form endospores to survive these kinds of conditions. An **endospore** is a thick, protective covering that forms around the DNA of a bacterium. The endospore protects the DNA from changes in the environment. When conditions are good, the endospores break open, and the bacteria begin to grow.

CLASSIFICATION

Scientists can classify bacteria by the way the bacteria get food. There are three ways for bacteria to get food: consume it, decompose it, or produce it.

- *Consumers* eat other organisms.
- *Decomposers* eat dead organisms or waste.
- *Producers* make their own food. Some bacteria can make food using the energy from sunlight.

Decomposers, such as the ones helping to decay this leaf, return nutrients to the soil. This allows other living things to use those nutrients.

CYANOBACTERIA

Cyanobacteria are producers. These bacteria have a green pigment called *chlorophyll*. Chlorophyll traps the energy from the sun. The cell uses this energy to make food.

Some scientists think that billions of years ago, bacteria similar to cyanobacteria began to live inside larger cells. According to this hypothesis, the bacteria made food for itself and the larger cells. In return, the larger cells protected the bacteria. This relationship may have led to the first plant cells on Earth.

STANDARDS CHECK

LS 4b Populations of organisms can be categorized by the functions they serve in an ecosystem. Plants and some microorganisms are producers—they make their own food. All animals, including humans, are consumers, which obtain their food by eating other organisms. Decomposers, primarily bacteria and fungi, are consumers that use waste materials and dead organisms for food. Food webs identify the relationship among producers, consumers, and decomposers in an ecosystem.

Word Help: categorized
to put into groups or classes

8. List Name three roles bacteria can play in an ecosystem.

_____

_____

_____

How Do Archaea Differ from Bacteria?

Like bacteria, archaea are prokaryotes. However, archaea are different from bacteria. For example, not all archaea have cell walls. When they do have them, the cell walls are made of different materials than the cell walls of bacteria. ☑

There are three types of archaea: heat lovers, salt lovers, and methane makers. *Heat lovers* live in hot ocean vents and hot springs. They usually live in water that is 60°C to 80°C. However, they have been found in living in water as hot as 250°C.

Salt lovers live where there are high levels of salt, such as the Dead Sea. *Methane makers* give off methane gas. Methane makers often live in swamps. They can also live inside animal intestines.

HARSH ENVIRONMENTS

Although bacteria can be found almost anywhere, archaea can live in places where even bacteria cannot survive. For example, many archaea live in places with little or no oxygen. Many can also survive very high temperatures and pressures.

Scientists have found archaea in the hot springs at Yellowstone National Park and beneath 430 m of ice in Antarctica. Archaea have even been found 8 km below the surface of the Earth! Even though they can be found in harsh environments, many archaea also live in more moderate environments, such as the ocean.

☑ **READING CHECK**

9. Identify What is one difference between archaea and bacteria?

_____

_____

_____

_____

TAKE A LOOK

10. Identify What type of archaea do you think would live in this swamp?

_____

_____

Critical Thinking

11. Infer What kind of prokaryote would most likely be found near vents at the bottom of the ocean with extremely high temperatures? Explain your answer.

_____

_____

_____

_____

Name _____ Class _____ Date _____

Section 1 Review

NSES   LS 1b, 1c, 2a, 3a, 3b, 4b

SECTION VOCABULARY

binary fission a form of asexual reproduction in single-celled organisms by which one cell divides into two cells of the same size	**prokaryote** an organism that consists of a single cell that does not have a nucleus
endospore a thick-walled protective spore that forms inside a bacterial cell and resists harsh conditions	

1. List What are the three domains that include all living things?

_____

2. Compare Fill in the Venn Diagram to compare bacteria and archaea.

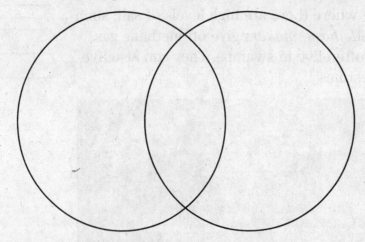

3. Describe How do some scientists think the first plants appeared on Earth?

_____

_____

_____

_____

4. List Name three kinds of archaea.

_____

_____

5. Infer Do you think it would be possible to find bacteria and archaea living in the same environment? Explain your answer.

_____

_____

_____

CHAPTER 12 Bacteria and Viruses

SECTION 2 Bacteria's Role in the World

National Science
Education Standards
LS 1f

BEFORE YOU READ

After you read this section, you should be able to answer these questions:

• How are some bacteria helpful?
• How are some bacteria harmful?

Are All Bacteria Harmful?

Bacteria are everywhere. They live in our water, our food, and our bodies. Some bacteria cause disease. However, most of the types of bacteria are helpful to organisms and the environment. Some bacteria move nitrogen throughout the environment. Other bacteria help recycle dead animals and plants. Still other bacteria are used to help scientists make medicines. Bacteria help us every day.

How Are Bacteria Helpful to Plants?

Plants need nitrogen to live. Although nitrogen makes up about 78% of the air, plants cannot use this nitrogen directly. They need to take in a different form of nitrogen. Bacteria in the soil take nitrogen from the air and change it into a form that plants can use. This process is called *nitrogen fixation*.

STUDY TIP

Outline As you read, make an outline of this section. Use the header questions to help you make your outline.

Say It

Discuss Does it surprise you to learn that not all bacteria are harmful? What were some things you used to believe about bacteria? With a partner, talk about how your view of bacteria has changed.

Bacteria's Role in the Nitrogen Cycle

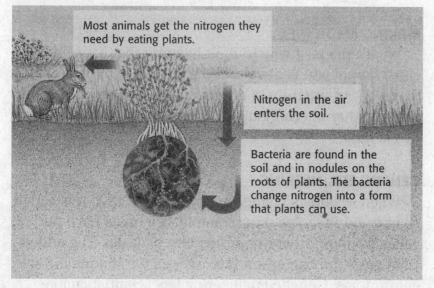

Most animals get the nitrogen they need by eating plants.

Nitrogen in the air enters the soil.

Bacteria are found in the soil and in nodules on the roots of plants. The bacteria change nitrogen into a form that plants can use.

TAKE A LOOK

1. **Explain** How do most animals get the nitrogen that they need?

_____

_____

_____

Name _____ Class _____ Date _____

How Are Bacteria Helpful to the Environment?

Bacteria are useful for recycling nutrients and cleaning up pollution. In nature, bacteria break down dead plants and animals. Breaking down dead matter helps recycle the nutrients so they can be used by other organisms.

Some bacteria are used to fight pollution. This process is called bioremediation. **Bioremediation** means using microorganisms, such as bacteria, to clean up harmful chemicals. Bioremediation is used to clean up oil spills and waste from cities, farms, and industries.

How Are Bacteria Helpful to People?

MAKING FOOD

Many of the foods you eat are made with the help of bacteria. Bacteria are commonly used in dairy products. Every time you eat cheese, yogurt, buttermilk, or sour cream, you are also eating bacteria. These products are made using milk and bacteria. Bacteria change the sugar in milk, called *lactose*, into *lactic acid*. Lactic acid adds flavor to the food and preserves it.

Bacteria are used to make many kinds of food.

MAKING MEDICINES

What's the best way to fight bacteria that cause disease? Would you believe that the answer is to use other bacteria? An **antibiotic** is a medicine that can kill bacteria and other microorganisms. Many antibiotics are made by bacteria.

GENETIC ENGINEERING

When scientists change an organism's DNA, it is called *genetic engineering*. For example, scientists have put genes from different organisms into bacteria. The added DNA gives the bacterium instructions for making different proteins. Genetic engineering lets scientists make products that are hard to find in nature. ☑

Math Focus

2. Calculate An ounce (1 oz) is equal to about 28 g. If 1 g of soil contains 25 billion bacteria, how many bacteria are in 1 oz of soil?

_____

 **READING CHECK**

3. Explain What does genetic engineering let scientists do?

_____

_____

_____

_____

Genes from the *Xenopus* frog were used to produce the first genetically engineered bacteria.

Scientists have used genetic engineering to produce insulin. The human body needs *insulin* to break down and use sugar. People who have a disease called *diabetes* cannot make enough insulin. In the 1970s, scientists found a way to put genes into bacteria so that the bacteria would produce human insulin.

How Can Bacteria Be Harmful?

Some bacteria can be harmful to people and other organisms. Bacteria that cause disease are called **pathogenic bacteria**. Pathogenic bacteria get inside a host organism and take nutrients from the organism's cells. Pathogenic bacteria can harm the organism. Today, people protect themselves from some bacterial diseases by vaccination. Many bacterial infections can be treated with antibiotics. ☑

Bacteria can cause disease in other organisms as well as people. Bacteria can rot or discolor a plant and its fruit. To stop this, plants are sometimes treated with antibiotics. Scientists also try to grow plants that have been genetically engineered to resist bacteria that cause disease.

Vaccines can protect you from bacterial diseases such as tetanus and diphtheria.

Critical Thinking

4. Infer Why do you think it is helpful to engineer bacteria to produce insulin?

_____

_____

_____

_____

_____

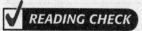
READING CHECK

5. Identify What can be used to treat bacterial infections?

_____

Section 2 Review

SECTION VOCABULARY

antibiotic medicine used to kill bacteria and other organisms **bioremediation** the biological treatment of hazardous waste by living organisms	**pathogenic bacteria** bacteria that cause disease

1. Define What is nitrogen fixation?

_____

_____

2. Complete Fill in the process chart for the nitrogen cycle. Be sure to describe what is happening during each step.

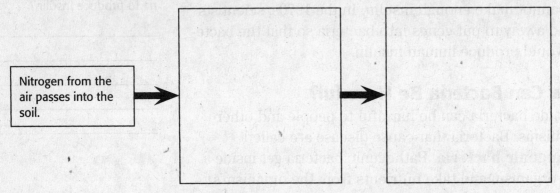

Nitrogen from the air passes into the soil.

3. Explain What are two ways that bacteria are helpful to other living things?

_____

_____

_____

4. Explain What is genetic engineering?

_____

_____

_____

5. Explain How do pathogenic bacteria harm an organism?

_____

_____

6. Identify Relationships Legumes, which include peanuts and beans, are good nitrogen-fixers. Legumes are also a good source of amino acids. What chemical element would you expect to find in amino acids?

CHAPTER 12 Bacteria and Viruses

SECTION
3 **Viruses**

BEFORE YOU READ

After you read this section, you should be able to answer these questions:

• What is a virus?

• How does a virus survive?

• How do viruses make more of themselves?

What Is a Virus?

Most people have either had chickenpox or seen someone with the disease. Chickenpox is caused by a virus. A **virus** is a tiny particle that gets inside a cell and usually kills it. Many viruses cause diseases, such as the common cold, the flu, and acquired immune deficiency syndrome (AIDS). Viruses are smaller than bacteria and can only be seen with a microscope. Viruses can also change quickly. These traits make it hard for scientists to fight viruses. ☑

STUDY TIP

Compare As you read, make a Venn Diagram to compare viruses and bacteria.

Are Viruses Living?

Like living things, viruses have protein and genetic material. However, viruses are not alive. A virus cannot eat, grow, or reproduce like a living thing. For a virus to function, it needs to get inside a living cell. Viruses use cells as hosts. A **host** is a living thing that a virus lives on or in. A virus uses a host cell to make more viruses.

READING CHECK

1. Explain Why is it difficult for scientists to fight viruses?

_____

_____

STANDARDS CHECK

LS 2a Reproduction is a characteristic of all living systems; because no living organism lives forever, reproduction is essential to the continuation of every species. Some organisms reproduce asexually. Others reproduce sexually.

2. List Give three reasons that viruses are not considered living things.

_____

_____

_____

_____

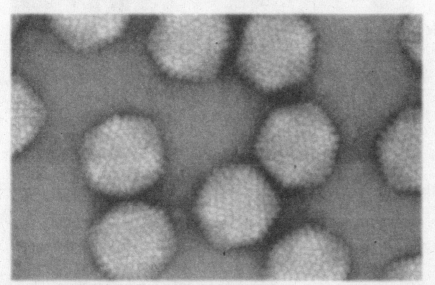

Viruses are not cells. They have genetic material, but they do not have cytoplasm or organelles.

What Is the Structure of a Virus?

Every virus is made up of genetic material inside a protein coat. The protein coat protects the genetic material and helps the virus enter a cell.

How Are Viruses Grouped?

Viruses can be grouped in several ways. These include shape and type of genetic material. The genetic material in viruses is either RNA or DNA. RNA is made of only one strand of nucleotides. DNA is made of two strands of nucleotides. The viruses that cause chickenpox and warts have DNA. The viruses that cause colds, flu, and AIDS have RNA. ☑

The Basic Shapes of Viruses

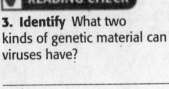 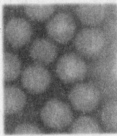

Crystals
The polio virus is shaped like the crystals shown here.

Spheres
Influenza viruses look like spheres. HIV, the virus that causes AIDS, also has this structure.

Cylinders
The tobacco mosaic virus is shaped like a cylinder and attacks tobacco plants.

Spacecraft
One group of viruses attacks only bacteria. Many of these look almost like spacecraft.

How Do Viruses Make More Viruses?

THE LYTIC CYCLE

Like living things, viruses make more of themselves. However, viruses cannot reproduce on their own. They attack living cells and turn them into virus factories. This process is called the *lytic cycle.* ☑

SECTION 3 Viruses *continued*

The Lytic Cycle

❶ The virus finds and joins itself to a host cell.

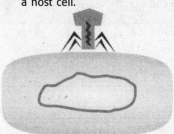

❷ The virus enters the cell, or the virus's genetic material is injected into the cell.

❸ Once the virus's genes are inside, they take control of the host cell and turn it into a virus factory.

❹ The host cell dies when the new viruses break out of it. The new viruses look for host cells, and the cycle continues.

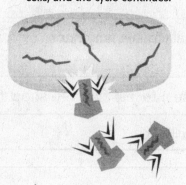

Math Focus
6. Calculate If you enlarged an average virus 600,000 times, it would be about the size of a small pea. How tall would you be if you were enlarged 600,000 times?

_____

TAKE A LOOK
7. Explain What happens to a host cell when the new viruses are released?

_____

THE LYSOGENIC CYCLE
Some viruses don't go right into the lytic cycle. These viruses put their genetic material inside a host cell, but new viruses are not made. As the host cells divides, the genetic material of the virus is passed to the new cells. This is called the *lysogenic cycle*. The host cells can carry the genes of the virus for a long time. When the genes do become active, the lytic cycle begins.

How Can You Stop Viruses?
Antibiotics cannot destroy viruses. However, scientists have made some medicines called *antiviral medications*. These medicines stop viruses from reproducing in their host. Most diseases caused by viruses do not have cures. It is best to try to stop a virus from entering your body. Washing your hands helps you avoid some viruses. The vaccinations some children get also help to prevent viral infections.

Critical Thinking
8. Infer Why shouldn't you take antibiotics to treat a cold?

_____

_____

_____

Section 3 Review

SECTION VOCABULARY

host an organism from which a parasite takes food or shelter	**virus** a microscopic particle that gets inside a cell and often destroys the cell

1. Compare How are viruses similar to living things?

_____

_____

2. Describe What is the structure of a virus?

_____

_____

3. List List four shapes that viruses may have.

_____

4. Summarize Complete the process chart to show the steps of the lytic cycle.

The virus attaches to a host cell.

↓

↓

↓

The new viruses break out of the host cell, and the host cell dies. The new viruses look for hosts.

5. Explain Why do viruses need hosts?

_____

_____

5. Compare How is the lysogenic cycle different from the lytic cycle?

_____

_____

_____

_____

6. List Name two ways to prevent a viral infection.

_____

CHAPTER 13 | Protists and Fungi

SECTION 1 | Protists

National Science
Education Standards
LS 1b, 1c, 1f, 2a, 4b, 5a

BEFORE YOU READ

After you read this section, you should be able to answer these questions:

• How are protists different from other organisms?

• How do protists get food?

• How do protists reproduce?

What Are Protists?

Protists are members of the kingdom Protista. They are a very diverse group of organisms. Most are single-celled, but some have many cells. Some are single-celled but live in colonies. Members of the kingdom Protista have many different ways of getting food, reproducing, and moving.

Protists are very different from plants, animals, and fungi. They are also very different from each other. Scientists group protists together because they do not fit into any other kingdom. However, they do share a few characteristics:

• They are *eukaryotic* (each of their cells has a nucleus).

• They are less complex than other eukaryotic organisms.

• They do not have specialized tissues.

STUDY TIP

List As you read, make a table showing the ways different protists can obtain food and reproduce.

Critical Thinking

1. Infer Are protists prokaryotes or eukaryotes? Explain your answer.

_____

_____

_____

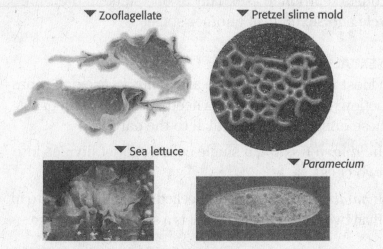

▼ Zooflagellate   ▼ Pretzel slime mold

▼ Sea lettuce   ▼ *Paramecium*

Protists have many different shapes.

How Do Protists Get Food?

Some protists can make their own food. Other protists need to get their food from other organisms or from the environment. Some protists use more than one method of getting food.

AUTOTROPHS

Some protists are producers because they make their own food. The cells of these protists have structures called *chloroplasts*, which capture energy from the sun. Protists use this energy to produce food in a process called *photosynthesis*. Plants use the same process to make their own food. Producers are also known as *autotrophs*. *Autos* is Greek for "self." ☑

✓ READING CHECK

2. Define What is an autotroph?

_____

_____

HETEROTROPHS

Some protists cannot make their own food. They are called **heterotrophs**. *Heteros* is Greek for "different." There are several ways that protist heterotrophs can obtain food.

- *Consumers* eat other organisms, such as bacteria, yeasts, or other protists.

- *Decomposers* break down dead organic matter.

- **Parasites** are organisms that feed on or invade other living things. The organism a parasite invades is called a **host**. Parasites usually harm their hosts. ☑

✓ READING CHECK

3. List What are the three kinds of protist heterotrophs?

_____

_____

_____

How Do Protists Reproduce?

Like all living things, protists reproduce. Some protists reproduce asexually, and some reproduce sexually. Some protists reproduce asexually at one stage in their life cycle and sexually at another stage.

ASEXUAL REPRODUCTION

Most protists reproduce asexually. In asexual reproduction, only one parent is needed to make offspring. These offspring are identical to the parent.

- In *binary fission*, a single-celled protist divides into two cells.

- In *multiple fission*, a single-celled protist divides into more than two cells. Each new cell is a single-celled protist.

Math Focus

4. Calculate If seven individuals of genus Euglena reproduce at one time, how many individuals will result?

_____

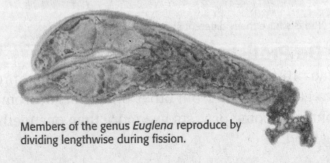

Members of the genus *Euglena* reproduce by dividing lengthwise during fission.

SEXUAL REPRODUCTION

Sexual reproduction requires two parents. The offspring have new combinations of genetic material. The two protists below are reproducing by conjugation. During *conjugation*, two organisms join and exchange genetic material. Then they divide to produce four new protists.

Protists of the genus *Paramecium* can reproduce by conjugation, a kind of sexual reproduction.

Some protists can reproduce both sexually and asexually. For example, some can reproduce asexually when environmental conditions are good. When conditions become difficult, such as when there is little food, the protists reproduce sexually. This allows the offspring to get new combinations of traits that might be helpful under difficult conditions.

What Is a Reproductive Cycle?

Some protists have complex reproductive cycles. These protists may change forms many times. For example, *Plasmodium vivax* is one of the protists that cause the disease malaria. *P. vivax* is a parasite. It needs both humans and mosquitoes to reproduce. Its reproductive cycle is shown below.

ⓐ An infected mosquito bites a human and releases *Plasmodium vivax* into the bloodstream.

ⓔ In the mosquito, the *P. vivax* matures into its original form. The cycle then repeats.

ⓑ *P. vivax* invades liver cells and reproduces. Then it enters the bloodstream again in a new form.

ⓒ *P. vivax* invades red blood cells and reproduces. When the red blood cells burst open, *P. vivax* enters the bloodstream in another new form.

ⓓ A mosquito bites a human and picks up *P. vivax*.

Critical Thinking

5. Apply Concepts Are the offspring produced by conjugation genetically identical to their parents? Explain your answer.

_____
_____
_____
_____
_____
_____

STANDARDS CHECK

LS 2a Reproduction is a characteristic of all living systems; because no living organism lives forever, reproduction is essential to the continuation of every species. Some organisms reproduce asexually. Others reproduce sexually.

6. Explain How does sexual reproduction during times of difficult conditions help protist offspring?

_____
_____
_____
_____
_____

TAKE A LOOK
7. Identify Inside the human body, where does *P. vivax* reproduce?

_____

Section 1 Review

NSES   LS 1b, 1c, 1f, 2a, 4b, 5a

SECTION VOCABULARY

heterotroph an organism that gets food by eating other organisms or their byproducts and that cannot make organic compounds from inorganic materials	**parasite** an organism that feeds on an organism of another species (the host) and that usually harms the host; the host never benefits from the presence of the parasite
host an organism from which a parasite takes food or shelter	**protist** an organism that belongs to the kingdom Protista

1. List Name three traits protists have in common.

_____

_____

_____

2. Identify What kind of reproduction results in offspring that are different from their parents? What kind of reproduction results in offspring that are identical to their parents?

_____

_____

3. Apply Concepts Why are mosquito control programs used to prevent the spread of malaria?

_____

_____

_____

4. List Name four ways protists obtain food.

_____

_____

_____

5. Explain Some scientists think kingdom Protista should be divided into several kingdoms. What do you think is the reason for this?

_____

_____

_____

CHAPTER 13 Protists and Fungi

SECTION 2 **Kinds of Protists**

After you read this section, you should be able to answer these questions:

• How are protists classified?

• What structures do protists use to move?

National Science Education Standards
LS 1a, 1b, 1f, 3a, 5a

How Do Scientists Classify Protists?

Protists are hard to classify, or group, because they are a very diverse group of organisms. Scientists are still learning how protists are related to each other. One way scientists classify protists is by their shared traits. Using this method, scientists classify protists into three groups: producers, heterotrophs that can move, and heterotrophs that cannot move. However, these groups do not show how protists are related to each other.

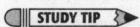

STUDY TIP

Organize As you read this section, make a table that describes each type of protist, how it gets food, and if and how it moves around.

Which Protists Are Producers?

Protist producers are known as **algae** (singular, *alga*). Like plants, algae have chloroplasts. Recall that chloroplasts contain a green pigment called chlorophyll that captures energy from the sun. During photosynthesis, protists use this energy, carbon dioxide, and water to make food. Most algae also have other pigments that give them different colors. ☑

Almost all algae live in water. Some are single-celled, and some are made of many cells. Many-celled algae are called *seaweeds*. They usually live in shallow water and can grow to be many meters long.

Free-floating, single-celled algae are called **phytoplankton**. They are too small to be seen without a microscope. They usually float near the surface of the water. They provide food and oxygen for other organisms.

READING CHECK

1. Identify What are protist producers called?

_____

_____

Critical Thinking

2. Compare What is the difference between seaweeds and phytoplankton?

_____

_____

_____

_____

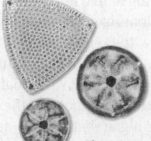

Diatoms are a kind of plankton.

SECTION 2 Kinds of Protists *continued*

RED ALGAE

Most of the world's seaweeds are red algae. Their cells contain chlorophyll, but a red pigment gives them their color. Most red algae live in tropical oceans. They can thrive in very deep water. Their red pigment allows them to absorb the light that filters into deep water. ☑

GREEN ALGAE

Green algae are the most diverse group of protist producers. They are green because chlorophyll is their main pigment. Most live in water or moist soil, but they can also live in melting snow, on tree trunks, and inside other organisms. Many are single-celled. Some have many cells and can grow to be eight meters long. Some individual cells live in groups called *colonies*.

Volvox is a green alga that grows in round colonies.

BROWN ALGAE

Most of the seaweeds found in cool climates are brown algae. These algae have chlorophyll and a yellow-brown pigment. Brown algae live in oceans. They attach to rocks or form large floating groups called *beds*. ☑

Brown algae can grow to be very large. Some grow 60 m in just one season. That is about as long as 20 cars! Only a brown alga's top is exposed to sunlight. This part of the alga makes food through photosynthesis. The food is carried to parts of the algae that are too deep in the water to get sunlight.

READING CHECK

3. Explain What is the function of red pigment in red algae?

_____

_____

_____

TAKE A LOOK

4. Infer Is *Volvox* single-celled or multicellular? Explain your answer.

_____

_____

_____

READING CHECK

5. Identify In what type of climate are brown algae likely to be found?

_____

DIATOMS

Diatoms make up a large percentage of phytoplankton. Most are single-celled protists found in both salt water and fresh water. They get their energy from photosynthesis. Diatoms are enclosed in thin, two-part shells. The shell is made of a glass-like substance called *silica*. ☑

DINOFLAGELLATES

Most dinoflagellates are single-celled. They generally live in salt water, but a few species live in fresh water, or even in snow. Dinoflagellates spin through the water using whiplike structures called *flagella* (singular, *flagellum*). Most dinoflagellates get their energy from photosynthesis. However, a few are consumers, decomposers, or parasites.

EUGLENOIDS

Euglenoids are single-celled protists. They move through the water using flagella. Most have two flagella, one long and one short. They cannot see, but they have eyespots that sense light. A special structure called a *contractile vacuole* removes excess water from the cell.

Euglenoids do not fit well into any one protist group because they can get their food in several ways. Many euglenoids are producers. When there is not enough light to make food, they can be heterotrophs. Some euglenoids are full-time consumers or decomposers. ☑

Structure of Euglenoids

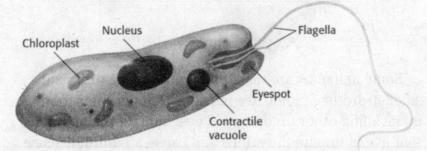

Chloroplast — Nucleus — Flagella — Eyespot — Contractile vacuole

☑ **READING CHECK**

6. Identify What substance makes up the thin shells of diatoms?

_____

_____

☑ **READING CHECK**

7. Explain Why don't euglenoids fit well into any one group of protists?

_____

_____

_____

Which Heterotrophs Can Move?

Heterotrophic protists that can move are sometimes called *protozoans*. They are usually single-celled consumers or parasites. They move using special structures, such as pseudopodia, flagella, or cilia. ☑

AMOEBAS

Amoebas are soft, jelly-like protozoans. Many amoebas are found in fresh water, salt water, and soil. Amoebas have contractile vacuoles to get rid of excess water. They move and catch food using pseudopodia (singular, *pseudopod*). *Pseudopodia* means "false feet." The figure below shows how an amoeba uses pseudopodia to move.

8. Identify What is another name for heterotrophic protists that can move?

_____

_____

TAKE A LOOK
9. Infer Are pseudopodia fixed structures? Explain your answer.

_____

_____

_____

_____

How Amoebas Move

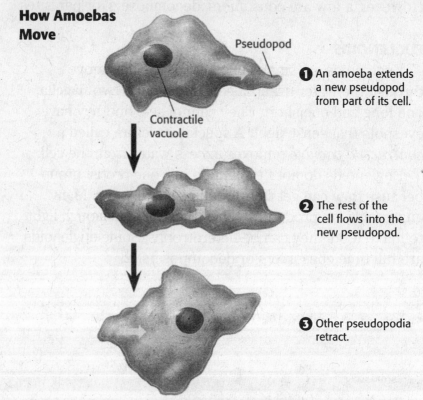

Pseudopod

Contractile vacuole

❶ An amoeba extends a new pseudopod from part of its cell.

❷ The rest of the cell flows into the new pseudopod.

❸ Other pseudopodia retract.

Some amoebas are consumers that eat bacteria and smaller protists. Some are parasites that get their food by invading other organisms. These include the amoebas that live in human intestines and cause a painful disease called *amoebic dysentery*.

Amoebas also use pseudopodia to catch food. When an amoeba senses food, it moves toward it using its pseudopodia. It surrounds the food with pseudopodia to form a *food vacuole*. Enzymes move into the vacuole to digest food. The digested food then passes into the amoeba.

An amoeba uses pseudopodia to catch food.

Food
vacuole

SHELLED AMOEBA-LIKE PROTISTS

Some amoeba-like protists have an outer shell. They
move by poking pseudopodia out of pores in the shells.
Some, such as *foraminiferans*, have snail-like shells.
Others, such as *radiolarian*, look like glass ornaments. ☑

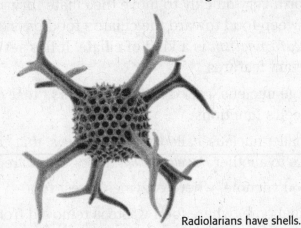

Radiolarians have shells.

✓ **READING CHECK**

10. Identify What are two
kinds of shelled amoeba-like
protists?

_____

_____

ZOOFLAGELLATES

Zooflagellates are heterotrophic protists that move by
waving flagella back and forth. Some live in water. Others
live in the bodies of other organisms.

Some zooflagellates are parasites that cause disease.
People who drink water containing the zooflagellate
Giardia lamblia can get severe stomach cramps.

Some zooflagellates live in mutualism with other
organisms. In *mutualism*, two different organisms live
closely together and help each other survive. The zooflagellate shown on the next page lives in the gut of ter-
mites. The organisms help each other. The zooflagellate
helps the termite digest wood, and the termite gives the
protist food and a place to live. ☑

✓ **READING CHECK**

11. Define What is
mutualism?

_____

_____

_____

_____

SECTION 2 Kinds of Protists *continued*

Structure of Flagellates

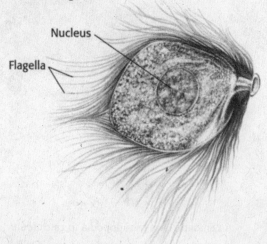

Nucleus

Flagella

TAKE A LOOK

12. Apply Concepts Is this organism a prokaryote or a eukaryote? Explain your answer.

_____

_____

CILIATES

Ciliates are complex protists. They have hundreds of tiny, hairlike structures called *cilia*. The cilia beat back and forth very quickly to move the ciliate forward. Cilia also sweep food toward the ciliate's food passageway.

A *Paramecium* is a kind of ciliate. It has several important features:

- a large nucleus, called a *macronucleus,* that controls the cell's functions

- a smaller nucleus, called a *micronucleus*, that can pass genes to another *Paramecium* during sexual reproduction

- a food vacuole, where enzymes digest food

- an anal pore, where food waste is removed from the cell

- a contractile vacuole to remove excess water ☑

READING CHECK

13. Identify What structure in a *Paramecium* is used to exchange genes with another *Paramecium* during sexual reproduction?

_____

_____

Structure of Paramecium

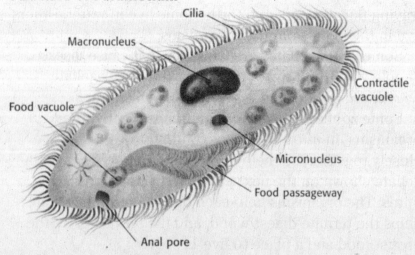

Cilia

Macronucleus

Contractile vacuole

Food vacuole

Micronucleus

Food passageway

Anal pore

Which Heterotrophs Cannot Move?

Not all protist heterotrophs can move. Some of these are parasites. Others in this group only move at certain phases in their life cycles.

SPORE-FORMING PROTISTS

Many spore-forming protists are parasites. They absorb nutrients from their hosts. Spore-forming protists have complicated life cycles that usually include two or more hosts. For example, the spore-forming protists that cause malaria use mosquitoes and humans as hosts.

WATER MOLDS

Most water molds are small, single-celled protists. They live in water, moist soil, or other organisms. Some water molds are decomposers. Many water molds are parasites. Their hosts can be living plants, animals, algae, or fungi.

SLIME MOLDS

Slime molds live in cool, moist places. They look like colorful globs of slime. At certain phases of their life cycles, slime molds can move using pseudopodia. Some live as a giant cell with many nuclei at one stage of life. Some live as single-celled organisms. Slime molds eat bacteria and yeast. They also surround bits of rotting matter and digest them.

When water or nutrients are hard to find, slime molds grow stalk-like structures with round knobs on top. The knobs are called *sporangia*, and they contain spores. *Spores* are small reproductive cells covered by a thick cell wall. The spores can survive for a long time without water or nutrients. When conditions improve, the spores develop into new slime molds. ☑

Sporangia

The sporangia of a slime mold contain spores.

Critical Thinking

14. Identify What is one characteristic that slime molds, at certain phases of their life cycles, share with amoebas?

_____

_____

_____

✓ READING CHECK

15. Explain How can spores help a species of slime mold survive difficult conditions?

_____

_____

_____

_____

Section 2 Review

SECTION VOCABULARY

algae eukaryotic organisms that convert the sun's energy into food through photosynthesis but that do not have roots, stems, or leaves (singular, alga)	**phytoplankton** the microscopic, photosynthetic organisms that float near the surface of marine or fresh water

1. List Name three kinds of protists that use flagella to move.

_____

_____

2. Organize Fill in the Venn Diagram below to organize the different kinds of protists based on how they get food. Remember that some protists can get food in more than one way.

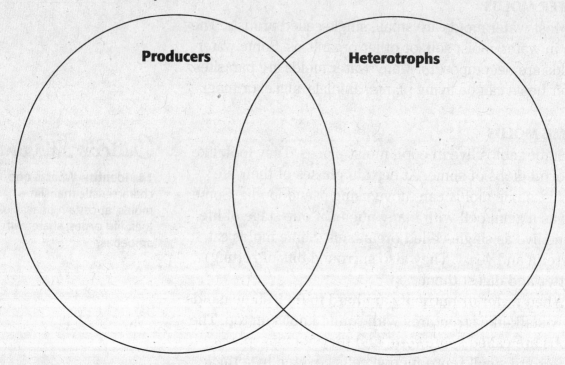

3. List Give two examples of each of the following: protist producers, heterotrophs that can move, and heterotrophs that cannot move.

_____

_____

_____

4. Identify What are two ways amoebas use pseudopodia?

_____

_____

CHAPTER 13 Protists and Fungi
SECTION 3 Fungi

National Science
Education Standards
LS 1a, 1b, 1d, 1f, 2a, 4b, 5a

BEFORE YOU READ

After you read this section, you should be able to answer these questions:

• What are the characteristics of fungi?

• What are the four groups of fungi?

What Are Fungi?

Fungi (singular, *fungus*) are everywhere. The mushrooms on pizza are a type of fungus. The yeast used to make bread is also a fungus. If you have ever had athlete's foot, you can thank a fungus for that, too.

Fungi are so different from other organisms that they are placed in their own kingdom. There are many different shapes, sizes, and colors of fungi, but they have several characteristics in common.

• They are eukaryotic. (Their cells have nuclei.)

• They are heterotrophs. (They cannot make their own food.)

• Their cells have rigid cell walls.

• They have no chlorophyll.

• They produce reproductive cells called spores.

STUDY TIP

Compare Make a chart showing how fungi in each group reproduce.

Critical Thinking

1. Compare What are two ways that fungi differ from plants?

_____

_____

_____

▼ Straight coral fungus

▲ Bird's nest fungus

Say It

Describe Where have you seen fungi before? What did they look like? With a partner, describe any fungi you have seen.

SECTION 3 Fungi *continued*

The mycelium of a fungus is underground. It is made up of hyphae.

Hyphae

STANDARDS CHECK

LS 4b Populations of organisms can be <u>categorized</u> by the <u>functions</u> they serve in an ecosystem. Plants and some microorganimsms are producers—they make their own food. All animals, including humans, are consumers, which obtain their food by eating other organisms. Decomposers, primarily bacteria and fungi, are consumers that use waste materials and dead organisms for food. Food webs identify the relationship among producers, consumers, and decomposers in an ecosystem.

Word Help: categorize
to put into groups or classes

Word Help: function
use or purpose

2. List What are three ways that fungi can get food?

_____

_____

_____

Some fungi are single cells, but most are made of many cells. Many-celled fungi are made up of chains of cells called *hyphae* (singular, *hypha*). A **hypha** is a thread-like filament. The cells in these filaments have openings in their cell walls. These openings allow cytoplasm to move between cells.

Most of the hyphae that make up a fungus grow together to form a twisted mass called the **mycelium**. The mycelium is generally the largest part of the fungus. It grows underground.

How Do Fungi Get Nutrients?

Fungi are heterotrophs, which means they feed on other organisms. Unlike other heterotrophs, however, fungi cannot catch or surround food. They must live on or near their food supply. They secrete digestive juices onto food and absorb the nutrients from the dissolved food.

Most fungi are decomposers. They feed on dead plant or animal matter. Some fungi are parasites. Some live in mutualism with other organisms. For example, some fungi grow on or in the roots of a plant. This relationship is called a *mycorrhiza*. The plant provides nutrients to the fungus. The fungus helps the plant absorb minerals and protects it from some diseases.

How Do Fungi Reproduce?

Reproduction in fungi can be asexual or sexual. Many fungi reproduce using spores. **Spores** are small reproductive cells or structures that can grow into a new fungus. They can be the formed by sexual or asexual reproduction. Many spores have thick cell walls that protect them from harsh environments.

ASEXUAL REPRODUCTION

Asexual reproduction can occur in several ways. Some fungi produce spores. The spores are light and easily spread by the wind. Asexual reproduction can also happen when hyphae break apart. Each new piece can become a new fungus. Single-celled fungi called yeasts reproduce through a process called *budding*. In budding, a new cell pinches off from an existing cell. ☑

SEXUAL REPRODUCTION

In sexual reproduction, special structures form to make sex cells. The sex cells join to produce sexual spores that grow into a new fungus.

This puffball is releasing sexual spores that can grow into new fungi.

☑ READING CHECK

3. List What are three types of asexual reproduction in fungi?

_____

_____

_____

What Are the Four Kinds of Fungi?

Scientists classify fungi based on the shape of the fungi and the way they reproduce. There are four main groups of fungi: threadlike fungi, sac fungi, club fungi, and imperfect fungi.

THREADLIKE FUNGI

Have you ever seen fuzzy black mold growing on bread? A **mold** is a shapeless, fuzzy fungus. Most thread-like fungi live in the soil and are decomposers. However, some are parasites.

Threadlike fungi can reproduce asexually. Structures called *sporangia* (singular, *sporangium*) produce spores. When sporangia break open, they release the spores into the air. ☑

Threadlike fungi can also reproduce sexually. Hyphae from two different individuals can join together and grow into specialized sporangia. These sporangia can survive in the cold and with little water. When conditions improve, these specialized sporangia release spores that can grow into new fungi.

☑ READING CHECK

4. Define What are sporangia?

_____

_____

_____

These groups of sporangia are magnified. Each tiny, round sporangium contains thousands of spores.

TAKE A LOOK

5. Circle On the figure, circle one sporangium.

SAC FUNGI

Sac fungi are the largest group of fungi. Members of this group include yeasts, mildew, truffles, and morels. Sac fungi can reproduce both asexually and sexually. Usually they use asexual reproduction. When they do reproduce sexually, they form a sac called an *ascus*. Sexual spores develop inside the ascus. ☑

Most sac fungi are multicellular. However, yeasts are single-celled sac fungi. When yeasts reproduce asexually, they use a process called budding. In budding, a new cell pinches off from an existing cell.

Yeasts are the only fungi that reproduce by budding.

READING CHECK

6. Identify How do sac fungi usually reproduce?

_____

_____

Yeasts use sugar as food and produce carbon dioxide and alcohol as waste. Because of this, humans use yeasts to make products such as alcohol and bread. Bread dough rises because of trapped bubbles of waste carbon dioxide from the yeast.

Some sac fungi are used to make vitamins and antibiotics. Others cause plant diseases. One of these diseases, Dutch elm disease, has killed millions of elm trees. Humans can use some sac fungi, such as morels and truffles, as food. ☑

READING CHECK

7. List Name three ways humans use sac fungi.

_____

_____

_____

Morels are only part of a larger sac fungus. They are the reproductive part of a fungus that lives under the soil.

CLUB FUNGI

Mushrooms are the most familiar club fungi. Other club fungi include bracket fungi, smuts, rusts, and puffballs. Bracket fungi grow on wood and form small shelves or brackets. Smuts and rusts are common plant parasites that can attack corn and wheat.

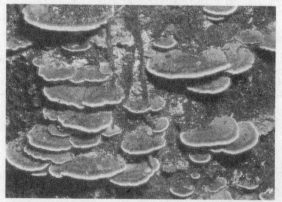
Bracket fungi look like shelves on trees. The underside of the bracket contains spores,

Club fungi reproduce sexually. They grow special hyphae that form clublike structures called *basidia* (singular, *basidium*). Sexual spores develop on the basidia. The spore-producing part of the organism is the only part of a mushroom that is above the ground. Most of the organism is underground. ☑

The most familiar mushrooms are called *gill fungi*. The basidia of these mushrooms develop in structures called *gills*, under the mushroom cap. Some gill fungi are the edible mushrooms sold in supermarket. However, other gill fungi are extremely poisonous.

Say It

Investigate Use your school's media center to research a bracket, rust, or smut club fungus. Learn about where the fungus lives and how it gets nutrients. Describe this fungus to your class.

READING CHECK

8. Identify What part of a mushroom grows above the ground?

_____

_____

TAKE A LOOK

9. Infer Does this picture show the largest part of the fungus? Explain your answer.

_____

_____

_____

Witch's hat fungus is a gill fungus.

IMPERFECT FUNGI

This group includes all of the fungi that do not fit into the other groups. Imperfect fungi do not reproduce sexually. This group includes some fungi that are harmful to humans and some that are useful.

Most imperfect fungi are parasites that cause diseases in plants and animals. One kind of imperfect fungus causes a skin disease called athlete's foot. Another kind of imperfect fungus produces aflatoxin. *Aflatoxin* is a poison that can cause cancer. Some imperfect fungi are used to make medicines, including the antibiotic penicillin. Other imperfect fungi are used to produce cheeses and soy sauce.

The fungus *Penicillium* produces a substance that kills certain bacteria.

STANDARDS CHECK

LS 1f Disease is the breakdown in structures or functions of an organism. Some diseases are the result of intrinsic failures of the system. Others are the result of damage by infection by other organisms.

10. Identify Name two diseases that can be caused by an imperfect fungus.

_____

_____

What Are Lichens?

A **lichen** is the combination of a fungus and an alga. The alga lives inside the cell walls of the fungus. This creates a mutualistic relationship. In a mututalistic relationship, both organisms benefit from living closely with another species. Even though lichens are made of two different organisms, scientists give lichens their own scientific names. This is because the two organisms together function like one organism. ☑

READING CHECK

11. Identify What two kinds of organisms make up a lichen?

_____

_____

▲ Wolf lichen

▼ Christmas lichen

British soldier ▲
lichen

These are some of the many types of lichens.

Critical Thinking

12. Infer How do you think a fungus and an alga evolved to live together?

_____

_____

_____

_____

_____

The fungus and alga that make up a lichen benefit each other. Unlike fungi alone, lichens are producers. The algae in the lichens produce food through photosynthesis. Algae alone would quickly dry out in a dry environment. Fungi, however, have protective walls that keep moisture inside the lichen. This allows lichens to live in dry environments.

Lichens need only air, light, and minerals to grow. They can grow in very dry and very cold environments. They can even grow on rocks. Lichens are important to other organisms because they can help form soil. As they grow on a rock, lichens produce acid. The acid breaks down the rock. Soil forms from bits of rock and dead lichens. Once soil forms, plants can move into the area. Animals that eat the plants can also move into the area. ☑

Lichens are sensitive to air pollution. Because of this, they are good *ecological indicators*. This means that if lichens start to die off in an area, there may be something wrong in the environment.

☑ **READING CHECK**

13. Explain How do lichens make soil?

_____

_____

_____

_____

_____

_____

_____

Section 3 Review

NSES LS 1a, 1b, 1d, 1f, 2a, 4b, 5a

SECTION VOCABULARY

fungus an organism whose cells have nuclei, rigid cell walls, and no chlorophyll and that belongs to the kingdom Fungi	**mold** in biology, a fungus that looks like wool or cotton
hypha a nonreproductive filament of a fungus	**mycelium** the mass of fungal filaments, or hyphae, that forms the body of a fungus
lichen a mass of fungal and algal cells that grow together in a symbiotic relationship and that are usually found on rocks or trees	**spore** a reproductive cell or multicellular structure that is resistant to stressful environmental conditions and that can develop into an adult without fusing with another cell

1. List What are the four groups of fungi?

_____

_____

2. Explain How does a mycorrhiza help both the plant and the fungus?

_____

_____

3. Identify Relationships How are a hypha and a mycelium related?

_____

4. Identify What part of a club fungus grows above the ground? What part grows below the ground?

_____

_____

_____

5. Explain What is the function of sporangia?

_____

6. Compare How are lichens different from fungi?

_____

_____

_____

7. Compare What is the difference between a lichen and a mycorrhiza?

_____

_____

8. Identify Which group of fungi forms basidia during sexual reproduction?

_____

CHAPTER 14 Introduction to Plants

SECTION 1

What Is a Plant?

National Science Education Standards
LS 1a, 2b, 4c, 5a

BEFORE YOU READ

After you read this section, you should be able to answer these questions:

- What characteristics do all plants share?
- What are two differences between plant cells and animal cells?

What Are the Characteristics of Plants?

A plant is an organism that uses sunlight to make food. Trees, grasses, ferns, cactuses, and dandelions are all types of plants. Plants can look very different, but they all share four characteristics.

STUDY TIP

Organize As you read, make a diagram to show the major groups of plants. Be sure to include the characteristics of each group.

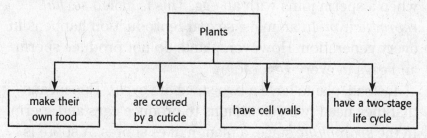

Plants
- make their own food
- are covered by a cuticle
- have cell walls
- have a two-stage life cycle

PHOTOSYNTHESIS

Plants make their own food from carbon dioxide, water, and energy from sunlight. This process is called *photosynthesis*. Photosynthesis takes place in special organelles called *chloroplasts*. Inside the chloroplasts, a green pigment called *chlorophyll* collects energy from the sun for photosynthesis. Chlorophyll is what makes most plants look green. Animal cells do not have chloroplasts. ☑

CUTICLES

Every plant has a cuticle that covers and protects it. A *cuticle* is a waxy layer that coats a plant's leaves and stem. The cuticle keeps plants from drying out by keeping water inside the plant.

CELL WALLS

How do plants stay upright? They do not have skeletons, as many animals do. Instead, each plant cell is surrounded by a stiff cell wall. The cell wall is outside the cell membrane. Cell walls support and protect the plant cell. Animal cells do not have cell walls.

READING CHECK

1. **Define** What is chlorophyll?

_____

_____

_____

_____

Structures in a Plant Cell

Large central vacuole A vacuole stores water and helps support the cell.

Chloroplast Chloroplasts contain chlorophyll. Chlorophyll captures energy from the sun. Plants use this energy to make food.

Cell wall The cell wall surrounds the cell membrane. It supports and protects the plant cell.

Cell membrane The cell membrane surrounds a plant cell and lies under the cell wall.

TAKE A LOOK
2. Identify What structure in a plant cell stores water?

_____

_____

3. Identify Where is chlorophyll found?

_____

TWO-STAGE LIFE CYCLE

Many organisms, including plants, produce offspring when a sperm joins with an egg. This is called *sexual reproduction*. In animals, sexual reproduction happens in every generation. However, plants do not produce sperm and eggs in every generation.

Instead, plants have a two-stage life cycle. This means that they need two generations to produce eggs and sperm. In the *sporophyte* stage, a plant makes spores. A spore is a cell that can divide and grow into a new plant. This new plant is called a *gametophyte*. In the gametophyte stage, the plants produce sperm and eggs. The sperm and eggs must join to produce a new sporophyte. ☑

READING CHECK
4. List What are the two stages of the plant life cycle?

_____

_____

What Are the Main Groups of Plants?

There are two main groups of plants: vascular and nonvascular. A **vascular plant** has specialized vascular tissues. *Vascular tissues* move water and nutrients from one part of a plant to another. A **nonvascular plant** does not have vascular tissues to move water and nutrients.

The Main Groups of Plants

Nonvascular Plants	Vascular Plants		
Mosses, liverworts, and hornworts	Seedless plants	Seed plants	
	Ferns, horsetails, and club mosses	Nonflowering	Flowering
		Gymnosperms	Angiosperms

SECTION 1 What Is a Plant? *continued*

NONVASCULAR PLANTS

Nonvascular plants depend on diffusion to move water and nutrients through the plant. In *diffusion*, water and nutrients move through a cell membrane and into a cell. Each cell must get water and nutrients from the environment or a cell that is close by.

Nonvascular plants can rely on diffusion because they are small. If a nonvascular plant were large, not all of its cells would get enough water and nutrients. Most nonvascular plants live in damp areas, so each of their cells is close to water.

VASCULAR PLANTS

Many of the plants we see in gardens and forests are vascular plants. Vascular plants are divided into two groups: seedless plants and seed plants. Seed plants are divided into two more groups—flowing and nonflowering. Nonflowering seed plants, such as pine trees, are called **gymnosperms**. Flowering seed plants, such as magnolias, are called **angiosperms**.

How Did Plants Evolve?

What would you see if you traveled back in time about 440 million years? The Earth would be a strange, bare place. There would be no plants on land. Where did plants come from?

The green alga in the figure below may look like a plant, but it is not. However, it does share some characteristics with plants. Both algae and plants have the same kind of chlorophyll and make their food by photosynthesis. Like plants, algae also have a two-stage life cycle. Scientists think these similarities mean that plants evolved from a species of green algae millions of years ago.

A modern green alga and plants, such as ferns, share several characteristics. Because of this, scientists think that both types of organisms shared an ancient ancestor.

Critical Thinking

5. Apply Concepts Do you think a sunflower is a gymnosperm or an angiosperm? Explain your answer.

_____

_____

_____

STANDARDS CHECK

LS 5a Millions of species of animals, plants, and microorganisms are alive today. Although different species might look dissimilar, the unity among organisms becomes apparent from an analysis of internal structures, the similarity of their chemical processes, and the evidence of common ancestry.

Word Help: evidence information showing whether an idea is true or valid

6. List Give three reasons scientists think plants evolved from an ancient green algae.

_____

_____

_____

_____

_____

_____

Name _____ Class _____ Date _____

Section 1 Review

NSES   LS 1a, 2b, 4c, 5a

SECTION VOCABULARY

angiosperm a flowering plant that produces seeds within a fruit	**nonvascular plant** the three groups of plants (liverworts, hornworts, and mosses) that lack specialized conducting tissues and true roots, stems, and leaves
gymnosperm a woody, vascular seed plant whose seeds are not enclosed by an ovary or fruit	**vascular plant** a plant that has specialized tissues that conduct materials from one part of the plant to another

1. Explain What are the two main differences between a plant cell and an animal cell?

_____

_____

_____

_____

2. Organize Fill in each box in the figure below with one of the main characteristics of plants.

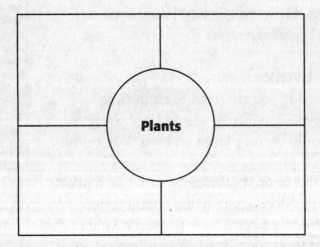

3. Predict What would happen to a plant if its chloroplasts stopped working? Explain your answer.

_____

_____

_____

4. Compare What is the main difference between vascular and nonvascular plants?

_____

_____

_____

BEFORE YOU READ

After you read this section, you should be able to answer these questions:

- What are the differences between seedless vascular plants and nonvascular plants?
- How can plants reproduce without seeds?

National Science Education Standards
LS 1a, 1c, 2b, 5c

What Are Seedless Plants?

When you think of plants, you probably think of plants that make seeds, such as flowers and trees. However, there are many plants that don't make seeds.

Remember that plants are divided into two main groups: nonvascular plants and vascular plants. All nonvascular plants are seedless, and some vascular plants are seedless, as well.

What Are the Features of Nonvascular Plants?

Mosses, liverworts, and hornworts are types of nonvascular plants. Remember that nonvascular plants do not have vascular tissue to deliver water and nutrients. Instead, each plant cell gets water and nutrients directly from the environment or from a nearby cell. Therefore, nonvascular plants usually live in places that are damp.

Nonvascular plants do not have true stems, roots, or leaves. However, they do have features that help them to get water and stay in place. For example, a **rhizoid** is a rootlike structure that holds some nonvascular plants in place. Rhizoids also help plants get water and nutrients. ☑

Nonvascular plants
• have no vascular tissue
• have no true roots, stems, leaves, or seeds
• are usually small
• live in damp places

REPRODUCTION IN NONVASCULAR PLANTS

Like all plants, nonvascular plants have a two-stage life cycle. They have a sporophyte generation, which produces spores, and a gametophyte generation, which produces eggs and sperm. Sperm from these plants need water so they can swim to the eggs. Nonvascular plants can also reproduce asexually, that is, without eggs and sperm.

STUDY TIP
Organize As you read this section, make a chart that compares vascular plants and nonvascular plants.

Critical Thinking
1. Apply Concepts Why wouldn't you expect to see nonvascular plants in the desert?
_____
_____
_____
_____

READING CHECK
2. List What are two functions of the rhizoid?
_____
_____

SECTION 2 Seedless Plants *continued*

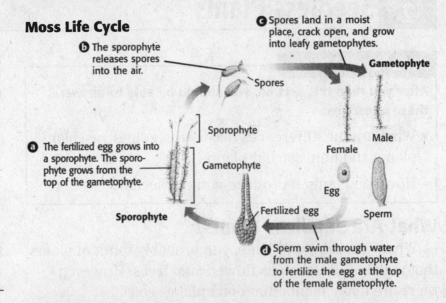

Moss Life Cycle

b The sporophyte releases spores into the air.

Spores

c Spores land in a moist place, crack open, and grow into leafy gametophytes.

Gametophyte

Sporophyte

Male

Female

a The fertilized egg grows into a sporophyte. The sporophyte grows from the top of the gametophyte.

Gametophyte

Egg

Sperm

Sporophyte

Fertilized egg

d Sperm swim through water from the male gametophyte to fertilize the egg at the top of the female gametophyte.

TAKE A LOOK
3. Identify Are the male and female gametophytes separate plants or part of the same plant?

_____

_____

Critical Thinking
4. Apply Concepts Why do you think nonvascular plants can be the first plants to grow in a new environment?

_____

_____

_____

_____

IMPORTANCE OF NONVASCULAR PLANTS

Nonvascular plants are usually the first plants to live in a new environment, such as newly exposed rock. When these plants die, they break down and help form a thin layer of soil. Then plants that need soil in order to grow can move into these areas.

Some nonvascular plants are important as food or nesting material for animals. A nonvascular plant called peat moss is important to humans. When it turns to peat, it can be burned as a fuel.

What Are the Features of Seedless Vascular Plants?

Vascular plants have specialized tissues that carry water and nutrients to all their cells. These tissues generally make seedless vascular plants larger than nonvascular plants. Because they have tissues to move water, vascular plants do not have to live in places that are damp. ☑

Many seedless vascular plants, such as ferns, have a structure called a rhizome. The **rhizome** is an underground stem that produces new leaves and roots.

☑ **READING CHECK**

5. Explain How do the cells of a seedless vascular plant get water?

_____

_____

_____

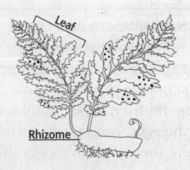

Leaf

Rhizome

SECTION 2 Seedless Plants *continued*

REPRODUCTION IN SEEDLESS VASCULAR PLANTS

The life cycles of vascular plants and nonvascular plants are similar. Sperm and eggs are produced in gametophytes. They join to form a sporophyte. The sporophyte produces spores. The spores are released. They grow into new gametophytes. ☑

Seedless vascular plants can also reproduce asexually in two ways. New plants can branch off from older plants. Pieces of a plant can fall off and begin to grow as new plants.

READING CHECK

6. Identify What do spores grow into?

_____

Fern Life Cycle

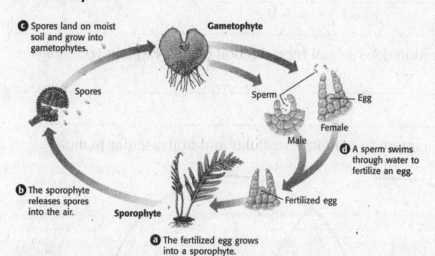

c Spores land on moist soil and grow into gametophytes.

Gametophyte

Spores

Sperm

Egg

Female

Male

d A sperm swims through water to fertilize an egg.

b The sporophyte releases spores into the air.

Sporophyte

Fertilized egg

a The fertilized egg grows into a sporophyte.

TAKE A LOOK
7. Apply Concepts Does this figure show sexual or asexual reproduction? Explain your answer.

_____

_____

_____

_____

IMPORTANCE OF SEEDLESS VASCULAR PLANTS

Did you know that seedless vascular plants that lived 300 million years ago are important to people today? After these ancient ferns, horsetails, and club mosses died, they formed coal and oil. Coal and oil are fossil fuels that people remove from Earth's crust to use for energy. They are called *fossil fuels* because they formed from plants (or animals) that lived long ago. ☑

Another way seedless vascular plants are important is they help make and preserve soil. Seedless vascular plants help form new soil when they die and break down. Their roots can make the soil deeper, which allows other plants to grow. Their roots also help prevent soil from washing away.

Many seedless vascular plants are used by humans. Ferns and some club mosses are popular houseplants. Horsetails are used in some shampoos and skincare products.

READING CHECK

8. Explain Where does coal come from?

_____

_____

_____

_____

_____

Section 2 Review

SECTION VOCABULARY

rhizoid a rootlike structure in nonvascular plants that holds the plants in place and helps plants get water and nutrients	rhizome a horizontal underground stem that produces new leaves, shoots, and roots

1. Compare What are two differences between a rhizoid and a rhizome?

_____

_____

_____

_____

2. Explain In which generation does sexual reproduction occur? Explain your answer.

_____

_____

3. Compare Use a Venn Diagram to compare vascular and nonvascular plants.

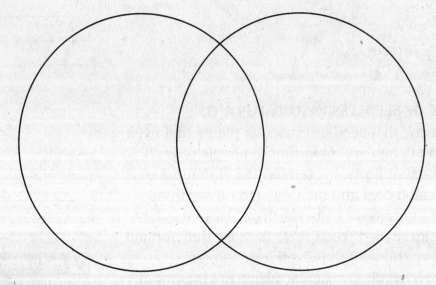

4. Describe What are two ways in which seedless nonvascular plants reproduce asexually?

_____

_____

5. Apply Concepts Nonvascular plants are usually very small. How does their structure limit their size?

_____

_____

CHAPTER 14 | Introduction to Plants
SECTION
3 **Seed Plants**

National Science
Education Standards
LS 1a, 1d, 2b, 2c, 2d, 4b, 4c,
4d, 5b

BEFORE YOU READ

After you read this section, you should be able to answer these questions:

• How are seed plants different from seedless plants?

• What are the parts of a seed?

• How do gymnosperms and angiosperms reproduce?

What Are Seed Plants?

Think about the seed plants that you use during the day. You probably use dozens of seed plants, including the food you eat and the paper you write on. Seed plants include trees, such as oaks and pine trees, as well as flowers, such as roses and dandelions. Seed plants are one of the two main groups of vascular plants.

Like all plants, seed plants have a two-stage life cycle. However, seed plants differ from seedless plants, as shown below.

Seedless plants	Seed plants
They do not produce seeds.	They produce seeds.
The gametophyte grows as an independent plant.	The gametophyte lives inside the sporophyte.
Sperm need water to swim to the eggs.	Sperm are carried to the eggs by pollen.

Seed plants do not depend on moist habitats for reproduction, the way seedless plants do. Because of this, seed plants can live in many more places than seedless plants can. Seed plants are the most common plants on Earth today.

What Is a Seed?

A *seed* is a structure that feeds and protects a young plant. It forms after fertilization, when a sperm and an egg join. A seed has the following three main parts:

• a young plant, or sporophyte

• *cotyledons*, early leaves that provide food for the young plant

• a seed coat that covers and protects the young plant ✓

STUDY TIP

Organize As you read this section, make cards showing the parts of the life cycle of seed plants. Practice arranging the cards in the correct sequence.

TAKE A LOOK
1. Compare How do the gametophytes of seedless plants differ from those of seed plants?

_____

_____

_____

_____

✓ **READING CHECK**

2. Identify What process must occur before a seed can develop?

_____

SECTION 3 Seed Plants *continued*

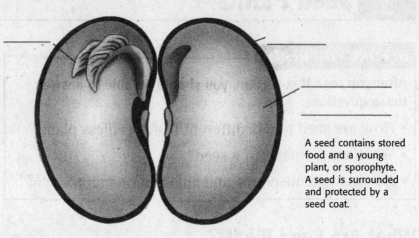

A seed contains stored food and a young plant, or sporophyte. A seed is surrounded and protected by a seed coat.

TAKE A LOOK
3. Label Label the parts of a seed with these terms: young plant, seed coat, cotyledon.

STANDARDS CHECK

LS 5b Biological evolution accounts for the <u>diversity</u> of species developed through gradual <u>processes</u> over many generations. Species acquire many of their unique characteristics through biological adaptation, which involves the selection of naturally occurring variations in populations. Biological adaptations include changes in structures, behaviors, or physiology that enhance <u>survival</u> and reproductive success in a particular environment.

Word Help: <u>diversity</u>
variety

Word Help: <u>process</u>
a set of steps, events, or changes

Word Help: <u>survival</u>
the continuing to live or exist

4. Identify What are two advantages seeds have over spores?

_____

_____

_____

_____

_____

_____

ADVANTAGES OF HAVING SEEDS

Seeds have some advantages over spores. For example, when the young plant inside a seed begins to grow, it uses the food stored in the seed. In contrast, the spores of seedless plants don't have stored food to help a new plant grow. Therefore, they will live only if they start growing when and where there are enough resources available.

Another advantage is that seeds can be spread by animals. The spores of seedless plants are usually spread by wind. Animals often spread seeds more efficiently than the wind spreads spores. Therefore, seeds that are spread by animals are more likely to find a good place to grow.

What Kinds of Plants Have Seeds?

Seed plants are divided into two main groups: gymnosperms and angiosperms. *Gymnosperms* are nonflowering plants, and *angiosperms* are flowering plants.

GYMNOSPERMS

Gymnosperms are seed plants that do not have flowers or fruits. They include plants such as pine trees and redwood trees. Many gymnosperms are evergreen, which means that they keep their leaves all year. Gymnosperm seeds usually develop in a cone, such as a pine cone.

Pine cone

Seeds

SECTION 3 Seed Plants *continued*

REPRODUCTION IN GYMNOSPERMS

The most well-known gymnosperms are the conifers. Conifers are evergreen trees and shrubs, such as pines, spruces, and firs, that make cones to reproduce. They have male cones and female cones. Spores in male cones develop into male gametophytes, and spores in female cones develop into female gametophytes. The gametophytes produce sperm and eggs.

A **pollen** grain contains the tiny male gametophyte. The wind carries pollen from the male cones to the female cones. This movement of pollen to the female cones is called **pollination**. Pollination is part of sexual reproduction in plants. ☑

After pollination, sperm fertilize the eggs in the female cones. A fertilized egg develops into a new sporophyte inside a seed. Eventually, the seeds fall from the cone. If the conditions are right, the seeds will grow into plants.

☑ **READING CHECK**
5. Explain How is gymnosperm pollen carried from one plant to another?

_____
_____

The Life Cycle of a Pine Tree

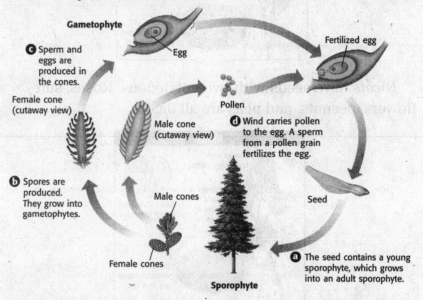

Gametophyte

c Sperm and eggs are produced in the cones.

Egg

Fertilized egg

Female cone (cutaway view)

Male cone (cutaway view)

Pollen

d Wind carries pollen to the egg. A sperm from a pollen grain fertilizes the egg.

b Spores are produced. They grow into gametophytes.

Male cones

Seed

Female cones

Sporophyte

a The seed contains a young sporophyte, which grows into an adult sporophyte.

TAKE A LOOK
6. Explain Does this picture show an example of sexual or asexual reproduction? Explain.

_____
_____
_____
_____

IMPORTANCE OF GYMNOSPERMS

Gymnosperms are used to make many products, such as medicines, building materials, and household products. Some conifers produce a drug used to fight cancer. Many trees are cut so that their wood can be used to build homes and furniture. Pine trees make a sticky substance called resin. Resin can be used to make soap, paint, and ink.

What Are Angiosperms?

Angiosperms are seed plants that produce flowers and fruit. Maple trees, daisies, and blackberries are all examples of angiosperms. There are more angiosperms on Earth than any other kind of plant. They can be found in almost every land ecosystem, including grasslands, deserts, and forests.

TWO KINDS OF ANGIOSPERMS

There are two kinds of angiosperms: monocots and dicots. These plants are grouped based on how many cotyledons, or seed leaves, the seeds have. *Monocots* have seeds with one cotyledon. Grasses, orchids, palms, and lilies are all monocots.

Math Focus
7. Calculate Percentages
More than 265,000 species of plants have been discovered. About 235,000 of those species are angiosperms. What percentage of plants are angiosperms?

_____

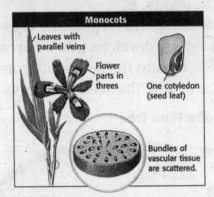

Dicots have seeds with two cotyledons. Roses, sunflowers, peanuts, and peas are all dicots.

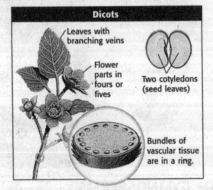

Monocots	Dicots
	flower parts in fours or fives
leaves with parallel veins	
	two cotyledons
bundles of vascular tissue scattered	

TAKE A LOOK
8. Complete Fill in the table to show the differences between monocots and dicots.

SECTION 3 Seed Plants *continued*

REPRODUCTION IN ANGIOSPERMS

In angiosperms, pollination takes place in flowers. Some angiosperms depend on the wind for pollination. Others rely on animals such as bees and birds to carry pollen from flower to flower.

Angiosperm seeds develop inside fruits. Some fruits and seeds, like those of a dandelion, are made to help the wind carry them. Other fruits, such as blackberries, attract animals that eat them. The animals drop the seeds in new places, where they can grow into plants. Some fruits, such as burrs, travel by sticking to animal fur. ☑

Each of the fluffy structures on this dandelion is actually a fruit. Each of the fruits contains a seed.

9. Identify Where do angiosperm seeds develop?

_____

_____

TAKE A LOOK

10. Identify How are the fruits of this dandelion spread?

_____

_____

IMPORTANCE OF ANGIOSPERMS

Like many other plants, flowering plants provide food for animals. A mouse that eats seeds and berries uses flowering plants directly as food. An owl that eats a field mouse uses flowering plants indirectly as food. Flowering plants can also provide food for the animals that pollinate them.

People use flowering plants, too. Major food crops, such as corn, wheat, and rice, come from flowering plants. Many flowering trees, such as oak trees, can be used for building materials. Plants such as cotton and flax are used to make clothing and rope. Flowering plants are also used to make medicines, rubber, and perfume oils.

 **Say It**

Describe Think of all the products you used today that came from angiosperms. Describe to the class five items you used in some way and what kind of angiosperm they came from.

Section 3 Review

NSES LS 1a, 1d, 2b, 2c, 2d, 4b, 4c, 4d, 5b

SECTION VOCABULARY

pollen the tiny granules that contain the male gametophyte of seed plants	**pollination** the transfer of pollen from the male reproductive structures to the female reproductive structures of seed plants

1. Compare How are the gametophytes of seed plants different from the gameto-phytes of seedless plants?

_____

_____

_____

2. Describe What happens during pollination?

_____

_____

_____

3. Compare Use a Venn Diagram to compare gymnosperms and angiosperms.

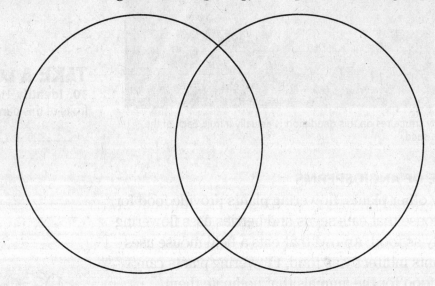

4. List What are the three main parts of a seed? What does each part do?

_____

_____

_____

5. Identify In what structure do gymnosperm seeds usually develop?

_____

6. Identify What two structures are unique to angiosperms?

_____

SECTION 4 Structures of Seed Plants

BEFORE YOU READ

After you read this section, you should be able to answer these questions:

- What are the functions of roots and stems?
- What is the function of leaves?
- What is the function of a flower?

National Science Education Standards
LS 1a, 1d, 2b, 3d, 4c, 5b

What Structures Are Found in a Seed Plant?

Remember that seed plants include trees, such as oaks and pine trees, as well as flowers, such as roses and dandelions. Seed plants are one of the two main groups of vascular plants.

You have different body systems that carry out many functions. Plants have systems too. Vascular plants have a root system, a shoot system, and a reproductive system. A plant's root and shoot systems help the plant to get water and nutrients. Roots are often found underground. Shoots include stems and leaves. They are usually found above ground. ☑

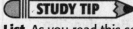

STUDY TIP

List As you read this section, make a chart listing the structures of seed plants and their functions.

READING CHECK

1. Identify What are the three main parts of a seed plant?

_____

_____

_____

Onion   **Dandelion**

Carrots

The roots of plants absorb and store water and nutrients.

SECTION 4 Structures of Seed Plants *continued*

VASCULAR TISSUE

Like all vascular plants, seed plants have specialized tissues that move water and nutrients through the plant. There are two kinds of vascular tissue: xylem and phloem. **Xylem** moves water and minerals from the roots to the shoots. **Phloem** moves food molecules to all parts of the plant. The vascular tissues in the roots and shoots are connected.

What Are Roots?

Roots are organs that have three main functions:

- to absorb water and nutrients from the soil
- to hold plants in the soil
- to store extra food made in the leaves

Roots have several structures that help them do these jobs. The *epidermis* is a layer of cells that covers the outside of the root, like skin. Some cells of the epidermis, called *root hairs*, stick out from the root. These hairs expose more cells to water and minerals in the soil. This helps the root absorb more of these materials.

Roots grow longer at their tips. A *root cap* is a group of cells found at the tip of a root. These cells produce a slimy substance. This helps the root push through the soil as it grows.

2. Describe What are the functions of xylem and phloem?

_____

_____

_____

_____

_____

Critical Thinking

3. Apply Concepts What do you think happens to water and minerals right after they are absorbed by roots?

_____

_____

_____

TAKE A LOOK
4. Identify Where is the vascular tissue located in this root?

_____

_____

The Parts of a Root

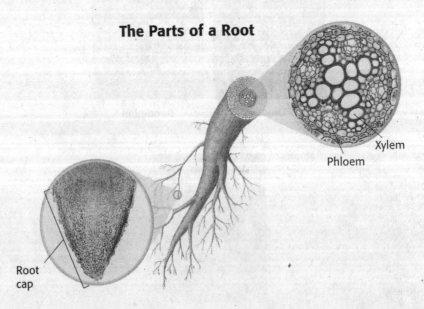

Xylem

Phloem

Root cap

TYPES OF ROOT SYSTEMS

There are two kinds of root systems: taproot systems and fibrous root systems. A *taproot system* has one main root, or taproot, that grows downward. Many smaller roots branch from the taproot. Taproots can reach water deep underground. Carrots are plants that have taproot systems.

A *fibrous root system* has several roots that spread out from the base of a plant's stem. The roots are usually the same size. Fibrous roots usually get water from near the soil surface. Many grasses have fibrous root systems.

What Are Stems?

A stem is an organ that connects a plant's roots to its leaves and reproductive structures. A stem does the following jobs: ☑

- Stems support the plant body. Leaves are arranged along stems so that each leaf can get sunlight.
- Stems hold up reproductive structures such as flowers. This helps bees and other pollinators find the flowers.
- Stems carry materials between the root system and the leaves and reproductive structures. Xylem carries water and minerals from the roots to the rest of the plant. Phloem carries the food made in the leaves to roots and other parts of the plant.
- Some stems store materials. For example, the stems of cactuses can store water.

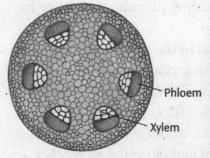

READING CHECK

5. Define What is a stem?

_____

_____

_____

HERBACEOUS STEMS

There are two different types of stems: herbaceous and woody. *Herbaceous* stems are thin, soft, and flexible. Flowers, such as daisies and clover, have herbaceous stems. Many crops, such as tomatoes, corn, and beans, also have herbaceous stems.

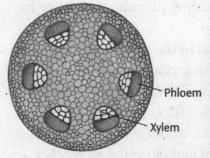

Phloem

Xylem

Herbaceous stems are thin and flexible

TAKE A LOOK

6. Compare Examine this figure and the pictures of woody stems on the next page. How are herbaceous and woody stems similar?

_____

_____

WOODY STEMS

Other plants have woody stems. *Woody* stems are stiff and are often covered by bark. Trees and shrubs have woody stems. The trunk of a tree is actually its stem!

Trees or shrubs that live in areas with cold winters grow mostly during the spring and summer. During the winter, these plants are *dormant*. This means they are not growing or reproducing. Plants that live in areas with wet and dry seasons are dormant during the dry season.

When a growing season starts, the plant produces large xylem cells. These large cells appear as a light-colored ring when the plant stem is cut. In the fall, right before the dormant period, the plant produces smaller xylem cells. The smaller cells produce a dark ring in the stem. A ring of dark cells surrounding a ring of light cells makes up a *growth ring*. The number of growth rings can show how old the tree is.

Critical Thinking

7. Infer How do you think growth rings can be used to tell how old a tree is?

_____

_____

_____

_____

_____

_____

_____

TAKE A LOOK

8. Compare How are herbaceous and woody stems different?

_____

_____

_____

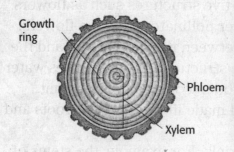

Growth ring

Phloem

Xylem

Woody stems are usually thick and stiff.

What Are Leaves?

FUNCTION OF LEAVES

Leaves are organs, too. The main function of leaves is to make food for the plant. The leaves are where most photosynthesis happens. Chloroplasts in the leaf cells trap energy from sunlight. The leaves also absorb carbon dioxide from the air. They use this energy, carbon dioxide, and water to make food. ☑

All leaf structures are related to the leaf's main job, photosynthesis. A *cuticle* covers the surfaces of the leaf. It prevents the leaf from losing water. The *epidermis* is a single layer of cells beneath the cuticle. Tiny openings in the epidermis, called *stomata* (singular, *stoma*), let carbon dioxide enter the leaf. *Guard cells* open and close the stomata.

9. Identify What is the main function of a leaf?

_____

_____

Structure of a Leaf

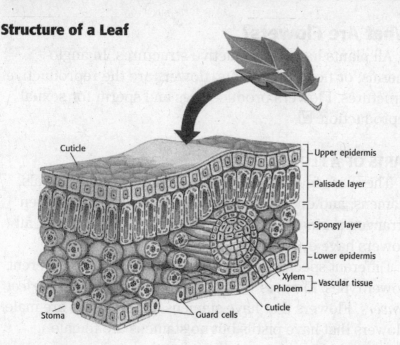

Cuticle

Upper epidermis
Palisade layer
Spongy layer
Lower epidermis
Xylem
Phloem — Vascular tissue
Cuticle
Guard cells
Stoma

TAKE A LOOK
10. Explain Is this plant vascular or nonvascular? Explain your answer.

_____

_____

_____

LEAF LAYERS

Most photosynthesis takes place in the two layers in the middle of the leaf. The upper layer, called the *palisade layer*, contains many chloroplasts. Sunlight is captured in this layer. The lower layer, called the *spongy layer*, has spaces between the cells, where carbon dioxide can move. The spongy layer also has the vascular tissues that bring water to the leaves and move food away.

LEAF SHAPES

Different kinds of plants can have different shaped leaves. Leaves may be round, narrow, heart-shaped, or fan-shaped. Leaves can also be different sizes. The raffia palm has leaves that may be six times longer than you are tall! Duckweed is a tiny plant that lives in water. Its leaves are so small that several of them could fit on your fingernail. Some leaves, such as those of poison ivy below, can be made of several leaflets.

This is one poison ivy leaf. It is made up of three leaflets

 **Say It**

Describe Some people are allergic to poison ivy. They can get a rash from touching its leaves. Some other plants can be poisonous to eat. Are there any other plants you know of that can be poisonous to touch or eat? Describe some of these plants to a partner.

What Are Flowers?

All plants have reproductive structures. In angiosperms, or flowering plants, flowers are the reproductive structures. Flowers produce eggs and sperm for sexual reproduction. ☑

PARTS OF A FLOWER

There are four basic parts of a flower: sepals, petals, stamens, and one or more pistils. These parts are often arranged in rings, one inside the other. However, not all flowers have every part.

Different species of flowering plants can have different flower types. Flowers with all four parts are called *perfect flowers*. Flowers that have stamens but no pistils are male. Flowers that have pistils but no stamens are female.

Parts of a Flower

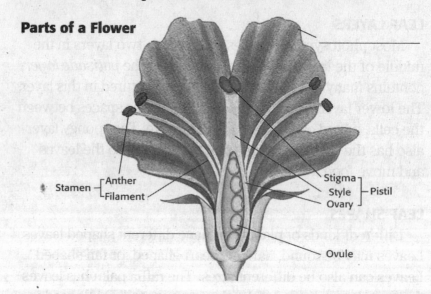

SEPALS

Sepals are leaves that make up the outer ring of flower parts. They are often green like leaves, but they may have other colors. Sepals protect and cover the flower while it is still a bud. When the flower begins to open, the sepals fold back, so the petals can be seen.

PETALS

Petals are leaflike parts of a flower. They make up the next ring inside of the sepals. Petals are sometimes brightly colored, like the petals of poppy flowers or roses. Many plants need animals to help spread their pollen. These colors help attract insects and other animals.

TAKE A LOOK
12. Label As you read, fill in the missing labels on the diagram.

13. Identify What two parts make up the stamen?

_____
_____

14. Identify What three parts make up the pistil?

_____
_____

STAMENS

A **stamen** is the male reproductive structure of a flower. Structures on the stamen called *anthers* produce pollen. Pollen contains the male gametophyte, which produces sperm. The anther rests on a thin stalk called a *filament*. ☑

PISTILS

A **pistil** is the female reproductive structure. The tip of the pistil is called the *stigma*. The long, thin part of the pistil is called the *style*. The rounded base of the pistil is called the **ovary**. The ovary contains one or more ovules. Each ovule contains an egg.

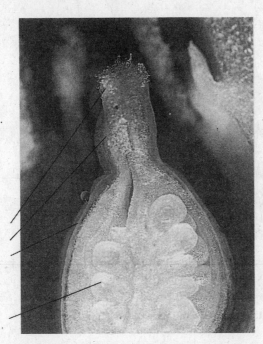

Pollinators brush pollen onto the style, and sperm from inside the pollen travel down the style to the ovary. One sperm can fertilize the egg of one ovule. After fertilization, an ovule develops into a seed. The ovary surrounding the ovule develops into a fruit.

IMPORTANCE OF FLOWERS

Flowers are important to plants because they help plants reproduce. They are also important to animals, such as insects and bats, that use parts of flowers for food. Humans also use flowers. Some flowers, such as broccoli and cauliflower, can be eaten. Others, such as chamomile, are used to make tea. Flowers are also used in perfumes, lotions, and shampoos.

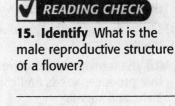

READING CHECK

15. Identify What is the male reproductive structure of a flower?

_____

_____

TAKE A LOOK

16. Label Label the female reproductive structures in this picture.

**Say It**

Discuss What is your favorite flower? Have you ever seen any unusual flowers in nature? In groups of two or three, discuss your experiences with flowers.

Section 4 Review

NSES LS 1a, 1d, 2b, 3d, 4c, 5b

SECTION VOCABULARY

ovary in flowering plants, the lower part of a pistil that produces eggs in ovules	**sepal** in a flower, one of the outermost rings of modified leaves that protect the flower bud
petal one of the usually brightly colored, leaf-shaped parts that make up one of the rings of a flower	**stamen** the male reproductive structure of a flower that produces pollen and consists of an anther at the tip of a filament
phloem the tissue that conducts food in vascular plants	**xylem** the type of tissue in vascular plants that provides support and conducts water and nutrients from the roots
pistil the female reproductive part of a flower that produces seeds and consists of an ovary, style, and stigma	

1. Label Label the parts of this flower.

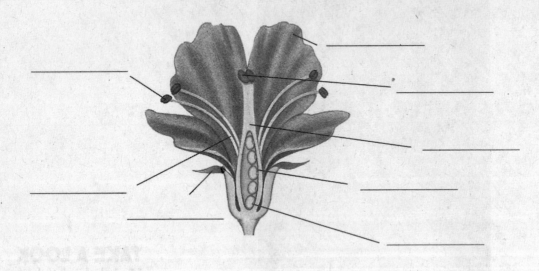

2. Compare How do taproot and fibrous root systems differ?

_____

_____

_____

3. Describe What are the three functions of a stem?

_____

_____

_____

4. List What are the four main organs of a flowering seed plant?

_____

_____

CHAPTER 15 Introduction to Animals

SECTION 1 # What Is an Animal?

After you read this section, you should be able to answer these questions:

• What is an animal?

• What are the seven basic characteristics of animals?

National Science Education Standards
LS 1a, 1b, 1d, 2a, 2b, 4b, 5a

What Is an Animal?

Animals come in many shapes and sizes, but they all share certain characteristics. An *animal* is an organism that is made up of many cells and must eat food to get energy. Animals cannot make their own food as plants do.

STUDY TIP

List As you read, make a list of the characteristics of animals.

What Are the Basic Characteristics of Animals?

ENERGY CONSUMPTION

All organisms need energy to survive. Unlike plants, animals must get energy from other organisms. They are consumers. A **consumer** is an organism that feeds on other organisms or parts of other organisms. Some animals, such as lions, eat other animals. Some, such as pandas, eat plants. Other animals, such as black bears, can eat both plants and animals. ☑

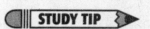

READING CHECK

1. Define What is a consumer?

_____

_____

_____

_____

MOVEMENT

Nearly all animals move during some part of their lives, often in order to find food or shelter. Muscle cells contract and relax to help the animal move. Groups of muscle cells work together.

MULTICELLULAR MAKEUP

Like all organisms, animals are made of cells. Unlike plant cells, animal cells do not have cell walls. Animal cells are surrounded only by cell membranes.

All animals are *multicellular*. That means they are made of many cells. All of an animal's cells work together to perform important functions, such as breathing, digesting food, and reproducing. ☑

READING CHECK

2. Define What does multicellular mean?

_____

_____

LEVELS OF ORGANIZATION

Animals have different levels of organization in their bodies. This means their bodies are organized into structures made of smaller structures. The levels are listed below in order from smallest to largest.

1. *Cells* are the first level of organization. Cells specialize to do specific jobs.
2. Groups of cells that are of the same kind and that work together form *tissues*. For example, muscle cells form muscle tissue.
3. Tissues work together to form an *organ*. The heart, lungs, and kidneys are all organs.
4. Organs work together to form an *organ system*. The organism may die if any part of an organ system stops working.

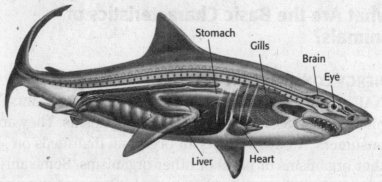

Like most other animals, sharks have organs for digestion, circulation, and sensing the environment.

CONTROLLED BODY TEMPERATURE

All animals need to keep their bodies within a range of temperatures. Birds and mammals do this by using the energy released by chemical reactions. These kinds of animals are called *endotherms*. Their body temperatures stay almost the same, even when the temperature of the environment changes.

The body temperatures of other animals change with the temperature of the environment. These kinds of animals are called *ectotherms*. Some of these animals have behaviors that help control their body temperatures. For example, some lizards sit in the sunlight to warm up.

BODY PLAN

A body plan is the general shape of an organism. One characteristic of a body plan is its symmetry. Animals can have three types of symmetry.

Say It

Discuss With a partner, choose 10 different kinds of animals, and describe what they eat. Make a list to share with your classmates.

TAKE A LOOK

3. List Name four organ systems you can see in this picture.

_____

_____

_____

_____

SECTION 1 What Is an Animal? *continued*

This tortoise has **bilateral symmetry**. The two sides of its body mirror each other.

This sea star has **radial symmetry**. Its body is organized around the center, like spokes on a wheel.

This sponge is **asymmetrical**. You cannot draw a straight line that divides its body into two or more equal parts. There is no center point that its body is organized around.

Another characteristic of a body plan is whether or not it has a coelom. A **coelom** is a body cavity, or space, that surrounds and protects many organs, such as the heart. Many animals have coeloms.

REPRODUCTION AND DEVELOPMENT

Animals make more animals through reproduction. In **asexual reproduction**, there is only one parent. All the offspring have the same genes as the parent. Some animals, such as the hydra, reproduce asexually by budding. In *budding*, part of an organism develops into a new organism. The new organism then breaks off from the parent. In *fragmentation*, parts of an organism break off and then develop into new organisms.

In **sexual reproduction**, offspring form when genetic information from two parents combines. The female parent produces sex cells called *eggs*. The male parent produces sex cells called *sperm*. The first cell of a new organism forms when an egg's nucleus and a sperm's nucleus join. This process is called *fertilization*.

A fertilized cell divides into many cells to form an *embryo*. As an animal embryo develops, its cells become specialized through differentiation. In **differentiation**, cells develop different structures to do specific jobs. ☑

Cells in the mouse embryo will differentiate as the mouse develops. These cells will produce skin, muscles, nerves, and all the other parts of the mouse's body.

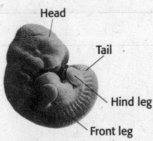

Head

Tail

Hind leg

Front leg

TAKE A LOOK

4. Demonstrate Draw a line or lines on two of the pictures to show how you can divide each body into like parts.

Critical Thinking

5. Apply Concepts What kind of symmetry do you have?

_____

_____

STANDARDS CHECK

LS 2a Reproduction is a characteristic of all living systems; because no living organism lives forever, reproduction is essential to the continuation of every species. Some organisms reproduce asexually. Others reproduce sexually.

Word Help: sexual having to do with sex

6. Identify What are two ways some animals can reproduce asexually?

_____

_____

_____

☑ **READING CHECK**

7. Explain What happens to the cells in an embryo as the embryo develops?

_____

_____

_____

281

Section 1 Review

NSES LS   1a, 1b, 1d, 2a, 2b, 4b, 5a

SECTION VOCABULARY

asexual reproduction reproduction that does not involve the union of sex cells and in which one parent produces offspring that are genetically identical to the parent	**differentiation** the process in which the structure and function of the parts of an organism change to enable specialization of those parts
coelom a body cavity that contains the internal organs	**sexual reproduction** reproduction in which the sex cells from two parents unite to produce offspring that share traits from both parents
consumer an organism that eats other organisms or organic matter	

1. Predict What would happen to an animal such as a shark if its heart failed?

_____

_____

_____

2. Compare What is the difference between an endotherm and an ectotherm?

_____

_____

_____

_____

3. Explain Why is differentiation important to multicellular organisms?

_____

_____

_____

_____

4. Explain Does fertilization happen in asexual reproduction?

_____

_____

_____

5. Compare How is fragmentation different from budding?

_____

_____

_____

_____

SECTION 2 The Animal Kingdom

After you read this section, you should be able to answer these questions:

• What is diversity?

• What are vertebrates?

• What are invertebrates?

What Is Diversity?

Insects, birds, and other animals look and are very different from one another. They also live in many different places. Scientists call this range of difference *diversity*.

Scientists have named more than 1 million species of animals. Some scientists estimate that more than 3 million species of animals live on Earth. That means many species that exist have not yet been discovered and named.

Animals that have been discovered and described are placed into groups. Grouping organisms makes it easier to study all of the different kinds of animals. The pie graph below shows the relative sizes of the main groups of animals in the animal kingdom.

STUDY TIP

Organize As you read, make combination notes about each group of animals. Write descriptions in the left column of the notes. Draw pictures in the right column that will help you remember what each type of animal looks like.

Makeup of the Animal Kingdom

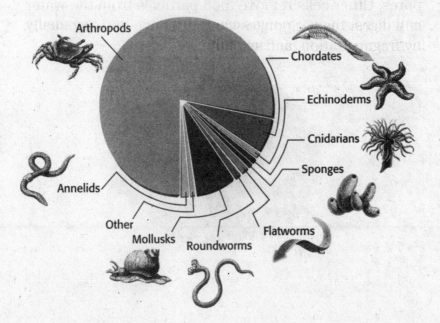

Arthropods
Chordates
Echinoderms
Cnidarians
Sponges
Annelids
Other
Mollusks
Roundworms
Flatworms

TAKE A LOOK
1. **Identify** Which group of animals has the most species?

_____

_____

How Do Scientists Classify Animals?

Scientists organize animals into several groups. These groups are based on the characteristics the animals share and how closely they are related.

In the past, scientists grouped animals on the basis of only their structure or appearance. Today, scientists also use DNA, or genetic material, to group animals. There are two general groups of animals: invertebrates and vertebrates. ☑

2. Identify What two types of information do scientists use to classify animals?

_____

_____

_____

3. Define What is an invertebrate?

_____

_____

What Are Invertebrates?

Most of the animals on Earth are invertebrates. An **invertebrate** is an animal that does not have a backbone. In fact, invertebrates do not have any bones. Insects, snails, and worms are all invertebrates. Invertebrates can be found living in every environment on Earth. ☑

SPONGES

Sponges are some of the simplest invertebrates. A sponge is a mass of specialized cells held together by a jelly-like material. Tiny, glassy structures in the sponge also provide support. Sponges are asymmetrical. Adult sponges generally do not move.

The body of a sponge has many tubes and thousands of small pores, or holes. Some cells sweep water into the pores. Other cells remove food particles from the water and digest them. Sponges can reproduce both asexually, by fragmentation, and sexually.

CNIDARIANS

Cnidarians include jellyfish, sea anemones, and corals. Most cnidarians live in the ocean. Their simple bodies have radial symmetry. They have specialized stinging cells called *cnidocysts* on their tentacles. These cells help the animals stun and catch the tiny animals they eat. ☑

Cnidarians have one of two radially symmetrical body plans: a medusa form or a polyp form. These two body plans are shown below. Many cnidarians reproduce sexually. Some cnidarians can also reproduce asexually by budding and fragmentation.

READING CHECK

4. Identify What specialized structures help cnidarians catch their food?

_____

_____

Jellyfish

The jellyfish has the medusa body form. A medusa is a bell-shaped body with tentacles.

As adults, sea anemones and corals are polyps. The polyp body form looks like a cup on a base. A polyp attaches the base of the cup to a hard surface.

TAKE A LOOK

5. Identify What two body forms can cnidarians have?

_____

_____

FLATWORMS

Flatworms are the simplest worms. Many flatworms live in water or damp soils. Other flatworms are parasites. A *parasite* is an organism that invades and feeds on the body of another organism.

Flatworms are more complex than sponges or cnidarians. They are bilaterally symmetrical. Flatworms have heads and eyespots, which are sensitive to light. Every flatworm is both male and female. They can reproduce both asexually, by fragmentation, and sexually.

READING CHECK

6. Identify What is one major difference in how flatworms and roundworms reproduce?

_____

_____

_____

_____

ROUNDWORMS

Roundworms are cylindrical like spaghetti. They live in freshwater, in damp soils, and as parasites in the tissues and body fluids of other animals. Like flatworms, roundworms have bilateral symmetry. Unlike flatworms, roundworms are either male or female. They reproduce sexually. ☑

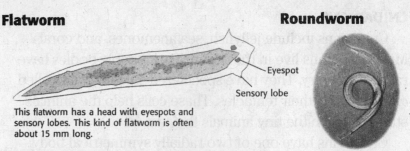

Flatworm   **Roundworm**

Eyespot

Sensory lobe

This flatworm has a head with eyespots and sensory lobes. This kind of flatworm is often about 15 mm long.

MOLLUSKS

Snails, slugs, clams, oysters, squids, and octopuses are mollusks. They live in water and on land. Mollusks have a specialized tissue called a *mantle*. The shells of snails, clams, and oysters are made by their mantles. A mollusk also has a muscular foot that helps it move. In some mollusks, the foot is modified into tentacles. ☑

Squids and octopuses use tentacles to capture prey, such as fish. Clams and oysters filter food from the water. Snails and slugs feed on plants and break down dead organisms. Mollusks reproduce sexually.

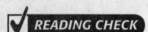

READING CHECK

7. Identify What specialized tissue makes a mollusk's shell?

_____

_____

Squid

ANNELIDS

Annelids are worms made of repeating body segments. For this reason, they are sometimes called *segmented worms*. Annelids have bilateral symmetry. Earthworms and leeches are types of annelids. ☑

Each annelid has both male and female sex organs. However, individuals must fertilize each other to reproduce sexually.

READING CHECK

8. Compare How are earthworms different from roundworms and flatworms?

_____

_____

_____

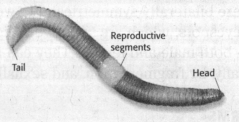

Reproductive segments

Tail

Head

SECTION 2 The Animal Kingdom *continued*

ARTHROPODS

Arthropods are the most diverse group in the animal kingdom. Arthropods have bilateral symmetry and a strong, external armor called an **exoskeleton**. The exoskeleton protects arthropods from predators. The exoskeleton also keeps the animal from drying out. ☑

Insects are a familiar group of arthropods because they are often seen on land. An insect's body is clearly divided into three segments called the *head, thorax,* and *abdomen*. Millipedes, centipedes, spiders, and scorpions are also arthropods. Arthropods that live in the water include crabs and shrimp. Most arthropods are either male or female and reproduce sexually.

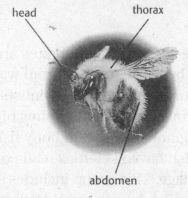

head   thorax

abdomen

ECHINODERMS

Echinoderms live in the ocean. This group includes sea stars, sea urchins, and sand dollars. The name echinoderm means "spiny skinned." Echinoderms, like the sea urchins in the picture below, have exoskeletons covered in bumps and spines.

Echinoderms have a system of water pumps and canals in their bodies. This system, called the *water vascular system*, helps the animal move, eat, breathe, and sense its environment. Echinoderms usually reproduce sexually. Males release sperm as females release eggs into the water, where fertilization takes place. ☑

Echinoderms have bilateral
symmetry as larvae and
radial symmetry as adults.

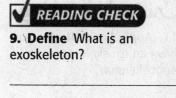

✓ READING CHECK

9. Define What is an exoskeleton?

_____

_____

📣 Say It

Describe There are more arthropods on Earth than any other kind of animal! Describe to the class the last time you saw an arthropod. Where were you? What did the animal look like? What kind of arthropod was it?

✓ READING CHECK

10. Define What is the water vascular system?

_____

_____

_____

_____

_____

_____

What Are Vertebrates?

Vertebrates are animals that have backbones. The backbone is a strong but flexible column made of individual units called *vertebrae* (singular, *vertebra*). The backbone is a part of the endoskeleton of a vertebrate. An **endoskeleton** is an internal skeleton. It supports the body of the animal. It also provides a place for muscles to attach so that the animal can move.

Less than 5% of the known animal species are vertebrates. Vertebrates are divided into five main groups: fishes, amphibians, reptiles, birds, and mammals. Vertebrates are either male or female, and they reproduce mainly by sexual reproduction.

FISH

Over half of the species of vertebrates are species of fish. Fish breathe by taking oxygen from water through specialized structures called gills. Scientists classify fishes into four groups: two small groups of jawless fishes plus cartilaginous fishes and bony fishes. ☑

Cartilaginous fish have skeletons made of a flexible tissue called cartilage. This group includes sharks and stingrays. All other fish have bony skeletons. Bony fish live in saltwater and freshwater environments all over the world. Trout, bass, and goldfish are all bony fishes.

AMPHIBIANS

Salamanders, toads, and frogs are all amphibians. Because many amphibians spend part of their lives on land and part in the water, scientists say they have a two-part life cycle. Most amphibians live near fresh water because their eggs and young need water to survive.

Critical Thinking

11. Apply Concepts Do you have an exoskeleton or an endoskeleton?

_____

READING CHECK

12. Identify What structure do fish use to breathe?

_____

Critical Thinking

13. Apply Concepts How is the development of a tadpole into an adult frog an example of differentiation?

_____

_____

_____

_____

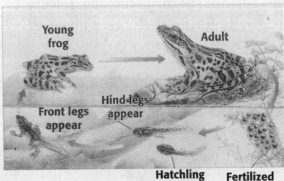

As adults, most amphibians still need water. Salamanders and frogs have thin skin that must stay moist.

Young amphibians live in water and breathe with gills. As they develop, they grow legs and lungs that help them live on land.

Young frog

Adult

Hind legs appear

Front legs appear

Hatchling tadpole

Fertilized eggs

SECTION 2 The Animal Kingdom *continued*

REPTILES

Reptiles include snakes, turtles, and alligators. Unlike amphibians, reptiles do not need to lay their eggs in water. Reptile eggs have membranes and a shell that protects them from drying out. Reptiles can live almost anywhere on land. They can also live in water. Reptiles generally reproduce sexually.

BIRDS

Birds share many characteristics with reptiles, such as sexual reproduction, eggs with shells, and similar feet structures. However, birds are the only animals on Earth today that have feathers. Feathers are lightweight structures that help birds stay warm. They also help shape the body and wings for flying.

Some kinds of birds no longer use their wings to fly. The penguin, for example, uses its wings to swim. Birds such as ostriches and emus do not fly, but they have unique characteristics that help them run.

Math Focus

14. Calculate A bird weighing 15 g eats 10 times its weight in food in a week. How much food does the bird eat in a day?

_____

MAMMALS

All mammals have hair at some time in their lives, and all female mammals can produce milk for their young. Mammals reproduce sexually.

Mammals are divided into three groups: monotremes, marsupials, and placental mammals.

Three Kinds of Mammals

Monotremes are mammals that lay eggs. The echidna is a monotreme that lives in Australia.

Marsupials are mammals with pouches. Kangaroos are marsupials. Young marsupials develop inside the mother's pouch.

TAKE A LOOK

15. List What are the three groups of mammals?

_____

_____

_____

Most mammals are **placental mammals**. The placenta is a special organ that lets nutrients and wastes be exchanged between the mother and unborn young.

Name _____ Class _____ Date _____

Section 2 Review

SECTION VOCABULARY

endoskeleton an internal skeleton made of bone and cartilage	**invertebrate** an animal that does not have a backbone
exoskeleton a hard, external, supporting structure	**vertebrate** an animal that has a backbone

1. Compare What are two main differences between a sponge and a roundworm?

_____

_____

_____

2. Describe Describe the two cnidarian body plans.

_____

_____

_____

3. Compare What is the difference between an exoskeleton and an endoskeleton?

_____

_____

_____

4. Explain Why do most adult amphibians need to live near water or in a moist habitat?

_____

_____

_____

5. Identify What are two characteristics of mammals?

_____

_____

_____

6. Explain What is the function of a placenta?

_____

_____

_____

CHAPTER 15 Introduction to Animals
SECTION
3 **Invertebrates**

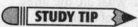

National Science
Education Standards

LS 1a, 1d, 1f, 2a, 3a, 3c, 5a

BEFORE YOU READ

After you read this section, you should be able to answer
these questions:

- What structures and systems perform basic life
  functions in invertebrates?

- How do invertebrates reproduce and develop?

What Are the Characteristics of Invertebrates?

Invertebrates are animals without backbones. They
can be found in almost every environment on Earth.
Invertebrates come in many different shapes and sizes.
Some invertebrates have heads, and others do not. Some
invertebrates eat food through their mouths. Others
absorb food particles through their tissues.

BODY SYMMETRY

Invertebrates have one of three basic body plans: irregular, radial, or bilateral. Sponges have irregular shapes.
They are asymmetrical. Sea anemones have radial symmetry. That means that body parts extend from a central
point. Animals with radial symmetry have only a top and
a bottom that are very different from each other.

Most invertebrates have bilateral symmetry. This
means the body can be divided into two mirror-image
halves by one straight line. Animals with bilateral symmetry have a top and bottom that differ, as well as a front
end and a back end that differ. The development of a
head is seen only in organisms with bilateral symmetry.

STUDY TIP

Underline As you read,
underline the characteristics
of invertebrates.

Critical Thinking

1. Predict Would you expect
a sea anemone to have a
head? Explain your answer.

_____

_____

_____

_____

TAKE A LOOK

2. Identify What type of
symmetry does the sea hare
have?

_____

_____

SEGMENTATION

The bodies of many animals are divided into **segments**, or sections. Segmentation in the body has many advantages. For example, each segment in an earthworm has a set of muscles that help the earthworm push through soil.

Segmentation in Invertebrate Bodies

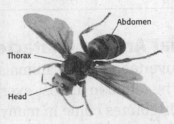

The body of a marine worm has many segments that are almost equal in size.

The body of an insect has three unequal segments: a head, a thorax, and an abdomen.

TAKE A LOOK
3. List What are the three segments of an insect's body?

_____

_____

_____

SUPPORT OF THE BODY

Invertebrate bodies need support and protection. Some invertebrates, like jellyfish and anemones, are supported by the water they live in. Others have structures in or on their bodies that support and protect them. The figure below shows the outer coverings of different invertebrates. Muscles attached to outer coverings in some invertebrates contract and relax to help the animals move.

Support in Invertebrate Bodies

TAKE A LOOK
4. List What are three types of support in invertebrate bodies?

_____

_____

_____

_____

A sponge is supported by jelly-like material and tiny glassy structures.

Some invertebrates, such as this roundworm, have thick skin as a tough outer covering.

Other invertebrates have tough outer coverings called exoskeletons.

RESPIRATION AND CIRCULATION

All animals take in oxygen and release carbon dioxide through respiration. Respiration is performed by the *respiratory system*. Different invertebrates have different structures for respiration. For example, lobsters have gills. Respiration in insects, however, is through a network of tubes, called *tracheae*, inside the body. ☑

Oxygen, carbon dioxide, and nutrients must circulate, or move around, within the body. The *circulatory system* moves these materials with blood through the body. Some invertebrates have an **open circulatory system**. In open circulatory systems, blood moves through open spaces in the body. Others have a **closed circulatory system**. In closed circulatory systems, blood moves through tubes that form a closed loop.

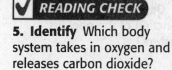

Trachea

This beetle moves air into and out of its body through small holes along the sides of its body.

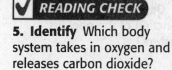
READING CHECK

5. Identify Which body system takes in oxygen and releases carbon dioxide?

_____

_____

DIGESTION AND EXCRETION

Animals get the energy they need by digesting food. Food is broken down and nutrients are absorbed by the *digestive system*. Invertebrates have relatively simple digestive systems. The mouth and anus form two ends of a tube called a *digestive tract*. Any material that is eaten but not digested is sent out of the body as waste. ☑

As cells in the body use up nutrients, another kind of waste forms. In many invertebrates, the digestive tract eliminates this kind of waste as well. Other invertebrates have a separate system, called the *excretory system*, that remove excess water and waste from cells.

READING CHECK

6. Identify Which body system breaks down food?

_____

_____

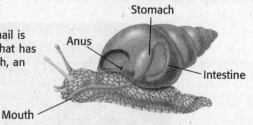

The digestive system in the snail is made up of a digestive tract that has four parts: a mouth, a stomach, an intestine, and an anus.

Stomach

Anus

Intestine

Mouth

NERVOUS SYSTEM

The nervous system receives and sends electrical signals that control the body. The figure below shows the nervous systems of three invertebrates. The simplest invertebrates have only nerve cells. More complex invertebrates have brains and sense organs, such as eyes. ☑

READING CHECK

7. Describe What is the function of the nervous system?

_____

_____

_____

_____

TAKE A LOOK

8. Compare How do the nervous systems of a hydra and a grasshopper differ?

_____

_____

_____

_____

_____

_____

Invertebrate Nervous Systems

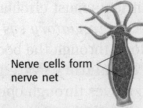

Nerve cells form nerve net

Hydra The simplest invertebrates have nervous systems made up of only nerve cells.

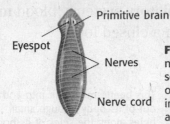

Primitive brain

Eyespot

Nerves

Nerve cord

Flatworm More complex nervous systems include sense organs, such as eyes or eyespots. They collect information such as sound and light.

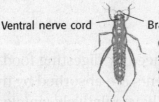

Ventral nerve cord

Brain

Grasshopper Many nervous systems have a specialized area called the brain. The brain acts as the control center.

Critical Thinking

9. Apply Concepts Are offspring produced in asexual reproduction genetically different from their parents? Explain your answer.

_____

_____

_____

_____

_____

REPRODUCTION

Many invertebrates can reproduce both asexually and sexually. A hydra, for example, can reproduce asexually by budding. Budding happens when a part of the parent organism develops into a new organism. Other invertebrates can reproduce asexually by fragmentation. In fragmentation, parts of an organism break off and then develop into new individuals.

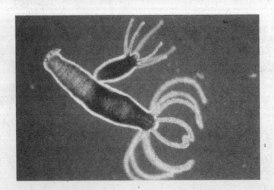

SECTION 3 Invertebrates *continued*

DEVELOPMENT

Some invertebrates, such as insects, change form as they develop. This change is called **metamorphosis**. Most insects, including butterflies, beetles, flies, bees, and ants, go through a complex change called *complete metamorphosis*.

Stages of Complete Metamorphosis

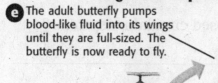

e The adult butterfly pumps blood-like fluid into its wings until they are full-sized. The butterfly is now ready to fly.

a An adult lays **eggs.** An embryo forms inside each egg.

d Adult body parts replace the larval body parts. The **adult** splits its chrysalis and emerges.

b A **larva** hatches from the egg. Butterfly and moth larvae are called *caterpillars.* The caterpillar eats leaves and grows rapidly. As the caterpillar grows, it sheds its outer layer several times. This process is called *molting.*

c After its final molt, the caterpillar makes a chrysalis and becomes a **pupa.** The pupal stage may last a few days or several months. During this stage, the insect is inactive.

Stages of Incomplete Metamorphosis

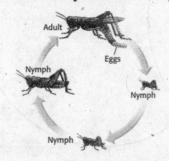

Adult

Eggs

Nymph

Nymph

Nymph

Some insects, such as grasshoppers and cockroaches, go through *incomplete metamorphosis*. Incomplete metamorphosis has three stages: egg, nymph, and adult.

Some nymphs shed their exoskeletons several times as they grow into adults. This shedding is called *molting*. In incomplete metamorphosis, nymphs look very much like small adults. ☑

TAKE A LOOK

10. List What are the four stages of complete metamorphosis?

_____

_____

_____

_____

TAKE A LOOK

11. Compare In which type of metamorphosis do more changes take place as the young develop?

_____

_____

✓ READING CHECK

12. List What are the three stages of incomplete metamorphosis?

_____

_____

_____

Section 3 Review

NSES LS 1a, 1d, 1f, 2a, 3a, 3c, 5a

SECTION VOCABULARY

closed circulatory system a circulatory system in which the heart circulates blood through a network of vessels that form a closed loop	**open circulatory system** a type of circulatory system in which the circulatory fluid is not contained entirely within vessels
metamorphosis a process in the life cycle of many animals during which a rapid change from the immature form of an organism to the adult form takes place	**segment** any part of a larger structure, such as the body of an organism, that is set off by natural arbitrary boundaries

1. Explain What is the difference between open and closed circulatory systems?

_____

_____

_____

2. Compare How is the life cycle of a butterfly different from the life cycle of a hydra?

_____

_____

_____

3. Compare Use a Venn Diagram to compare complete metamorphosis and incomplete metamorphosis.

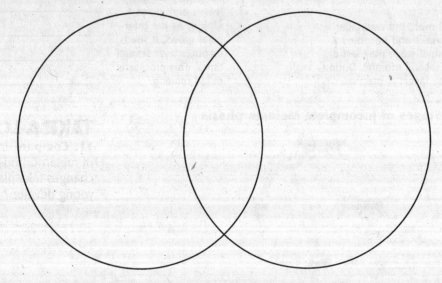

4. Explain Why are outer coverings important for movement in many invertebrates?

_____

_____

_____

CHAPTER 15 | Introduction to Animals

SECTION 4 Vertebrates

National Science Education Standards

LS 1a, 1d, 2a, 3a, 3b, 3c, 5a, 5c

BEFORE YOU READ

After you read this section, you should be able to answer these questions:

- How are vertebrates different from invertebrates?
- How do vertebrate organ systems work?
- How do vertebrate embryos develop?

What Are the Characteristics of Vertebrates?

All vertebrates have a structure called a backbone. The backbone is part of a skeleton that is made of bone. Bone is a type of very hard tissue found only in vertebrates.

All vertebrates also have a head protected by a skull. The skull is made of either cartilage or bone. **Cartilage** is a flexible material made of cells and proteins. All vertebrate embryos have skeletons made of cartilage. However, as most vertebrates grow, bone replaces the cartilage. ☑

BODY COVERINGS

The bodies of vertebrates are covered with skin. Skin protects the body from the environment. The structure of skin is different in different vertebrates.

Body Coverings in Vertebrates

Scales Reptiles and fish are covered with thin, small plates called scales.

Feathers Feathers on birds, like hairs on mammals, help keep the body temperature stable.

Fur Some body coverings have colors and patterns that help vertebrates hide from prey or predators.

Skin Skin protects the body from the environment.

STUDY TIP

Summarize As you read, make an outline of the characteristics of vertebrates.

READING CHECK

1. List Name two characteristics of vertebrates.

_____

_____

_____

TAKE A LOOK

2. List Name three different types of body coverings that protect vertebrate bodies.

_____

_____

_____

SECTION 4 Vertebrates *continued*

BODY SYMMETRY

All vertebrates have bilateral symmetry. A bilaterally symmetrical body has four main parts. (Though we are vertebrates, it helps to think of bilateral symmetry in an animal that walks on four legs.) The upper surface, or back, is the *dorsal* side. The lower surface, or belly, is the *ventral* side. The head is in the front of the body, or *anterior*. The tail is in the back of the body, or *posterior*. ☑

SUPPORT OF THE BODY

The body of a vertebrate is supported by an endoskeleton. An endoskeleton has three main parts: a skull, a backbone, and limb bones. Vertebrates need large bones and muscles for support and movement if they don't live in the water.

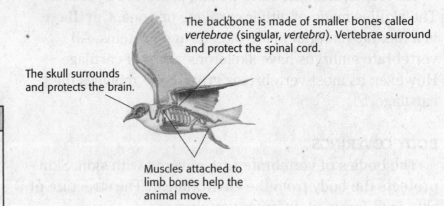

The backbone is made of smaller bones called *vertebrae* (singular, *vertebra*). Vertebrae surround and protect the spinal cord.

The skull surrounds and protects the brain.

Muscles attached to limb bones help the animal move.

RESPIRATORY SYSTEM

The respiratory system in vertebrates brings oxygen into the body and takes carbon dioxide out. The main respiratory organs in vertebrates are either lungs or gills.

Vertebrates that breathe air, rather than water, have their respiratory organs inside the body. This protects them from drying out.

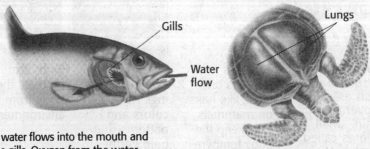

Gills

Water flow

Lungs

In fish, water flows into the mouth and over the gills. Oxygen from the water moves across the gills and into the blood. Carbon dioxide moves from the blood, across the gills, and into the water.

The inside surfaces of lungs have many small pockets. These pockets make more surface area for the exchange of oxygen and carbon dioxide.

✓ **READING CHECK**

3. Identify List the four main parts of an animal that has bilateral symnmetry.

_____

_____

_____

_____

STANDARDS CHECK

LS 1a Living organisms at all levels of organization demonstrate the complementary nature of <u>structure</u> and <u>function</u>. Important levels of organization for structure and function include cells, organs, tissues, organ systems, whole organisms, and ecosystems.

Word Help: <u>structure</u> the arrangement of the parts of a whole

Word Help: <u>function</u> use or purpose

4. Compare How are the functions of gills and lungs similar and different?

_____

_____

_____

_____

_____

_____

CIRCULATORY SYSTEM

The circulatory system moves nutrients and other substances around the body. Vertebrates have closed circulatory systems made up of blood, blood vessels, and a heart. Arteries are vessels that carry blood away from the heart. Veins are vessels that carry blood to the heart. Tiny vessels called capillaries connect veins and arteries. ☑

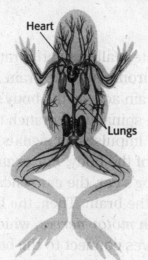

Heart

Lungs

☑ READING CHECK

5. Identify What kind of circulatory system do vertebrates have?

_____

_____

_____

DIGESTIVE SYSTEM

Vertebrates have digestive systems to break down food and absorb nutrients. The digestive system is made up of a long tube called the *digestive tract*. Food moves through the digestive tract from the mouth to the anus. ☑

☑ READING CHECK

6. Identify What is the beginning and what is the end of the vertebrate digestive tract?

_____

_____

Food passes from the mouth and into the stomach.

↓

Chemicals in the stomach turn the food into a liquid.

↓

The liquid moves into the **small intestine**, which absorbs nutrients.

↓

Undigested material moves into the **large intestine**, which turns this material into feces.

↓

The feces leave the body through the anus.

SECTION 4 Vertebrates *continued*

EXCRETORY SYSTEM

The digestive system produces waste from the food that animals eat. However, cells also produce waste. In vertebrates, these wastes are removed by the *excretory system*. One of these wastes is ammonia. In mammals, the liver turns ammonia into urea. Then, the kidneys filter the urea from the blood. Urea combines with water to form urine.

NERVOUS SYSTEM

The nervous system allows vertebrates to sense and respond to the environment. The brain is part of the spinal cord. The brain acts as the body's control center.

Nerves from the spinal cord branch throughout the body. Nerves carry impulses, or signals, between the brain and the rest of the body. For example, when a sound reaches a dog's ear, the ear sends a signal through *sensory nerves* to the brain. Then, the brain sends signals to the body through *motor nerves*, which cause the body to react. Some nerves connect to the body's muscles. Signals sent to these muscles cause them to contract.

Brain size is very different in different kinds of vertebrates. Although all vertebrates use instinct to react, those with larger brains depend more on learning. An *instinct* is a behavior or reaction that an animal is born with. *Learning* is a behavior in which new experiences change the way an animal reacts.

Nervous Systems in Vertebrates

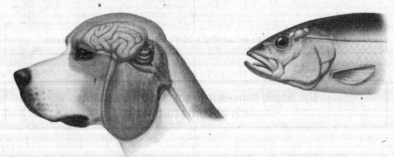

REPRODUCTION

Most vertebrates reproduce sexually. Fertilization happens when the nucleus of a sperm cell joins with the nucleus of an egg cell. A fertilized egg cell divides many times. It becomes a multicellular embryo. As an embryo develops, its cells differentiate. That is, the cells develop different structures so that they can perform different functions.

Say It

Discuss Work with a partner to discuss how the five senses—touch, taste, sight, hearing, and smell—are all important to animal survival. Prepare a short presentation to tell your classmates how an animal uses each sense.

TAKE A LOOK

7. Explain Which of these animals probably relies more on learning than instinct? Explain your answer.

_____

_____

_____

_____

DEVELOPMENT

Most fish and amphibians have a larval stage in their life cycles. A *larva* (plural, *larvae*) is a newly hatched animal that must go through metamorphosis to become an adult. The larvae of fish and amphibians usually hatch in the water and live on their own. Over time, larvae develop new structures or lose old structures to become adults.

Reptiles, birds, and mammals do not have a larval stage in their life cycles. These animals make eggs that are protected by special membranes. The eggs of reptiles, birds, and some mammals also have a shell. Shelled eggs are laid on land. However, most mammals do not lay eggs. Their embryos develop inside the female until the offspring are born. ☑

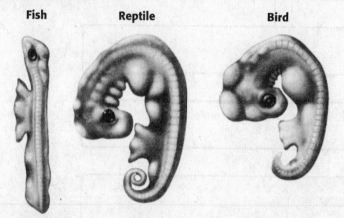

Fish   Reptile   Bird

Embryos of different species look similar at early stages of development. Embryos begin to look more like their own species as they develop.

☑ **READING CHECK**

8. Identify How do most vertebrates reproduce?

_____

_____

PARENTAL CARE

Human babies need a great deal of care from their parents for many years. However, not all vertebrates need as much parental care. Many fish, for example, simply lay their eggs and leave. These animals lay so many eggs that at least a few survive. Some fish and reptiles guard a nest until the young hatch. Usually, once they hatch, the young are on their own.

Birds and mammals generally show more parental care than other vertebrates. They have fewer offspring but spend more time feeding and protecting them. More parental care gives each offspring a better chance of survival.

Critical Thinking

9. Infer Why are birds more likely to care for their young than frogs are?

_____

_____

_____

_____

Name _____ Class _____ Date _____

Section 4 Review

NSES LS 1a, 1d, 2a, 3a, 3b, 3c, 5a, 5c

SECTION VOCABULARY

cartilage a flexible and strong connective tissue **large intestine** the wider and shorter portion of the intestine that removes water from mostly digested food and that turns the waste into semisolid feces, or stool	**small intestine** the organ between the stomach and the large intestine where most of the breakdown of food happens and most of the nutrients from food are absorbed

1. Summarize Complete the Flow Chart to show how food passes through the digestive system.

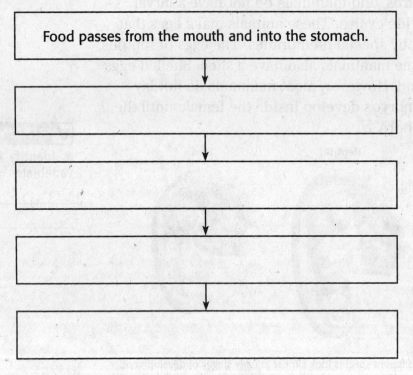

Food passes from the mouth and into the stomach.

2. List What are the three main parts of an endoskeleton?

_____

_____

_____

3. Explain Why do cells in a developing embryo differentiate?

_____

_____

4. Compare How does parental care differ in fish and mammals?

_____

_____

_____

CHAPTER 16 Interactions of Living Things

SECTION 1 Everything Is Connected

After you read this section, you should be able to answer these questions:

- What do organisms in an ecosystem depend on for survival?
- What are biotic and abiotic factors?
- What are the levels of organization in the environment?

National Science Education Standards
LS 4a, 4b, 4c, 4d

What Is the Web of Life?

All organisms, or living things, are linked together in the web of life. In this web, energy and resources pass between organisms and their surroundings. The study of how different organisms interact with one another and their environment is **ecology**.

An alligator may hunt along the edge of a river. It may catch a fish, such as a gar, that swims by too closely. As it hunts, the alligator is interacting with its environment. Its environment includes other organisms living in the area. The alligator depends on other organisms to survive, and other organisms depend on the alligator.

However, one organism eating another is not the only way living things interact. For example, when it gets too hot, the alligator may dig a hole in the mud under water. When the alligator no longer uses the hole, fish and other organisms can use it. They may live in the hole when the water level in the rest of the river is low.

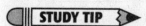

STUDY TIP

Underline As you read, underline any new science terms. Find their definitions in the section review or a dictionary. Make sure you learn what each term means before you move to the next section.

Living things in an environment interact.

**Say It**

Discuss With a partner, talk about the organisms in this picture. How do you think each type of organism interacts with the others? What kind of things do you think each of these organisms needs to survive?

SECTION 1 Everything Is Connected *continued*

What Are the Two Parts of an Environment?

An organism's environment is made up of biotic and abiotic parts. **Biotic** describes the living parts of the environment, such as fish. **Abiotic** describes the nonliving parts of the environment, such as rivers. Organisms need both biotic and abiotic parts of the environment to live. ☑

How Is the Environment Organized?

The environment can be organized into five levels. Individual organisms are at the first level. The higher levels include more and more parts of the environment. The highest level is called the biosphere. It is the largest level, and includes all the other levels.

1. An *individual* is a single organism.

Organism

2. A **population** is a group of individuals of the same species in the same area. For example, all the alligators in the same river make a population. The whole population uses the same area for food and shelter.

Population

3. A **community** is made up of all the different populations that live and interact in the same area. The different populations in a community depend on each other. For example, alligators eat other animals, including fish. Alligators create water-filled holes where fish and other organisms in the river can live during dry seasons.

Community

Critical Thinking

2. Identify Relationships How are the first two levels of organization related?

_____

_____

_____

_____

Critical Thinking

3. Infer Could a community be made up of only one population of organisms? Explain.

_____

_____

_____

_____

SECTION 1 Everything Is Connected *continued*

4. An **ecosystem** is made up of a community and its abiotic environment. The abiotic factors provide resources for all the organisms and energy for some. A river, for example, can provide water for river plants and many animals, and shelter for water insects. It can provide nutrients for plants, as well as food for fish and alligators.

Ecosystem

TAKE A LOOK
4. Identify Use colored pencils to make circles on the picture.
Circle an individual in red.
Circle a population in blue.
Circle a community in brown.
Circle an ecosystem in green.

5. The **biosphere** is the part of Earth where life exists. The biosphere is the largest environmental level. It reaches from the bottom of the ocean and the Earth's crust to high in the sky. Scientists study the biosphere to learn how organisms interact with abiotic parts of the environment. These abiotic parts include Earth's atmosphere, water, soil, and rock.

Math Focus
5. Calculate From sea level, the biosphere goes up about 9 km and down about 19 km. What is the thickness of the biosphere in meters?

_____

Biosphere

Name ______________________ Class ______________ Date ______________

Section 1 Review

NSES LS 4a, 4b, 4c, 4d

SECTION VOCABULARY

abiotic describes the nonliving part of the environment, including water, rocks, light, and temperature

biosphere the part of Earth where life exists

biotic describes living factors in the environment

community all of the populations of species that live in the same habitat and interact with each other

ecology the study of the interactions of living organisms with one another and with their environment

ecosystem a community of organisms and their abiotic, or nonliving, environment

population a group of organisms of the same species that live in a specific geographical area

1. Compare What is the difference between a community and an ecosystem?

____

____

____

2. Organize Complete the chart below to describe the five levels of the environment, from smallest to largest.

Level	Description
	a single organism
Population	
	all of the populations of species that live in the same habitat and interact with one another
Ecosystem	
Biosphere	

3. Identify What two kinds of factors does an organism depend on for survival?

____

____

4. Infer Would all the birds in an area make up a population? Explain your answer.

____

____

____

CHAPTER 16 | Interactions of Living Things

SECTION 2 **Living Things Need Energy**

National Science Education Standards
LS 4a, 4b, 4c, 4d

BEFORE YOU READ

After you read this section, you should be able to answer these questions:

- How do producers, consumers, and decomposers get energy?
- What is a food web?

How Do Organisms Get Energy?

Eating gives organisms two things they cannot live without—energy and nutrients. Prairie dogs, for example, eat grasses and seeds to get their energy and nutrients. Like all organisms, prairie dogs need energy to live.

Organisms in any community can be separated into three groups based on how they get energy: producers, consumers, and decomposers.

PRODUCERS

Producers are organisms that use the energy from sunlight to make their own food. This process is called *photosynthesis*. Most producers are green plants, such as grasses on the prairie and trees in a forest. Some bacteria and algae also photosynthesize to make food. ☑

CONSUMERS

Consumers cannot make their own food. They need to eat other organisms to obtain energy and nutrients. Consumers can be put into four groups based on how they get energy: herbivores, carnivores, omnivores, and scavengers.

STUDY TIP

Circle Choose different colored pencils for producers, primary consumers, secondary consumers, and decomposers. As you read, circle these terms in the text with the colors you chose. Use the same colors to circle animals in any figures that are examples of each group.

READING CHECK

1. **Explain** Why is sunlight important to producers?

_____

_____

_____

Critical Thinking

2. Apply Concepts What types of consumers are the following organisms?

tigers _____

deer _____

humans _____

Herbivore   Carnivore

Omnivore   Scavenger

3. Define What is the role of decomposers in an ecosystem?

_____

_____

_____

_____

An **herbivore** is a consumer that eats only plants. Prairie dogs and bison are herbivores. A **carnivore** is a consumer that eats other animals. Eagles and cougars are carnivores. An **omnivore** is a consumer that eats both plants and animals. Bears and raccoons are omnivores.

A *scavenger* is a consumer that eats dead plants and animals. Turkey vultures are scavengers. They will eat animals and plants that have been dead for days. They will also eat what is left over after a carnivore has had a meal.

DECOMPOSERS

Decomposers recycle nature's resources. They get energy by breaking down dead organisms into simple materials. These materials, such as carbon dioxide and water, can then be used by other organisms. Many bacteria and fungi are decomposers.

What Is a Food Chain?

When an organism eats, it gets energy from its food. If that organism is then eaten, the energy stored in its body is passed to the organism eating it. A **food chain** is the path energy takes from one organism to another. Producers form the beginning of the food chain. Energy passes through the rest of the chain as one organism eats another.

SECTION 2 Living Things Need Energy *continued*

A Prairie Ecosystem Food Chain

TAKE A LOOK
4. Identify Label the food chain diagram with the following terms: energy, producer, primary consumer, secondary consumer, tertiary consumer, decomposer.

In a food chain:

- Producers are eaten by *primary consumers*.
- Primary consumers are eaten by *secondary consumers*.
- Secondary consumers are eaten by *tertiary consumers*.

In the food chain above, the grasses are the producers. The grasses are eaten by prairie dogs, which are the primary consumers. The prairie dogs are eaten by coyotes, which are the secondary consumers. When coyotes die, they are eaten by turkey vultures, which are the tertiary consumers. The tertiary consumer is usually the end of the food chain.

What Is a Food Web?

In most ecosystems, organisms eat more than one thing. Feeding relationships in an ecosystem are shown more completely by a food web. A **food web** is a system of many connected food chains in an ecosystem. Organisms in different food chains may feed upon one another. ☑

As in a food chain, in a food web, energy moves from one organism to the next in one direction. The energy in an organism that is eaten goes into the body of the organism that eats it.

✓ READING CHECK
5. Explain Why does a food web show feeding relationships better than a food chain?

_____

_____

_____

Critical Thinking

6. Predict What do you think would happen if all of the plants were taken out of this food web?

_____

_____

_____

_____

_____

Simple Food Web

MANY AT THE BASE

An organism uses much of the energy from its food for life processes such as growing or reproducing. When this organism is eaten, only a small amount of energy passes to the next consumer in the chain. Because of this, many more organisms have to be at the base, or bottom, of the food chain than at the top. For example, in a prairie community, there is more grass than prairie dogs. There are more prairie dogs than coyotes.

What Is an Energy Pyramid?

Energy is lost as it passes through a food chain. An **energy pyramid** is a diagram that shows this energy loss. Each level of the pyramid represents a link in the food chain. The bottom of the pyramid is larger than the top. There is less energy for use at the top of the pyramid than at the bottom. This is because most of the energy is used up at the lower levels. Only about 10% of the energy at each level of the energy pyramid passes on to the next level.

Math Focus

7. Calculate How much energy is lost at each level of the energy pyramid?

_____

_____

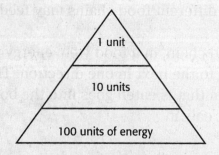

Energy Pyramid

8. Explain In which level of this energy pyramid do you think deer would belong? Explain your answer.

_____

_____

_____

EFFECT OF ONE SPECIES

A single species can change the flow of energy in an ecosystem. For example, gray wolves are at the top of their food chains. They eat a lot of different organisms but are usually not eaten by any other animal. By eating other organisms, wolves help control the size of those populations.

At one time, wolves were found across the United States. As settlers moved west, many wolves were killed. With few wolves left to feed on the primary consumers, such as elk, those populations began to grow. The elk ate all the grass, and there was none left for the smaller herbivores, such as hares. As these small herbivores died, there was less food for the secondary consumers. When wolves were removed from the food web, the whole ecosystem was affected. ☑

9. Summarize Why did a change in the wolf population affect the other organisms in the community?

_____

_____

_____

_____

_____

When wolves were removed from the ecosystem, other organisms were affected.

Name _____ Class _____ Date _____

Section 2 Review

NSES LS 4a, 4b, 4c, 4d

SECTION VOCABULARY

carnivore an organism that eats animals

energy pyramid a triangular diagram that shows an ecosystem's loss of energy, which results as energy passes through the ecosystem's food chain

food chain the pathway of energy transfer through various stages as a result of the feeding patterns of a series of organisms

food web a diagram that shows the feeding relationships between organisms in an ecosystem

herbivore an organism that eats only plants

omnivore an organism that eats both plants and animals

1. **Explain** Why are producers important in an ecosystem?

_____

_____

_____

2. **Connect** Make a food chain using the following organisms: mouse, snake, grass, hawk. Draw arrows showing how energy flows through the chain. Identify each organism as a producer, primary consumer, secondary consumer, or tertiary consumer.

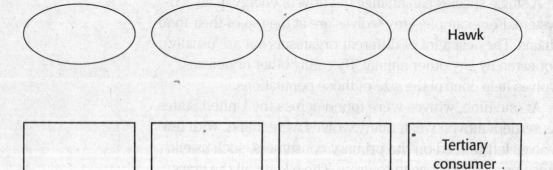

Hawk

Tertiary consumer

3. **Apply Concepts** Organisms can be part of more than one food chain. Make a food chain that includes one of the organisms above.

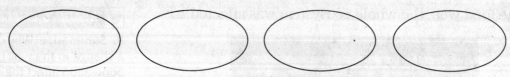

4. **Infer** Do you think you could find a food chain that had 10 organisms? Explain.

_____

_____

_____

CHAPTER 16 Interactions of Living Things

SECTION 3 Types of Interactions

National Science
Education Standards
LS 3a, 3c, 4b, 4d

BEFORE YOU READ

After you read this section, you should be able to answer these questions:

- What determines an area's carrying capacity?
- Why does competition occur?
- How do organisms avoid being eaten?
- What are three kinds of symbiotic relationships?

How Does the Environment Control Population Sizes?

Most living things have more offspring than will survive. A female frog, for example, may lay hundreds of eggs in a small pond. If all of the eggs became frogs, the pond would soon become very crowded. There would not be enough food for the frogs or other organisms in the pond. But in nature, this usually does not happen. The biotic and abiotic factors in the pond control the frog population so that it does not get too large.

Populations cannot grow without stopping because the environment has only a certain amount of food, water, space, and other resources. A resource that keeps a population from growing forever is called a *limiting factor*. Food is often a limiting factor in an ecosystem.

STUDY TIP

Make a List As you read this section, write down any questions you may have. Work with a partner to find the answers to your questions.

STANDARDS CHECK

LS 4d The number of organisms an ecosystem can support depends on the <u>resources</u> available and abiotic factors, such as quantity of light and water, range of temperatures, and soil composition. Given adequate biotic and abiotic resources and no disease or predators, populations (including humans) increase at rapid rates. Lack of resources and other factors, such as predation and climate, limit the growth of populations in specific niches in the ecosystem.

Word Help: resource
anything that can be used to take care of a need

1. Define What is a limiting factor?

_____

_____

_____

All plants need sunlight. In this forest, sunlight may be a limiting factor. Not all plants can get the same amount of light.

What Is Carrying Capacity?

The largest number of organisms that can live in an environment is called the **carrying capacity**. When a population grows beyond the carrying capacity, limiting factors will cause some individuals to leave the area or to die. As individuals die or leave, the population decreases.

The carrying capacity of an area can change if the amount of the limiting factor changes. For example, the carrying capacity of an area will be higher in seasons when more food is available. ☑

READING CHECK

2. Explain Why can the carrying capacity of an area change?

_____

_____

_____

How Do Organisms Interact in an Ecosystem?

Populations are made of individuals of the same species. Communities are made of different populations that interact. There are four main ways that individuals and populations affect one another in an ecosystem: in competition, as predator and prey, through symbiosis, and coevolution. ☑

READING CHECK

3. List What are four ways that organisms in an ecosystem interact?

_____

_____

_____

Why Do Organisms Compete?

Competition happens when more than one individual or population tries to use the same resource. There may not be enough resources, such as food, water, shelter, or sunlight, for all the organisms in an environment. When one individual or population uses a resource, there is less for others to use.

Competition can happen between organisms in the same population. For example, in Yellowstone National Park, elk compete with one another for the same plants. In the winter, when there are not many plants, competition is much higher. Some elk will die because there is not enough food. In spring, when many plants grow, there is more food for the elk, and competition is lower.

Competition can also happen between populations. In a forest, different types of trees compete to grow in the same area. All of the plant populations must compete for the same resources: sunlight, space, water, and nutrients.

How Do Predators and Prey Interact?

Another way organisms interact is when one organism eats another to get energy. The organism that is eaten is called the **prey**. The organism that eats the prey is called the **predator**. When a bird eats a worm, for example, the bird is the predator, and the worm is the prey.

PREDATORS

Predators have traits or skills that help them catch and kill their prey. Different types of predators have different skills and traits. For example, a cheetah uses its speed to catch prey. On the other hand, tigers have colors that let them blend with the environment so that prey cannot see them easily. ☑

Critical Thinking

4. Predict In a prairie ecosystem, which two of the following organisms most likely compete for the same food source: elk, coyotes, prairie dogs, vultures?

_____

_____

✓ READING CHECK

5. Identify What are two traits different predators may have to help them catch prey?

_____

_____

_____

SECTION 3 Types of Interactions *continued*

PREY

Prey generally have some way to protect themselves from being eaten. Different types of organisms protect themselves in different ways:

1. Run Away When a rabbit is in danger, it runs.

2. Travel in Groups Some animals, such as musk oxen, travel in herds, or groups. Many fishes, such as anchovies, travel in schools. All the animals in these groups can help one another by watching for predators.

When musk oxen sense danger, they move close together to protect their young.

3. Show Warning Colors Some organisms have bright colors that act as a warning. The colors warn predators that the prey might be poisonous. A brightly colored fire salamander, for example, sprays a poison that burns.

Critical Thinking

6. Infer Why do you think it would be difficult for predators to attack animals in a herd?

_____

_____

_____

_____

_____

SECTION 3 Types of Interactions *continued*

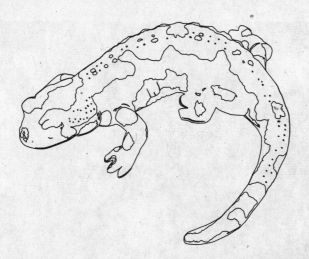

TAKE A LOOK
7. Color A fire salamander has a black body with bright orange or yellow spots. Use colored pencils to give this salamander its warning colors.

4. Use Camouflage Some organisms can hide from predators by blending in with the background. This is called *camouflage*. A rabbit's natural colors, for example, may help it blend in with dead leaves or shrubs so that it cannot be seen. Some animals may look like twigs, stone, or bark.

What Is Symbiosis?

Some species have very close interactions with other species. A close association between two or more species is called **symbiosis**. Each individual in a symbiotic relationship may be helped, hurt, or not affected by another individual. Often, one species lives on or in another species. Most symbiotic relationships can be divided into three types: mutualism, commensalism, and parasitism. ☑

READING CHECK
8. List List the three types of symbiotic relationships.

_____

_____

_____

MUTUALISM

When both individuals in a symbiotic relationship are helped, it is called **mutualism**. You can see mutualism in the relationship between a bee and a flower.

Organism hurt?	Organism helped?	Example
No one	both organisms	A bee transfers pollen for a flower; a flower provides nectar to a bee.

SECTION 3 Types of Interactions *continued*

In a mutualistic relationship, both species benefit.

COMMENSALISM

When one individual in a symbiotic relationship is helped but the other is not affected, this is called **commensalism**.

Critical Thinking

9. Compare How does mutualism differ from commensalism?

_____
_____
_____
_____
_____
_____

Organism hurt?	Organism helped?	Example
No one	one of the organisms	A fish called a remora attaches to a shark and eats the shark's leftovers.

The remoras get a free meal, but the shark is not harmed.

PARASITISM

A symbiotic relationship in which one individual is hurt and the other is helped is called **parasitism**. The organism that is helped is called the parasite. The organism that is hurt is called the *host*. ☑

READING CHECK

10. Define In parasitism, is the host helped or hurt?

_____

Organism hurt?	Organism helped?	Example
Host	parasite	A flea is a parasite on a dog.

SECTION 3 Types of Interactions *continued*

This tomato hornworm is being parasitized by young wasps. Their cocoons are on the caterpillar's back.

What Is Coevolution?

Relationships between organisms change over time. Interactions can even be one reason that organisms change. When a long-term change happens in two species because of their close interactions, the change is called **coevolution**.

One example of coevolution can be seen in some flowers and the organisms that pollinate them. A *pollinator* is an organism, such as a bird, insect, or bat, that carries pollen from one flower to another. Flowers need to attract pollinators to help them reproduce. Different flowers have evolved different ways to attract pollinators. Some use colors or odors. Others use nectar as a food reward for the pollinator.

Some plants can use a variety of pollinators. Others have coevolved with certain pollinators. For example, the bat in the picture below has a long sticky tongue. It uses its tongue to get nectar from deep inside the flower. Only an organism with a way to reach the nectar could be a pollinator for this flower.

TAKE A LOOK
11. Infer How do you think the caterpillar helps the wasps?

_____

_____

_____

Say It

Investigate With a partner, look up the meaning of the suffix *co-*. Discuss how the meaning of this suffix can help you remember what *coevolution* means. Think of some other words that have *co-*.

Section 3 Review

NSES LS 3a, 3c, 4b, 4d

SECTION VOCABULARY

carrying capacity the largest population that an environment can support at any given time	**parasitism** a relationship between two species in which one species, the parasite, benefits from the other species, the host, which is harmed
coevolution the evolution of two species that is due to mutual influence, often in a way that makes the relationship more beneficial to both species	**predator** an organism that kills and eats all or part of another organism
commensalism a relationship between two organisms in which one organism benefits and the other is unaffected	**prey** an organism that is killed and eaten by another organism
mutualism a relationship between two species in which both species benefit	**symbiosis** a relationship in which two different organisms live in close association with each other

1. Identify What are two resources for which organisms are likely to compete?

_____

2. Explain What happens to a population when it grows larger than its carrying capacity?

_____

_____

3. Infer Do you think the carrying capacity is the same for all species in an ecosystem? Explain your answer.

_____

_____

4. Summarize Complete the chart below to describe the different kinds of symbiotic relationships.

Example organisms	Type of symbiosis	Organism(s) helped	Organism(s) hurt
Flea and dog			host (dog)
Bee and flower	mutualism		
Remora and shark			none

5. Apply Concepts The flowers of many plants provide a food reward, such as nectar, to pollinators. Some plants, however, attract pollinators but provide no reward. What type of symbiosis best describes this relationship? Explain your answer.

_____

_____

CHAPTER 17 Cycles in Nature

SECTION 1 **The Cycles of Matter**

BEFORE YOU READ

After you read this section, you should be able to answer these questions:

• Why does matter need to be recycled?

• How are water, carbon, and nitrogen recycled?

Why Is Matter Recycled on Earth?

The matter in your body has been on Earth since the planet formed billions of years ago. Matter on Earth is limited, so it must be used over and over again. Each kind of matter has its own cycles. In these cycles, matter moves between the environment and living things.

What Is the Water Cycle?

Without water there would be no life on Earth. All living things are made mostly of water. Water carries other nutrients to cells and carries wastes away from them. It also helps living things regulate their temperatures. Like all matter, water is limited on Earth. The water cycle lets living things use water over and over.

In the environment, water moves between the oceans, atmosphere, land, and living things. Eventually, all the water taken in by organisms returns to the environment. The movement of water is known as the *water cycle*. The parts of the water cycle are explained in the figure below.

STUDY TIP

Mnemonic As you read, create a mnemonic device, or memory trick, to help you remember the parts of the water cycle.

Say It

Identify Describe to the class all the things you and your family do in a day that use water. Can you think of any ways you might be able to use less water?

Precipitation is rain, snow, sleet, or hail that falls from clouds to Earth's surface. Most precipitation falls into the ocean. It never touches the land.

Condensation happens when water vapor cools and changes into drops of liquid water. The water drops form clouds in the atmosphere.

Evaporation happens when liquid water on Earth's surface changes into water vapor. Energy from the sun makes water evaporate.

Groundwater is water that flows under the ground. Gravity can make water that falls on the land move into rocks underground.

Runoff is water that flows over the land into streams and rivers. Most of the water ends up in the oceans.

Transpiration happens when plants give off water vapor from tiny holes in their leaves.

TAKE A LOOK

1. Describe How do clouds form?

_____

_____

_____

_____

_____

2. Analyze Explain the role of photosynthesis in the carbon cycle.

_____

_____

_____

_____

_____

_____

_____

_____

What Is the Carbon Cycle?

Besides water, the most common molecules in living things are *organic molecules.* These are molecules that contain carbon, such as sugar. Carbon moves between the environment and living things in the *carbon cycle*.

PHOTOSYNTHESIS AND RESPIRATION

Plants are producers. This means they make their own food. They use water, carbon dioxide, and sunlight to make sugar. This process is called *photosynthesis*. Photosynthesis is the basis of the carbon cycle.

Animals are consumers. This means they have to consume other organisms to get energy. Most animals get the carbon and energy they need by eating plants. How does this carbon return to the environment? It returns when cells break down sugar molecules to release energy. This process is called *respiration*.

DECOMPOSITION AND COMBUSTION

Fungi and some bacteria get their energy by breaking down wastes and dead organisms. This process is called **decomposition**. When organisms decompose organic matter, they return carbon dioxide and water to the environment.

When organic molecules, such as those in wood or fossil fuels, are burned, it is called **combustion**. Combustion releases the carbon stored in these organic molecules back into the atmosphere.

TAKE A LOOK

3. Complete Carbon dioxide in the air is used for _____

_____.

4. List What three processes release carbon dioxide into the environment?

_____

_____

The Carbon Cycle

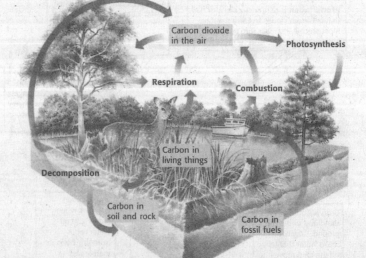

Carbon dioxide in the air

Photosynthesis

Respiration

Combustion

Carbon in living things

Decomposition

Carbon in soil and rock

Carbon in fossil fuels

What Is the Nitrogen Cycle?

Nitrogen is also important to living things. Organisms need nitrogen to build proteins and DNA for new cells. Like water and carbon, nitrogen cycles through living things and the environment. This is called the *nitrogen cycle*.

NITROGEN FIXATION

About 78% of Earth's atmosphere is nitrogen gas. Most organisms cannot use nitrogen gas directly. Bacteria in soil can change nitrogen gas into forms that plants can use. This is called *nitrogen fixation*. Other organisms can get the nitrogen they need by eating plants or by eating organisms that eat plants.

The Nitrogen Cycle

- Nitrogen in the air
- Lightning causes some nitrogen fixation.
- Animals get nitrogen from plants.
- Plant roots take up nitrogen from soil.
- Dead animals and plants
- Bacteria in soil convert nitrogen back to gas.
- Decomposition releases nitrogen into soil.
- Bacteria in soil and plant roots perform most nitrogen fixation.

How Are the Cycles of Matter Connected?

Other forms of matter on Earth also cycle through the environment. These include many minerals that living cells need, such as calcium and phosphorus. When an organism dies, every substance in its body will be recycled in the environment or reused by other organisms. ☑

All of the cycles of matter are connected. For example, water carries some forms of carbon and nitrogen through the environment. Many nutrients pass from soil to plants to animals and back. Living things play a part in each of the cycles.

Critical Thinking
5. Apply Concepts How is nitrogen fixation important to animals?

TAKE A LOOK
6. Identify What process releases nitrogen into the soil?

READING CHECK
7. Explain What happens to the substances in an organism's body when the organism dies?

Name _____ Class _____ Date _____

Section 1 Review

NSES LS 1c, 4b, 4c, 5a

SECTION VOCABULARY

combustion the burning of a substance	**evaporation** the change of state from a liquid to a gas
condensation the change of state from a gas to a liquid	**precipitation** any form of water that falls to Earth's surface from the clouds
decomposition the breakdown of substances into simpler molecular substances	

1. Identify In the water cycle, what makes water evaporate?

_____

2. Summarize Draw arrows to show how carbon cycles through the environment and living things.

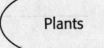

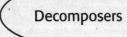

Decomposers

3. Explain Why does matter need to be recycled?

_____

4. Explain Why is water so important to life on Earth?

_____
_____
_____

5. Define What is nitrogen fixation?

_____
_____

6. Define What are organic molecules?

_____

Cycles in Nature

CHAPTER 17 Cycles in Nature

SECTION 2 # Ecological Succession

> **BEFORE YOU READ**
>
> After you read this section, you should be able to answer these questions:
>
> • How do communities of living things form?
>
> • Why do the type of organisms in a community change over time?

National Science Education Standards
LS 1a, 4d

What Is Succession?

In the spring of 1988, much of Yellowstone National Park was a forest. The trees grew close together. Large areas were in shade, and few plants grew under the trees.

That summer, fires burned much of the forest and left a blanket of gray ash on the forest floor. Most of the trees were dead, though some of them were still standing.

The following spring, the forest floor was green. Some of the dead trees had fallen over, and many small, green plants, such as grasses, were growing.

STUDY TIP
Organize As you read, make a table comparing primary succession and secondary succession.

Math Focus

1. Calculate Percentages The fires in Yellowstone National Park in 1988 burned 739,000 acres. The park has 2.2 million acres total. What percentage of the park burned?

_____

Why were grasses the first things to grow? After the fire, the forest floor was sunny and empty. Nonliving parts of ecosystems, such as water, light, and space, are called *abiotic factors*. When the trees were dead, grasses had the abiotic factors they needed, and their populations grew quickly.

In a few years, larger plants began growing in some areas, and the grasses could not grow without sunlight. Within 10 years, the trees were starting to grow back. The trees began to shade out those plants.

When one type of community replaces another type of community, this is called **succession**. The grasses and other species that are the first to live or grow in an area are called **pioneer species**. ☑

READING CHECK

2. Define What is a pioneer species?

_____

_____

_____

SECTION 2 Ecological Succession *continued*

PRIMARY SUCCESSION

Sometimes, a small community starts to grow in an area where living things have never grown before. The area is only bare rock and there is no soil. Over a very long time, a community can develop. The change from bare rock to a community of organisms is called *primary succession*.

Lichens are pioneer species on bare rock. A lichen's structure allows it to function on bare rock. Lichens don't have roots, and they get their water from the air. This means they do not need soil. Most other organisms, however, cannot move into the area without soil.

Lichens produce acid that breaks down the rock they are living on. The rock particles, mixed with the remains of dead lichens, become the first soil.

After many years, there is enough soil for mosses to grow. The mosses eventually replace the lichens. Tiny organisms and insects begin to live there. When they die, their remains add to the soil.

Over time, the soil gets deeper, and ferns replace mosses. The ferns may be replaced later by grasses and wildflowers. If there is enough soil, shrubs and small trees may grow. After hundreds of years, the soil may be deep enough and rich enough to support a forest community.

Critical Thinking

3. Analyze What makes lichens good pioneer species?

_____

_____

_____

_____

_____

TAKE A LOOK

4. Identify Which kind of plants are generally the last to appear in an area going through primary succession?

_____

Succession of Lichen and Plant Species in a Forest

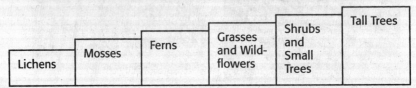

Remember that a community is made up of all the living things in an area. It includes the plants that can live with the abiotic factors there at the time. It also includes the animals that can use the resources there at the time.

When the abiotic factors and resources change, so does the community. For example, a population of cottontail rabbits will get bigger as more small plants grow in the soil over the rock. Later, there will be fewer small plants, when more trees grow and block the sun. Then, there will be fewer rabbits. However, the populations of animals that need trees, such as squirrels, will increase.

SECTION 2 Ecological Succession *continued*

SECONDARY SUCCESSION

Sometimes, a community is destroyed by a natural disaster, such as a flood or fire. Sometimes, humans or animals alter an environment. For example, a farmer may stop growing crops in a field. In either case, if there is soil and the area is left alone, the natural community can grow back. The plant species change in a series of stages called *secondary succession*. Secondary succession happens in areas where living things already exist. ☑

The figure below shows secondary succession in a farm field that used to be a forest.

First Year Weeds start to grow.

Second Year New weeds appear. Their seeds may have been blown to the field by the wind, or insects may have carried them.

In 5 to 15 Years Small conifer trees, such as pines and firs, grow among the weeds. After about 100 years, the weeds are gone and a forest has formed.

After 100 Years or More As older conifer trees die, they may be replaced by hardwood trees. Oak and maple will grow if the temperature and precipitation are right.

READING CHECK

5. Describe Where does secondary succession happen?

_____

_____

_____

TAKE A LOOK

6. Identify In this example, what are the first kind of plants to grow in secondary succession?

_____

7. Identify What are the first kind of trees that may grow in an area?

_____

MATURE COMMUNITIES AND BIODIVERSITY

As succession goes on, a community can end up having one well-adapted plant species. This is called a *climax species*. However, in many places, a community is more likely to include many species. The variety of species that live in an area is called its *biodiversity*.

Section 2 Review

NSES LS 1a, 4d

SECTION VOCABULARY

1. Define What are abiotic factors? Give three examples.

_____

_____

2. Compare What is the difference between primary and secondary succession?

_____

_____

_____

3. Apply Concepts Secondary succession generally happens faster than primary succession. Why do you think this happens?

_____

_____

_____

4. Apply Ideas Consider a species of animal that eats grass and a species of animal that eats nuts. Which species do you think would have a larger population in a mature forest? Explain your answer.

_____

_____

_____

5. Analyze Why, in general, can't tall trees be pioneer species?

_____

_____

_____

6. Define What is biodiversity?

_____

7. Describe When soil first forms over bare rock, what is it made of?

_____

CHAPTER 18 | Properties and States of Matter)

SECTION
1 | **Properties of Matter**

National Science
Education Standards
PS 1a, 1b, 1c

BEFORE YOU READ

After you read this section, you should be able to answer these questions:

- What is matter?
- How do scientists classify matter?
- How do scientists describe matter?

What Is Matter?

Look around you. Almost everything you see is matter. A chair, a desk, a book, and a pencil are all examples of matter. **Matter** is anything that has mass and volume. *Mass* consists of all the particles that something has. *Volume* is the amount of space that something takes up.

It can be confusing to tell the difference between matter and non-matter. For example, you can see light. However, light is not matter because it does not have mass or volume. On the other hand, air is invisible, but air is matter because it has mass and volume. ☑

To understand that air has mass and volume, think about blowing up a balloon. Imagine that you use a balance to measure the mass of the balloon both before and after you blow it up. You would find that the inflated balloon has more mass than the empty balloon. Therefore, air must have mass. The inflated balloon also takes up more space than the empty balloon did. Therefore, air must have volume. Because it has both mass and volume, air is matter.

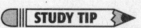

STUDY TIP

Organize Information Make a list of 10 different objects or substances. As you read, describe each in terms of the type of matter it is, and the physical and chemical properties it has.

READING CHECK

1. Explain Why is air considered to be matter even though it is invisible?

_____

_____

_____

Air has mass and volume as you can see by comparing the masses and volumes of a balloon before and after it is inflated with air.

TAKE A LOOK
2. Identify List three examples of matter shown in the photograph.

_____

_____

ATOMS

All types of matter are made up of tiny particles called atoms. **Atoms** are the building blocks of matter. Matter can consist of just one type of atom. Matter can also consist of different types of atoms that combine or mix with one another. ☑

How Do Scientists Classify Matter?

To *classify* something means to place it into a specific group. All members in the group share certain features. For example, the CDs in a music store are classified into various groups such as rock, classical, and jazz. Because the CDs are classified into different groups, customers can easily find what they want. Like CDs, matter is classified into different groups.

In most cases, there is more than one way to classify things. For example, the same CD can be classified in both the rock and best-seller sections. Similarly, matter can be classified in more than one way. One useful way is based on the way the atoms in a sample of matter are arranged.

ELEMENTS

Have you ever seen someone inflate a balloon with helium gas from a tank? Helium is an example of an element. An **element** is a substance that is composed of a single kind of atom. Helium gas is composed only of individual helium atoms. Like all atoms, a helium atom cannot be broken down into a simpler substance.

There are more than 100 elements. Elements can be further classified into three major categories. The table below describes these categories.

Category	Examples	Some characteristics
Metals	lead, copper, tin	conduct electricity and heat
Nonmetals	sulfur, neon, iodine	brittle; crumble when struck
Metalloids	boron, antimony, silicon	have a mixture of properties of metals and nonmetals

Some elements are made of molecules of a single type of atom. *Molecules* are made of two or more atoms that are combined in a definite ratio. For example, oxygen gas exists as molecules of two oxygen atoms that are joined to each other.

READING CHECK

3. Identify What is all matter made of?

_____

Say It

Brainstorm In groups of three or four, choose a group of objects (for example, plants) and discuss different ways that they can be classified.

TAKE A LOOK
4. Compare Name two things that all of these substances have in common?

_____
_____
_____
_____

COMPOUNDS

You could also inflate a balloon with air from your lungs. Air contains many different elements that are in the form of gases, including oxygen and nitrogen. Air is mostly nitrogen gas. A molecule of nitrogen consists of two nitrogen atoms joined together.

Air also contains other gases, such as carbon dioxide and water vapor. Both carbon dioxide and water are examples of compounds. A **compound** is a substance that forms when atoms of two or more different elements are joined, or *bonded*, to each other.

Table salt, also known as sodium chloride, is another example of a compound. Salt forms when sodium atoms bond with chlorine atoms. As you can see in the figure below, the properties of the compound sodium chloride differ from the properties of the elements sodium and chlorine. In fact, the properties of every compound are different from the properties of the elements that form it.

Critical Thinking
5. Compare How are compounds different from elements?
_____
_____
_____

TAKE A LOOK
6. Compare Describe two ways that sodium chloride is different from sodium or chlorine.
_____
_____
_____
_____

Sodium is a soft, silvery white metal that reacts violently with water.

Chlorine is a poisonous greenish-yellow gas.

Sodium chloride, or table salt, is a white solid. It dissolves easily in water and is safe to eat.

Both elements and compounds are pure substances. A **pure substance** consists of only one type of atom or one type of molecule. Oxygen gas, water vapor, and table salt are all examples of pure substances. Air, however, is not a pure substance.

MIXTURES

Air is an example of a mixture. A **mixture** is a combination of two or more pure substances that are not bonded with one another. Air is a mixture of several gases, shown in the table below. Notice that the gases in air consist of both elements and compounds.

Gas	Percent of air
Nitrogen	78%
Oxygen	21%
Carbon dioxide	0.03%
Other gases	less than 1%

TYPES OF MIXTURES

There are two types of mixtures: homogeneous mixtures and heterogeneous mixtures. A *homogeneous mixture* is one in which the substances that make up the mixture are spread evenly throughout the mixture. Apple juice is an example of a homogeneous mixture. The water, sugars, and other dissolved substances in apple juice are mixed together uniformly. You do not have to shake a container of apple juice before you drink it. ☑

Orange juice, especially one that contains pulp, is an example of a heterogeneous mixture. A *heterogeneous mixture* is one in which the substances are not mixed uniformly. For example, the pulp in orange juice slowly settles to the bottom of the container.

You may have to shake a container of orange juice before you drink it. Shaking the container mixes up the orange juice so that it temporarily becomes a homogenous mixture. As a result, when you pour a glass, you will get some of the pulp evenly mixed with the other ingredients in the orange juice.

SOLUTIONS

If you add a little bit of table salt to water and stir, the sodium chloride will dissolve. The molecules of salt actually break apart, and particles of sodium and chlorine move around separately in the water. These particles will remain distributed evenly throughout the water. What you have made is a type of mixture called a solution. A *solution* is made by dissolving one or more substances in another substance.

All solutions are homogeneous mixtures because the dissolved substances are distributed evenly throughout the solution. Many solutions, such as saltwater, are made by dissolving a solid in a liquid. However, not all solutions consist of solids dissolved in liquids. For example, some solutions consist of one solid dissolved in another solid. For example, brass is a solution of zinc in copper.

✓ **READING CHECK**

7. Explain Why is it unnecessary to shake apple juice before you drink it?

_____

_____

_____

_____

TAKE A LOOK

8. Compare How would the atoms of zinc and copper be distributed differently if the mixture was heterogeneous?

_____

_____

Copper atom

Zinc atom

Brass

SUSPENSIONS

If you add sand to water, the particles will slowly settle to the bottom. You can stir or shake the mixture of sand and water, but the particles will eventually separate and settle to the bottom of the container. What you have made is a type of mixture called a suspension. A *suspension* is a mixture in which the substances eventually separate from each other.

All suspensions are heterogeneous mixtures. Some suspensions, such as sand and water, are made by mixing a solid in a liquid. There are other kinds of suspensions as well. For example, you may have seen light strike dust particles suspended in air. The solid dust particles and gaseous air make up a suspension.

COLLOIDS

If you drink a glass of milk, you are drinking a type of mixture known as a colloid. A *colloid* is a mixture that has particles that are smaller than those in suspensions but are larger than those in solutions.

Milk appears to be a solution. However, if you looked at a drop of milk under a microscope, you would see small clumps of fats and proteins. These clumps in milk do not settle to the bottom like the particles in a suspension. These clumps also do not dissolve like the particles in a solution. Instead, the clumps remain evenly distributed throughout the liquid. Milk is a colloid that is a homogeneous mixture. ☑

Some colloids are heterogeneous mixtures. For example, clouds consist of tiny water droplets or ice crystals that are suspended in the air. Certain areas of a cloud can contain more water droplets or ice crystals than other areas. Clouds are colloids that are heterogeneous mixtures.

Clouds are colloids that form when water droplets or ice crystals become suspended in the air.

Critical Thinking

9. Analyze Do you think ocean water is a homogeneous mixture or a heterogeneous mixture? Explain.

_____
_____
_____
_____
_____
_____
_____
_____

✓ **READING CHECK**

10. Compare How does a colloid differ from a suspension?

_____
_____

How Do Scientists Describe Matter?

How would you describe baking soda? You might say that it is a white powder. How would you describe white vinegar? You might say that it is a clear liquid. How would you describe what happens when baking soda and vinegar are mixed, as shown in the figure below?

Gas bubbles form when baking soda is mixed with vinegar.

PHYSICAL PROPERTIES

"White powder" and "clear liquid" are descriptions of physical properties. A **physical property** is a characteristic that can be observed or measured without changing the identity of the substance. The baking soda and vinegar are still baking soda and vinegar after you have observed them and described their physical properties. ☑

Physical properties include color, texture, shape, and odor. The table below describes some other physical properties.

Physical property	Description	Example
Mass	the amount of matter in an object	A baseball has a mass of 145 g.
Weight	a measure of the force of gravity on an object	On Earth, a baseball weighs 1.5 N. On the moon, it would weigh 0.25 N.
Volume	the amount of space matter takes up	The volume of a baseball is 199 cm³.
Density	the mass of an object per unit volume (mass ÷ volume)	A baseball has a density of 0.729 g/cm³.
Electrical conductivity	the ability of a material to allow an electrical current to flow through it	Copper is used for wiring because it has a high conductivity.
Boiling point	the temperature at which a substance changes from a liquid into a gas	The boiling point of water is 100°C.

TAKE A LOOK

12. Explain Why do you think a baseball weighs less on the moon than on Earth, even if it has the same mass?

_____

_____

_____

_____

SECTION 1 Properties of Matter *continued*

CHEMICAL PROPERTIES

What happens when baking soda and vinegar are mixed depends on their chemical properties. A **chemical property** is a property that is based on how a substance reacts with other substances.

For example, a chemical property of baking soda and vinegar is their *reactivity*. Baking soda and vinegar react to produce gas bubbles. This reaction produces new substances. In other words, the identities of both the baking soda and vinegar have been changed. In fact, it is not possible to observe the chemical properties of a substance without destroying or changing it. ☑

Another example of a chemical property is the flammability of wood. Wood burns by reacting with oxygen. As it reacts with oxygen, new substances such as ash and smoke form. These new substances have different properties from the wood and oxygen.

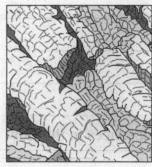

The ash that forms as wood burns has new properties. For example, ash is powdery and nonflammable while wood is hard and flammable.

13. Identify Name one chemical property of vinegar.

_____

_____

PROPERTIES OF PURE SUBSTANCES AND MIXTURES

All samples of a pure substance have exactly the same physical and chemical properties. For example, a pinch of salt has the same taste as a teaspoon of salt.

In contrast, the physical and chemical properties of a mixture can vary. Think about a mixture of popcorn and melted butter. The flavor of each mouthful of popcorn depends on how much butter and salt are on the kernels.

STANDARDS CHECK

PS 1a A substance has characteristic properties, such as density, a boiling point, and solubility, all of which are independent of the amount of the sample. A mixture of substances often can be separated into the original substances using one or more of the characteristic properties.

14. List Name three physical properties and one chemical property of popcorn.

_____

_____

_____

_____

Section 1 Review

NSES PS 1a, 1b, 1c

SECTION VOCABULARY

atom the smallest unit of an element that maintains the properties of that element	**matter** anything that has mass and takes up space
chemical property a property of matter that describes a substance's ability to participate in chemical reactions	**mixture** a combination of two or more substances that are not chemically combined
compound a substance made up of atoms of two or more different elements joined by chemical bonds	**physical property** a characteristic of a substance that does not involve a chemical change, such as density, color, or hardness
element a substance that cannot be separated or broken down into simpler substances by chemical means	**pure substance** a sample of matter, either a single element or a single compound, that has definite chemical and physical properties

1. Identify List two types of pure substances and give one example of each.

_____

_____

2. Describe What are two differences between compounds and mixtures?

_____

_____

_____

_____

3. Complete Fill in the blank spaces in the flow chart below.

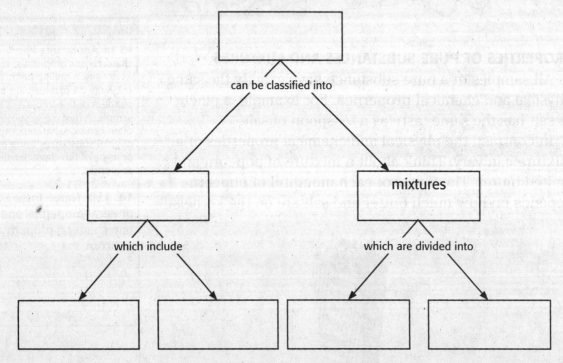

SECTION 2 | Physical and Chemical Changes

National Science
Education Standards
PS 1b, 3a, 3e

BEFORE YOU READ

After you read this section, you should be able to answer these questions:

- What happens to mass when matter changes?
- What is a physical change?
- What is a chemical change?

What Is the Law of Conservation of Mass?

Have you ever toasted marshmallows around a campfire? As you might guess, the matter that makes up a marshmallow changes as it is toasted by the heat from the fire. As it is heated, the marshmallow darkens and gets bigger. As the wood burns, fire and smoke rise into the air. Once all the wood has burned, only ashes remain in the fire pit.

It may seem like the marshmallow gains mass as it gets bigger. It might seem like the wood loses mass as it burns because the ashes take up much less space than the wood. However, mass is never created or destroyed when matter changes. This is known as the **law of conservation of mass**. ☑

The marshmallow might have changed in size, but it still has the same mass. As the wood reacts with oxygen and burns, it turns into different substances. These new substances have the same mass as the wood and oxygen. Matter can change form, be rearranged, and can move from one place to another. However, the total amount of matter will always stay the same.

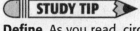

STUDY TIP

Define As you read, circle the words that you don't understand. Look them up or discuss them with a partner. Write down the meaning of each word in the margin. Then reread the paragraph to make sure you understand it completely.

READING CHECK

1. Explain What happens to the total amount of matter when matter changes?

_____

_____

TAKE A LOOK

2. Infer What do you think happens to the matter in the marshmallow after someone eats it?

_____

_____

_____

What Is a Physical Change?

Water on Earth is changing continuously. The water that falls from the sky as large droplets was once much smaller droplets in a cloud. Before that, the water existed as water vapor – invisible gas in the sky. That water vapor came from liquid water that evaporated from the oceans or on land. Some of the liquid water formed when frozen water in snow and ice melted. No matter where the water is or what form it is in, it is still water.

These changes that water goes through as it changes between solid, liquid, and gas are examples of physical changes. A **physical change** is any change in which one or more physical properties of a substance change, but the substance's identity does not change. During a physical change, the size, shape, temperature, or density of a material might change. ☑

PHYSICAL CHANGES IN NATURE AND INDUSTRY

Physical changes are common in nature. Examples include water freezing, rocks breaking apart, and sticks breaking in two.

Physical changes are also common in industry. People make physical changes to substances in order to transform them into products that are more useful. For example, wood is changed into lumber to make furniture, musical instruments, and toothpicks. The wood has changed form, from being part of a tree in a forest to becoming part of a consumer product in a home. However, the chemical identity of the wood has not changed.

The flowchart below shows how sugar cane is turned into raw table sugar through physical changes.

READING CHECK

3. Explain What happens to the identity of a substance when it goes through a physical change?

_____

_____

_____

TAKE A LOOK
4. Identify List three physical changes that happen to sugar before it gets to the grocery store.

_____

_____

_____

1. Sugar cane is grown in large fields. The stem of the plant contains sugar dissolved in liquid.

2. Sugar cane is cut, washed, shredded, and squeezed to press out the sugary juice.

3. Water evaporates from the sugary juice, leaving raw sugar crystals that we can use to sweeten our food.

What Is a Chemical Change?

Wood looks very different after it has burned. Burning wood changes some of its physical properties, such as shape and size. However, burning wood also changes some of its chemical properties. For example, ashes do not burn as easily as wood.

Burning wood is an example of a chemical change. A **chemical change** involves the formation of new substances. The identity and properties of the new substances differ from the identity and properties of the original substance. ☑

There are certain clues that indicate a chemical change has happened. The following events are evidence that a chemical change may have taken place.

- production of gas bubbles
- change in color or odor
- release of light, heat, or sound

Unlike a physical change, a chemical change involves breaking bonds between atoms. The atoms are then rearranged, and new bonds are formed. As a result, new substances are produced. This rearrangement of atoms is also called a *chemical reaction*. The figure below shows hydrogen reacting with oxygen to form water.

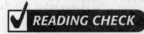

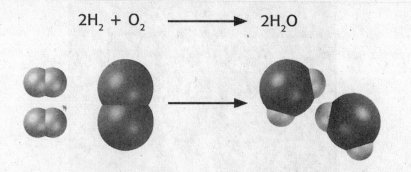

$$2H_2 + O_2 \longrightarrow 2H_2O$$

Before the reaction, each hydrogen atom is bonded to one other hydrogen atom. Each oxygen atom is bonded to one other oxygen atom. During the reaction, bonds are broken and reformed. After the reaction, each oxygen atom is bonded to two hydrogen atoms.

Chemical changes, or reactions, are constantly taking place naturally, both around and inside you. For example, chemical changes occur when fruit ripens, when tree leaves change color in the fall, and when you digest food.

READING CHECK

5. Explain What happens during a chemical change?

_____

_____

TAKE A LOOK

6. Explain Give two reasons why the production of water from hydrogen and oxygen is a chemical change rather than a physical change.

_____

_____

_____

_____

CHEMICAL CHANGES IN NATURE

One of the most important chemical changes happens when green plants make food. This process is called *photosynthesis*. Plants use energy from sunlight to carry out this chemical change. Photosynthesis is a chemical reaction that changes carbon dioxide (from the air), and water (from the air or ground), into oxygen and sugars. The chemical equation for photosynthesis is shown below.

carbon dioxide + water + energy = oxygen + sugars

In animals, chemical changes happen during eating and digestion. Digestion frees nutrients from foods. Chemical changes also happen when organisms use oxygen to obtain energy from foods. This process is known as *respiration*. The chemical equation for respiration is shown below.

sugars + oxygen = carbon dioxide + water + energy

Chemical changes also occur in living things as they grow, age, die, and decay. The spoonbill shown below has distinctive pink feathers. The pigments that color the spoonbill come from the bodies of the shrimp that it eats. During digestion, chemical changes release the pigments, giving the spoonbill its pink color.

STANDARDS CHECK

PS 1b Substances react chemically in characteristic ways with other substances to form new substances (compounds) with different characteristic properties. In <u>chemical reactions</u>, the total mass is conserved. Substances often are placed in categories or groups if they react in similar ways; metals is an example of such a group.

Word Help: <u>chemical</u>
of or having to do with the properties or actions of substances

Word Help: <u>reaction</u>
a response or change

7. Infer If the total mass of oxygen and sugars produced during photosynthesis is 100 g, what is the total mass of carbon dioxide and water that is used during photosynthesis?

_____

TAKE A LOOK
8. Infer What do you think might happen to the spoonbill if its diet changes?

_____

_____

_____

_____

Chemical changes can change an animal's appearance.

CHEMICAL CHANGES IN INDUSTRY

Like physical changes, chemical changes play an important role in industry. Think about how a bicycle is made. The metal, plastic, and rubber in bicycles come from natural materials that have been changed chemically. Metal comes from minerals in rocks. The minerals have undergone chemical changes to free the metals from them. Plastic comes from oil that has been changed chemically. Rubber comes from a type of tree sap called latex that has gone through chemical changes.

TAKE A LOOK
9. Describe What might be some *physical* changes that the metal, plastic, and rubber went through as they were turned into the pieces of a bike?

_____

_____

_____

Are Physical and Chemical Changes Reversible?

Most physical changes are reversible. That is, the change can be "undone." For example, if you dissolve salt in water, you can evaporate the water. The salt is left behind. However, in some cases, a physical change greatly alters the form of matter. As a result, it is impossible to reverse the physical change. Imagine trying to change a dining room table back into a tree!

Some chemical changes are also reversible. However, most chemical changes are not reversible. For example, the rubber used to make bicycle tires cannot be changed back into the tree sap used in the process. Other chemical changes that are not reversible include the burning of wood, the digestion of food, and the rotting of an apple.☑

☑ READING CHECK
10. Explain Can physical and chemical changes be reversed?

_____

_____

_____

_____

Section 2 Review

NSES PS 1b, 3a, 3e

SECTION VOCABULARY

chemical change a change that occurs when one or more substances change into entirely new substances with different properties **law of conservation of mass** the law that states that mass cannot be created or destroyed in ordinary chemical and physical changes	**physical change** a change in matter from one form to another without a change in chemical properties

1. Explain Why are both photosynthesis and respiration considered chemical changes?

_____

_____

_____

2. Compare Dissolving sugar is a physical change, while digesting sugar is a chemical change. How is this possible?

_____

_____

_____

3. Describe Some iron filings are mixed with sugar. Is this an example of a physical change or a chemical change? Is this change reversible? Explain your answers.

_____

_____

_____

_____

4. Explain What happens to mass in any physical or chemical change?

_____

_____

5. Infer Charcoal is made of carbon. It reacts with oxygen to produce heat and carbon dioxide gas. If 12 grams of carbon react with 32 grams of oxygen, how many grams of carbon dioxide are produced? Show your work.

CHAPTER 18 Properties and States of Matter

SECTION
3 **States of Matter**

National Science
Education Standards
PS 1a, 3a

BEFORE YOU READ

After you read this section, you should be able to answer these questions:

• What are the four states of matter?

• How are the particles of matter arranged in each state?

• How do temperature and pressure affect gases?

What Are the States of Matter?

Recall that matter can be classified into elements, compounds, and mixtures based on how the atoms are arranged. Matter can also be classified in other ways. For example, matter can be classified based on the movement and energy of its particles. This method classifies matter into different *states*, or physical forms.

There are four states of matter: solid, liquid, gas, and plasma. You are probably most familiar with three of these states—solid, liquid, and gas. Although it is probably the least familiar to you, 99% of the matter in the universe exists as plasma. On Earth, however, matter is usually found as a solid, liquid, or gas. For example, the water on Earth exists as a solid (ice), liquid (water), and gas (vapor or steam). The figure below shows the solid state of the element bromine.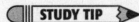

STUDY TIP

Make a Table As you read this section, make a table comparing solids, liquids, gases, and plasma. Include examples of each state of matter.

READING CHECK

1. List What are the four states of matter?

_____

_____

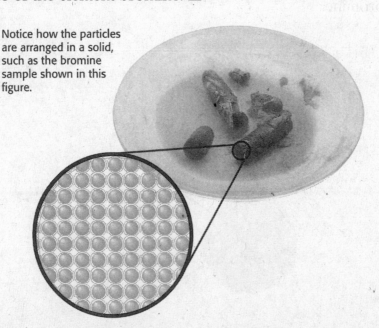

Notice how the particles are arranged in a solid, such as the bromine sample shown in this figure.

TAKE A LOOK

2. Describe Describe the arrangement of atoms in bromine.

_____

_____

_____

What Is a Solid?

Notice that the particles of solid bromine shown in the figure are highly organized. The particles are close together and form a pattern. These particles also do not move around. Instead, they vibrate in place. ☑

The state of matter in which particles are fixed in a rigid structure is a **solid**. A solid does not change shape or size because its particles do not move around. A solid has a definite shape and a definite volume.

MELTING POINT

If heat is applied to a solid, its particles absorb energy and start to vibrate faster. If the particles absorb enough energy, they start to move around. The more energy they absorb, the greater their motion.

The particles may gain enough energy that they can free themselves from their fixed positions. At this point, the solid starts to *melt* as the particles move apart from each other.

The temperature at which a solid melts is called the *melting point* of the solid. For example, the melting point of ice is 0°C. In contrast, the melting point of sodium chloride, or table salt, is more than 800°C. This high melting point indicates that sodium chloride requires more energy than water does to get its particles moving fast enough to turn it into a liquid.

The figure below shows the liquid state of the element bromine.

READING CHECK

3. Describe In what way do the particles in a solid move?

_____

_____

STANDARDS CHECK

PS 3a Energy is a property of many substances and is associated with heat, light, electricity, mechanical motion, sound, nuclei, and the nature of a chemical. Energy is transferred in many ways.

4. Describe What happens to the particles in a solid when they gain energy?

_____

_____

_____

TAKE A LOOK
5. Compare How is the arrangement of particles in the liquid bromine different from the arrangement in solid bromine?

_____

_____

_____

_____

What Is a Liquid?

Notice that the particles in the figure on the previous page are farther apart than those in the solid. The particles in a liquid are in contact with each other but they are not in fixed positions. Instead, they can move around and flow past each other.

As the particles in a liquid move around, the liquid takes the shape of its container. For example, notice in the figure on the previous page that the bromine liquid takes the rounded shape of its container. ☑

A **liquid** has a definite volume but no definite shape. Think what happens if you pour lemonade from a pitcher into a glass. The volume in the glass is the same as it was in the pitcher. However, the shape of the lemonade is different in the glass.

BOILING POINT

If you heat a liquid, its particles absorb energy and start to move faster. If the particles absorb enough energy, they start to move far apart from each other. The more energy the particles absorb, the farther apart they move.

The particles may gain enough energy that they are very far apart from each other. When they are far enough apart, the particles no longer attract each other. At this point, the liquid starts to *evaporate* or *boil*. The temperature at which a liquid boils is called the *boiling point* of the liquid. At this temperature, the liquid turns into a gas. The figure below shows a sample of bromine in the gas state. ☑

READING CHECK

6. Explain Why don't liquids have specific shapes?

_____

_____

_____

_____

READING CHECK

7. Explain What happens when a liquid reaches its boiling point?

_____

_____

Notice how the particles are arranged in a gas, such as the bromine sample shown in this figure.

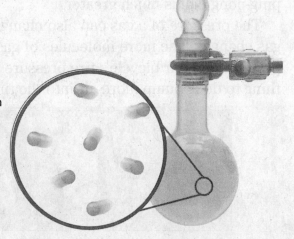

SECTION 3 States of Matter *continued*

What Is a Gas?

Notice that the particles in the figure on the previous page are farther apart than those in the liquid. The particles in a gas are so far apart that they are free of one another. Each particle moves freely inside the container.

As the particles in a gas move around, they fill up whatever volume is available to them. For example, notice in the figure on the previous page that the bromine gas fills up the entire container.

A **gas** has no definite volume and no definite shape. Think what would happen if you opened the valve on a tank of helium gas that is used to fill balloons. All the gas inside the container would escape. The gas particles would spread out to fill up the room. If a window were open, the gas particles would even spread outdoors.

GAS PRESSURE, TEMPERATURE, AND VOLUME

When particles of a gas hit a surface, they press on the surface. **Pressure** is the measure of the amount of force per unit area of surface. For example, the air pressure inside a tire is measured in pounds per square inch (lb/in^2). The pressure of a gas depends on the amount of space the gas has to occupy, the number of molecules of the gas, and the temperature of the gas. ☑

Unlike solids and liquids, gases can be *compressed*, or squeezed, to fit into a smaller space. If you increase the pressure on a gas, its volume decreases. The inverse happens as well. If you decrease the amount of space that a gas has available to fill, the pressure it exerts increases. For example, if you put the same amount of air in a ping-pong ball and in a basketball, the pressure inside the ping-pong ball is much greater.

The pressure of a gas can also change if the amount of gas changes. The more molecules of gas, the higher the pressure. If your bicycle's tire pressure is low, the easiest thing to do is pump more air into the tire.

Critical Thinking

8. Compare When water changes from a liquid into a gas (water vapor), how might its shape and volume change?

_____

_____

_____

_____

☑ **READING CHECK**

9. List Name two things that affect the pressure of a gas.

_____

_____

TAKE A LOOK

10. Explain How is this person increasing the pressure in the bike tire?

_____

_____

The temperature of a gas also affects its pressure and the amount of space it can take up. For example, less gas is needed to fill up parade balloons on a hot day than on a cold day. When gas is warm, its particles have more energy than when it is cold. They move faster, hit the inside walls of the balloon harder, and can fill up more space.

CHANGES IN STATE

Recall that adding heat changes a solid to a liquid, or a liquid to a gas. Changes from one state to another are called *changes of state*. Changes of state are physical changes because the identity of the substance does not change. Only its physical state changes. ☑

Removing heat can also cause changes in state. If a gas is cooled, the particles slow down and get closer to each other. The gas can *condense*, or change into a liquid if the particles lose enough energy.

If the liquid is cooled, the particles slow down even more and get still closer to each other. The liquid can *freeze*, or turn into a solid if the particles lose enough energy.

What Is Plasma?

If enough energy is added to a gas, the particles will break apart into electrically charged particles called electrons and ions. The state of matter that consists of electrons and ions is called a **plasma**.

Like gas, plasma does not have a fixed shape or volume. Plasma makes up most of the matter in the universe, including the sun and other stars. On Earth, plasma forms during lightning strikes and when electricity passes through fluorescent light bulbs. ☑

Critical Thinking
11. Infer If parade balloons are not filled completely in cold weather, and then the temperature suddenly increases, what might happen? Why?

_____

_____

_____

_____

_____

✔ READING CHECK
12. Explain Why is a change of state a physical change?

_____

_____

✔ READING CHECK
13. Define What is plasma made of?

_____

_____

Name _____ Class _____ Date _____

Section 3 Review

NSES PS 1a, 3a

SECTION VOCABULARY

gas a form of matter that does not have a definite volume or shape	**pressure** the amount of force exerted per unit area of a surface
liquid the state of matter that has a definite volume but not a definite shape	**solid** the state of matter in which the volume and shape of a substance are fixed
plasma the state of matter that starts as a gas and then becomes ionized; it consists of free-moving ions and electrons, it takes on an electric charge, and its properties differ from those of a solid, liquid, or gas	

1. Explain What is the relationship between gas and pressure?

_____

_____

2. Compare Fill in the table below

	Solid	Liquid	Gas	Plasma
Particle movement	vibrate in place			move freely
Shape	definite			not definite; takes the shape of the container
Volume			not fixed	
Compressible?			yes	yes
Example		water		

3. Describe Explain what happens to water molecules when energy is added to ice.

_____

_____

_____

4. Predict A big parade with huge, helium-filled balloons is held every year in New York City to celebrate Thanksgiving. What advice would you give to the organizers of this parade about filling the balloons if the parade were held to celebrate the Fourth of July?

_____

_____

_____

CHAPTER 19 | Matter in Motion
SECTION 1 | Measuring Motion

National Science
Education Standards
PS 2a

BEFORE YOU READ

After you read this section, you should be able to answer these questions:

• What is motion?

• How is motion shown by a graph?

• What are speed and velocity?

• What is acceleration?

What Is Motion?

Look around the room for a moment. What objects are in motion? Are students writing with pencils in their notebooks? Is the teacher writing on the board? Motion is all around you, even when you can't see it. Blood is circulating throughout your body. Earth orbits around the sun. Air particles shift in the wind.

When you watch an object move, you are watching it in relation to what is around it. Sometimes the objects around the object you are watching are at rest. An object that seems to stay in one place is called a *reference point*. When an object changes position over time in relation to a reference point, the object is in **motion**. ☑

You can use *standard reference directions* (such as north, south, east, west, right, and left) to describe an object's motion. You can also use features on Earth's surface, such as buildings or trees, as reference points. The figure below shows how a mountain can be used as a reference point to show the motion of a hot-air balloon.

STUDY TIP

Describe Study each graph carefully. In the margin next to the graph, write a sentence or two explaining what the graph shows.

READING CHECK

1. Describe What is the purpose of a reference point?

_____

_____

The hot-air balloon changed position relative to a reference point.

TAKE A LOOK

2. Identify What is the fixed reference point in the photos?

_____

How Can Motion Be Shown?

In the figure below, a sign-up sheet is being passed around a classroom. You can follow its path. The paper begins its journey at the reference point, the origin.

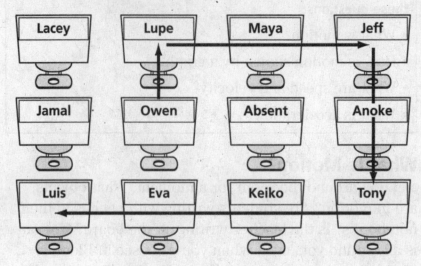

The path taken by a field trip sign-up sheet.

TAKE A LOOK

3. Identify What is the origin, or reference point, of the paper?

_____

_____

The figure below shows a graph of the position of the sign-up sheet as it is passed around the class. The paper moves in this order:

1. One positive unit on the *y*-axis
2. Two positive units on the *x*-axis
3. Two negative units on the *y*-axis
4. Three negative units on the *x*-axis

The graph provides a method of using standard reference directions to show motion.

4. Identify What is the shortest path that the paper could take to return to Owen's desk? The paper cannot move diagonally.

_____

_____

_____

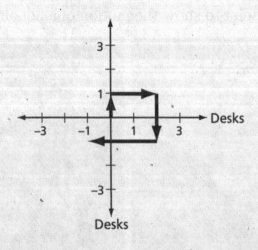

The position of the sign-up sheet as it moves through the classroom.

SECTION 1 Measuring Motion *continued*

What Is Speed?

Speed is the rate at which an object moves. It is the distance traveled divided by the time taken to travel that distance. Most of the time, objects do not travel at a constant speed. For example, when running a race, you might begin slowly but then sprint across the finish line.

So, it is useful to calculate *average speed*. We use the following equation:

$$average\ speed = \frac{total\ distance}{total\ time}$$

Suppose that it takes you 2 s to walk 4 m down a hallway. You can use the equation above to find your average speed:

$$average\ speed = \frac{4\ m}{2\ s} = 2\ m/s$$

Your speed is 2 m/s. Units for speed include meters per second (m/s), kilometers per hour (km/h), feet per second (ft/s), and miles per hour (mi/h).

How Can You Show Speed on a Graph?

You can show speed on a graph by showing how the position of an object changes over time. The *x*-axis shows the time it takes to move from place to place. The *y*-axis shows distance from the reference point.

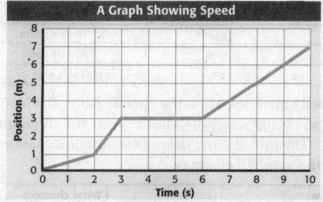

A Graph Showing Speed

A graph of position versus time also shows the dog's speed during his walk. The more slanted the line, the faster the dog walked.

Suppose you watched a dog walk beside a fence. The graph above shows the total distance the dog walked in 10 s. The line is not straight because the dog did not walk the same distance in each second. The dog walked slowly for 2 s and then quickly for 1 s. From 3 s to 5 s, the dog did not move.

Copyright © by Holt, Rinehart and Winston. All rights reserved.

Critical Thinking

5. Explain The average flight speed of a bald eagle is about 50 km/h. A scientist has measured an eagle flying 80 km/h. How is this possible?

_____

_____

_____

_____

Math Focus

6. Calculate Suppose you walk 10 m down a hallway in 2.5 s. What is your average speed? Show your work.

TAKE A LOOK

7. Apply Concepts Suppose the dog walks at a constant speed the whole way. On the graph, draw a line showing that the dog walks at a constant speed during the walk.

The average speed of the dog is:

$$average\ speed = \frac{total\ distance\ walked}{total\ time} = \frac{7\ m}{10\ s} = .07\ m/s$$

What Is Velocity?

Suppose that two birds leave the same tree at the same time. They both fly at 10 km/h for 5 min, then 5 km/h for 10 min. However, they don't end up in the same place. Why not?

The birds did not end up in the same place because they flew in different directions. Their speeds were the same, but because they flew in different directions, their velocities were different. **Velocity** is the speed of an object and its direction. ☑

The velocity of an object is constant as long as both speed and direction are constant. If a bus driving at 15 m/s south speeds up to 20 m/s south, its velocity changes. If the bus keeps moving at the same speed but changes direction from south to east, its velocity also changes. If the bus brakes to a stop, the velocity of the bus changes again.

The table below shows that velocity is a combination of both the speed of an object and its direction.

Speed	Direction	Velocity
15 m/s	south	15 m/s south
20 m/s	south	20 m/s south
20 m/s	east	20 m/s east
0 m/s	east	0 m/s east

Velocity changes when the speed changes, when the direction changes, or when both speed and direction change. The table below describes various situations in which the velocity changes.

Situation	What changes
Raindrop falling faster and faster	
Runner going around a turn on a track	direction
Car taking an exit off a highway	speed and direction
Train arriving at a station	speed
Baseball being caught by a catcher	speed
Baseball hit by a batter	
	speed and direction

READING CHECK

8. Analyze Someone tells you that the velocity of a car is 55 mi/h. Is this correct? Explain your answer.

_____

_____

Say It

Share Experiences Have you ever experienced a change in velocity on an amusement park ride? In pairs, share an experience. Explain how the velocity changed—was it a change in speed, direction, or both?

TAKE A LOOK
9. Identify Fill in the empty boxes in the table.

1 m/s   2 m/s   3 m/s   4 m/s   5 m/s

This cyclist moves faster and faster as he peddles his bike south.

TAKE A LOOK
10. Identify Is the cyclist accelerating? How do you know?

_____

_____

What Is Acceleration?

Acceleration is how quickly velocity changes. An object accelerates if its speed changes, its direction changes, or both its speed and direction change.

The units for acceleration are the units for velocity divided by a unit for time. The resulting unit is often meters per second per second (m/s/s or m/s^2).

Looking at the figure above, you can see that the speed increases by 1 m/s during each second. This means that the cyclist is accelerating at 1 m/s^2.

An increase in speed is referred to as *positive acceleration*. A decrease in speed is referred to as *negative acceleration* or *deceleration*. ☑

Acceleration can be shown on a graph of speed versus time. Suppose you are operating a remote control car. You push the lever on the remote to move the car forward. The graph below shows the car's acceleration as the car moves east. For the first 5 s, the car increases in speed. The car's acceleration is positive because the speed increases as time passes.

For the next 2 s, the speed of the car is constant. This means the car is no longer accelerating. Then the speed of the car begins to decrease. The car's acceleration is then negative because the speed decreases over time.

☑ READING CHECK
11. Explain What happens to an object when it has negative acceleration?

_____

_____

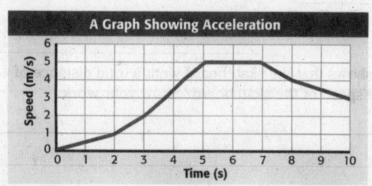

A Graph Showing Acceleration

The graph of speed versus time also shows that the acceleration of the car was positive and negative. Between 5 s and 7 s, it had no acceleration.

Math Focus
12. Interpret Graphs Is the slope positive or negative when the car's speed increases? Is the slope positive or negative when the car's speed decreases?

_____

_____

_____

Section 1 Review

SECTION VOCABULARY

acceleration the rate at which the velocity changes over time; an object accelerates if its speed, direction, or both change	**motion** an object's change in position relative to a reference point
speed the distance traveled divided by the time interval during which the motion occurred	**velocity** the speed of an object in a particular direction

1. Identify What is the difference between speed and velocity?

_____

_____

_____

2. Complete a Graphic Organizer Fill in the graphic organizer for a car that starts from one stop sign and approaches the next stop sign. Use the following terms: constant *velocity, positive acceleration, deceleration,* and *at rest.*

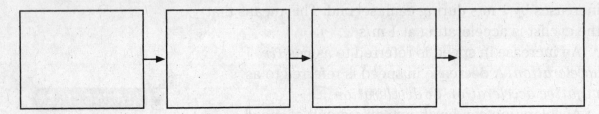

3. Interpret a Graph Describe the motion of the skateboard using the graph below. Write what the skateboard does from time = 0 s to time = 40 s.

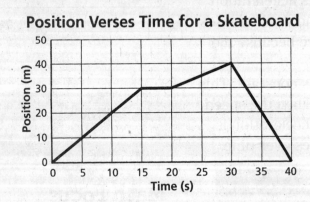

Position Verses Time for a Skateboard

_____

_____

_____

_____

_____

_____

_____

4. Calculate The graph above shows that the skateboard went a total distance of 80 m. What was the average speed of the skateboard? Show your work.

_____

_____

_____

| CHAPTER 19 | Matter in Motion |

SECTION 2 What Is a Force?

National Science
Education Standards
PS 2b, PS 2c

BEFORE YOU READ

After you read this section, you should be able to answer these questions:

- What is a force?
- How do forces combine?
- What is a balanced force?
- What is an unbalanced force?

What Is a Force?

You probably hear people talk about force often. You may hear someone say, "That storm had a lot of force" or "Mrs. Larsen is the force behind the school dance." But what exactly is a force in science?

In science, a **force** is a push or a pull. All forces have two properties: direction and size. A **newton** (N) is the unit that describes the size of a force. ☑

Forces act on the objects around us in ways that we can see. If you kick a ball, the ball receives a push from you. If you drag your backpack across the floor, the backpack is pulled by you.

Forces also act on objects around us in ways that we cannot see. For example, in the figure below, a student is sitting on a chair. What are the forces acting on the chair?

The student is pushing down on the chair, but the chair does not move. Why? The floor is balancing the force by pushing up on the chair. When the forces on an object are *balanced*, the object does not move.

STUDY TIP

Brainstorm As you read, think about objects you see every day. What kinds of forces are affecting them? How do the forces affect them?

☑ **READING CHECK**

1. **List** What two properties do all forces have?

_____

_____

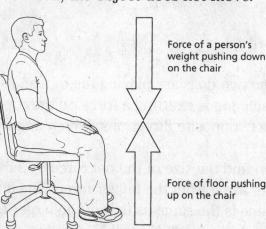

Force of a person's weight pushing down on the chair

Force of floor pushing up on the chair

A person sitting on a chair.

TAKE A LOOK

2. **Explain** Since the chair is not moving, what kind of forces are acting on it?

_____

_____

How Do Forces Combine?

As you saw in the previous example, more than one force often acts on an object. When all of the forces acting on an object are added together, you determine the **net force** on the object. An object with a net force more than 0 N acting on it will change its state of motion.

FORCES IN THE SAME DIRECTION

Suppose your music teacher asks you and a friend to move a piano, as shown in the figure above. You push the piano from one end and your friend pulls the piano from the other end. You and your friend are applying forces in the same direction. Adding the two forces gives you the size of the net force. The direction of the net force is the same as the direction of the forces.

$$125 \text{ N} + 120 \text{ N} = 245 \text{ N}$$
net force = 245 N to the right

FORCES IN DIFFERENT DIRECTIONS

Suppose two dogs are playing tug of war, as shown above. Each dog is exerting a force on the rope. Here, the forces are in opposite directions. Which dog will win the tug of war?

You can find the size of the net force by subtracting the smaller force from the bigger force. The direction of the net force is the same as that of the larger force:

$$120 \text{ N} - 80 \text{ N} = 40 \text{ N}$$
net force = 40 N to the right

What Happens When Forces Are Balanced or Unbalanced?

Knowing the net force on an object lets you determine its effect on the motion of the object. Why? The net force tells you whether the forces on the object are balanced or unbalanced.

BALANCED FORCES

When the forces on an object produce a net force of 0 N, the forces are *balanced*. There is no change in the motion of the object. For example, a light hanging from the ceiling does not move. This is because the force of gravity pulls down on the light while the force of the cord pulls upward. ☑

The soccer ball moves because the players exert an unbalanced force on the ball each time they kick it.

UNBALANCED FORCES

When the net force on an object is not 0 N, the forces on the object are *unbalanced*. Unbalanced forces produce a change in motion of an object. Think about a soccer game. Players kick the ball to each other. When a player kicks the ball, the kick is an unbalanced force. It sends the ball in a new direction with a new speed.

An object can continue to move when the unbalanced forces are removed. For example, when it is kicked, a soccer ball receives an unbalanced force. The ball continues to roll on the ground after the ball was kicked until an unbalanced force changes its motion.

✓ **READING CHECK**

5. Describe What happens to the motion of an object if the net force acting on it is 0 N?

_____

_____

STANDARDS CHECK

PS 2c If more than one force acts on an object along a straight line, then the forces reinforce or cancel one another, depending on their direction and magnitude. Unbalanced forces will cause changes in the speed or direction of an object's motion.

6. Describe What will happen to an object that has an unbalanced force acting on it?

_____

_____

Section 2 Review

SECTION VOCABULARY

force a push or a pull exerted on an object in order to change the motion of the object; force has size and direction	**newton** the SI unit for force (symbol, N)
net force the combination of all the forces acting on an object	

1. Explain If there are many forces acting on an object, how can the net force be 0?

_____

_____

_____

2. Apply Concepts Identify three forces acting on a bicycle when you ride it.

_____

_____

3. Calculate Determine the net force on each of the objects shown below. Don't forget to give the direction of the force.

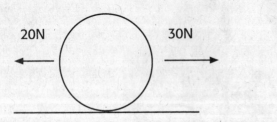

net force = _____   *net force =* _____

4. Explain How will the net force affect the motion of each object shown above?

_____

_____

_____

5. Describe What is the difference between balanced and unbalanced forces?

_____

_____

_____

_____

SECTION 3
Friction: A Force That Opposes Motion

National Science Education Standards
PS 2c

BEFORE YOU READ

After you read this section, you should be able to answer these questions:

• What is friction?

• How does friction affect motion?

• What are the types of friction?

• How can friction be changed?

What Causes Friction?

Suppose you are playing soccer and you kick the ball far from you. You know that the ball will slow down and eventually stop. This means that the velocity of the ball will decrease to 0. You also know that an unbalanced force is needed to change the velocity of objects. So, what force is stopping the ball?

Friction is the force that opposes the motion between two surfaces that touch. Friction causes the ball to slow down and then stop. ☑

What causes friction? The surface of any object is rough. Even an object that feels smooth is covered with very tiny hills and valleys. When two surfaces touch, the hills and valleys of one surface stick to the hills and valleys of the other. This contact between surfaces causes friction.

If a force pushes two surfaces together even harder, the hills and valleys come closer together. This increases the friction between the surfaces.

STUDY TIP

Imagine As you read, think about the ways that friction affects your life. Make a list of things that might happen, or not happen, if friction did not exist.

READING CHECK

1. **Describe** What is friction?

_____

_____

_____

_____

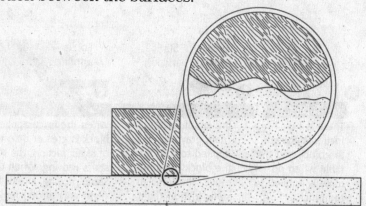

When the hills and valleys of one surface stick to the hills and valleys of another surface, friction is created.

TAKE A LOOK

2. **Explain** Why can't you see the hills and valleys without a close-up view of the objects?

_____

_____

3. Identify Two identical balls begin rolling next to each other at the same velocity. One is on a smooth surface and one is on a rough surface. Which ball will stop first? Why?

_____
_____
_____

What Affects the Amount of Friction?

Imagine that a ball is rolled over a carpeted floor and another ball is rolled over a wood floor. Which surface affects a ball more?

The smoothness of the surfaces of the objects affects how much friction exists. Friction is usually greater between materials that have rough surfaces than materials that have smooth surfaces. The carpet has greater friction than the wood floor, so the ball on the carpet stops first.

What Types of Friction Exist?

There are two types of friction: kinetic friction and static friction. *Kinetic friction* occurs when force is applied to an object and the object moves. When a cat slides along a countertop, the friction between the cat and the countertop is kinetic friction. The word kinetic means "moving."

The amount of kinetic friction between moving surfaces depends partly on how the surfaces move. In some cases, the surfaces slide past each other like pushing a box on the floor. In others, one surface rolls over another like a moving car on a road. There is usually less friction between surfaces that roll than between surfaces that slide.

Static friction occurs when force applied to an object does not cause the object to move. When you try to push a piece of furniture that will not move, the friction observed is static friction.

TAKE A LOOK
4. Describe When does static friction become kinetic friction?

_____
_____
_____

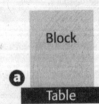

There is no friction between the block and the table when no force is applied to the block.

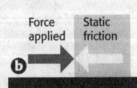

If a small force (dark gray arrow) is applied to the block, the block does not move. The force of static friction (light gray arrow) balances the force applied.

When the force applied to the block is greater than the force of static friction, the block starts moving. When the block starts moving, kinetic friction (light gray arrow) replaces all of the static friction and opposes the force applied.

How Can Friction Be Decreased?

To reduce the amount of friction, you can apply a lubricant between two surfaces. A *lubricant* is a substance that reduces the friction between surfaces. Motor oil, wax, and grease are examples of lubricants.

You can also reduce friction by rolling, rather than sliding, an object. A refrigerator on rollers is much easier to move than one that just slides.

Another way of reducing friction is to smooth the surfaces that rub against each other. Skiers have their skis sanded down to make them smoother. This makes it easier for the skis to slide over the snow.

If you work on a bicycle, you may get dirty from the chain oil. This lubricant reduces friction between sections of the chain.

How Can Friction Be Increased?

Increasing the amount of friction between surfaces can be very important. For example, when the tires of a car grip the road better, the car stops and turns corners much better. Friction causes the tires to grip the road. Without friction, a car could not start moving or stop.

On icy roads, sand can be used to make the road surface rougher. Friction increases as surfaces are made rougher.

You can also increase friction by increasing the force between the two objects. Have you ever cleaned a dirty pan in the kitchen sink? You may have found that cleaning the pan with more force allows you to increase the amount of friction. This makes it easier to clean the pan. ☑

Critical Thinking

5. Infer How does a lubricant reduce the amount of friction?

_____

_____

_____

TAKE A LOOK

6. Identify Why is it important to put oil on a bicycle chain?

_____

☑ READING CHECK

7. Identify Name two things that can be done to increase the friction between surfaces.

_____

_____

_____

_____

Section 3 Review

SECTION VOCABULARY

friction a force that opposes the motion between two surfaces that are in contact	

1. Describe What effect does friction have when you are trying to move an object at rest?

_____

2. Compare Explain the difference between static friction and kinetic friction. Give an example of each.

_____

_____

_____

3. Compare The figure on the left shows two surfaces up close. On the right, draw a sketch. Show what the surfaces of two objects that have less friction between them might look like.

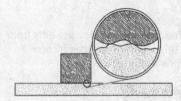

4. Analyze Name three common lubricants and describe why they are used.

_____

_____

5. Analyze In what direction does friction always act?

_____

6. Identify A car is driving on a flat road. When the driver hits the brakes, the car slows down and stops. What would happen if there were no friction between the tires and the road? Explain your answer.

_____

_____

_____

CHAPTER 19 | Matter in Motion

SECTION 4 | # Gravity: A Force of Attraction

BEFORE YOU READ

After you read this section, you should be able to answer these questions:

• What is gravity?

• How are weight and mass different?

How Does Gravity Affect Matter?

Have you ever seen a video of astronauts on the moon? The astronauts bounce around like beach balls even though the space suits weighed 180 pounds on Earth. See the figure below. Why is it easier for a person to move on the moon than on Earth? The reason is that the moon has less gravity than Earth. **Gravity** is a force of attraction, or a pull, between objects. It is caused by their masses. ☑

All matter has mass. Gravity is a result of mass. Therefore all matter has gravity. This means that all objects attract all other objects in the universe! The force of gravity pulls objects toward each other. For example, gravity between the objects in the solar system holds the solar system together. Gravity holds you to Earth.

Small objects also have gravity. You have gravity. This book has gravity. Why don't you notice the book pulling on you or you pulling on the book? The reason is that the book's mass and your mass are both small. The force of gravity caused by small mass is not large enough to move either you or the book.

STUDY TIP

Discuss Ideas Take turns reading this section out loud with a partner. Stop to discuss ideas that seem confusing.

✓ READING CHECK

1. Describe What is gravity?

_____

_____

_____

Critical Thinking

2. Infer Why can't you see two soccer balls attracting each other?

_____

_____

_____

Because the moon has less gravity than Earth does, walking on the moon's surface was a very bouncy experience for the Apollo astronauts.

What Is the Law of Universal Gravitation?

According to a story, Sir Isaac Newton, while sitting under an apple tree, watched an apple fall. This gave him a bright idea. Newton realized that an unbalanced force on the apple made it fall.

He then thought about the moon's orbit. Like many others, Newton had wondered what kept the planets in the sky. He realized that an unbalanced force on the moon kept it moving around Earth. Newton said that these forces are both gravity.

Newton's ideas are known as the *law of universal gravitation*. Newton said that all objects in the universe attract each other because of gravitational force. ☑

This law says that gravitational force depends on two things:

1. the masses of the objects
2. the distance between the objects

The word "universal" means that the law applies to all objects. See the figure below. ☑

Sir Isaac Newton said that the same unbalanced force caused the motions of the apple and the moon.

✓ **READING CHECK**

3. Describe What is the law of universal gravitation?

_____

_____

_____

_____

✓ **READING CHECK**

4. Identify What two things determine gravitational force?

_____

_____

STANDARDS CHECK

PS 2c If more than one force acts on an object along a straight line, then the forces reinforce or cancel one another, depending on their direction and magnitude. Unbalanced forces will cause changes in the speed or direction of an object's motion.

5. Identify What is the unbalanced force that affects the motions of a falling apple and the moon?

_____

How Does Mass Affect Gravity?

Imagine an elephant and a cat. Because the elephant has a larger mass than the cat does, gravity between the elephant and Earth is larger. So, the cat is much easier to pick up than the elephant. The gravitational force between objects depends on the masses of the objects. See the figure below.

mass = 100 kg   mass = 100 kg

● Gravitational force is small between objects that have small masses.

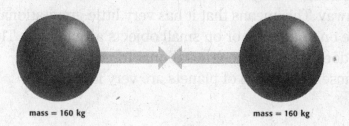

mass = 160 kg   mass = 160 kg

● If the mass of one or both objects increases, the gravitational force pulling them together increases.

The arrows indicate the gravitational force between two objects. The length of the arrows indicates the magnitude of the force.

TAKE A LOOK
6. Compare Is there more gravitational force between objects with small masses or objects with large masses?

_____

_____

_____

Mass also explains why an astronaut on the moon can jump around so easily. The moon has less mass than Earth does. This gives the moon a weaker pull on objects than the pull of Earth. The astronaut is not being pulled toward the moon as much as he is by Earth. So the astronaut can jump higher and more easily on the moon.

The universal law of gravitation can let us predict what happens to gravity when mass changes. According to the universal law of gravitation, suppose there is a 5 N force of gravity between two objects. If the mass of one object doubles and the other stays the same, the force of gravity also doubles.

Let's try a problem. The force due to gravity between two objects is 3 N. If the mass of one object triples and the other stays the same, what is the new force of gravity?

Solution: Since the mass of one object tripled and the other stayed the same, the force of gravity also triples. It is 9 N.

Math Focus
7. Infer Two objects of equal mass have a force of gravity of 6 N between them. Imagine the mass of one is cut in half and the other stays the same, what is the force due to gravity?

_____

SECTION 4 Gravity: A Force of Attraction *continued*

How Does Distance Affect Gravity?

The mass of the sun is 300,000 times bigger than that of Earth. However, if you jump up, you return to Earth every time you jump rather than flying toward the sun. If the sun has more mass, then why doesn't it have a larger gravitational pull on you?

This is because the gravitational force also depends on the distance between the objects. As the distance between two objects gets larger, the force of gravity gets much smaller. And as the distance between objects gets smaller, the force of gravity gets much bigger. This is shown in the figure below.

Although the sun has tremendous mass, it is also very far away. This means that it has very little gravitational force on your body or on small objects around you. The sun does have a large gravitational force on planets because the masses of planets are very large.

Critical Thinking

8. Analyze The sun is much more massive than Earth. Why is the force of gravity between you and the sun so much less than Earth's gravity and you?

_____

_____

_____

_____

● Gravitational force is large when the distance between two objects is small.

TAKE A LOOK
9. Describe Use the diagram to describe the effect of distance on gravitational force.

_____

_____

_____

_____

_____

● If the distance between two objects increases, the gravitational force pulling them together decreases rapidly.

The length of the arrows indicates the magnitude of the gravitational force between two objects.

What Is the Difference Between Mass and Weight?

You have learned that gravity is a force of attraction between objects. **Weight** is a measure of the gravitational force on an object. The SI unit for weight is the newton (N).

Mass is a measure of the amount of matter in an object. This seems similar to weight, but it is not the same. An object's mass does not change when gravitational forces change, but its weight does. Mass is usually expressed in kilograms (kg) or grams (g). ☑

In the figure below, you can see the difference between mass and weight. Compare the astronaut's mass and weight on Earth to his mass and weight on the moon.

10. Contrast How is mass different from weight?

_____

_____

_____

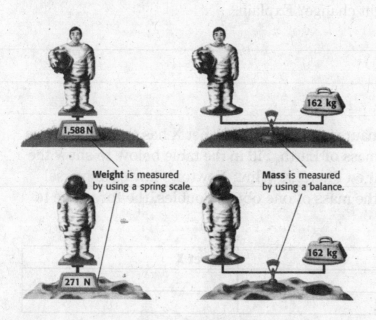

Weight is measured by using a spring scale.

Mass is measured by using a balance.

1,588 N

271 N

162 kg

162 kg

TAKE A LOOK

11. Identify What is the weight of the astronaut on Earth? What is the weight of the astronaut on the moon?

_____

_____

Gravity can cause objects to move because it is a type of force. But gravity also acts on objects that are not moving, or *static*. Earth's gravity pulls objects downward. However, not all objects move downward. Suppose a framed picture hangs from a wire. Gravity pulls the picture downward, but tension (the force in the wire) pulls the picture upward. The forces are balanced so that framed picture does not move.

Critical Thinking

12. Contrast What forces act on a framed picture on a shelf?

_____

_____

Section 4 Review

SECTION VOCABULARY

gravity a force of attraction between objects that is due to their masses	**weight** a measure of the gravitational force exerted on an object; its value can change with the location of the object in the universe
mass a measure of the amount of matter in an object	

1. Identify What is gravity? What determines the gravitational force between objects?

_____

_____

2. Describe A spacecraft is moving toward Mars. Its rocket engines are turned off. As the spacecraft nears the planet, what will happen to the pull of Mars's gravity?

_____

3. Summarize An astronaut travels from Earth to the moon. How does his mass change? How does his weight change? Explain.

_____

_____

_____

4. Applying Concepts An astronaut visits Planet X. Planet X has the same radius as Earth but has twice the mass of Earth. Fill in the table below to show the astronaut's mass and weight on Planet X. (Hint: Newton's law of universal gravitation says that when the mass of one object doubles, the force due to gravity also doubles.)

	Earth	Planet X
Mass of astronaut	80 kg	
Weight of astronaut	784 N	

5. Select Each of the spheres shown below is made of iron. Circle the pair of spheres that would have the greatest gravitational force between them. Below the spheres, explain the reason for your choice.

_____

_____

CHAPTER 20 Forces and Motion
SECTION
1 **Gravity and Motion**

BEFORE YOU READ

After you read this section, you should be able to answer
these questions:

• How does gravity affect objects?

• How does air resistance affect falling objects?

• What is free fall?

• Why does an object that is thrown horizontally follow
a curved path?

How Does Gravity Affect Falling Objects?

In ancient Greece, a great thinker named Aristotle
said that heavy objects fall faster than light objects. For
almost 2,000 years, people thought this was true. Then, in
the late 1500s, an Italian scientist named Galileo Galilei
proved that heavy and light objects actually fall at the
same rate.

It has been said that Galileo proved this by dropping
two cannonballs from the top of a tower at the same
time. The cannonballs were the same size, but one was
much heavier than the other. The people watching saw
both cannonballs hit the ground at the same time. ☑

Why don't heavy objects fall faster than light objects?
Gravity pulls on heavy objects more than it pulls on light
objects. However, heavy objects are harder to move than
light objects. So, the extra force from gravity on the heavy
object is balanced by how much harder it is to move.

Falling Balls

Time 1 — Golf ball
— Table tennis ball
Time 2

Time 3

Time 4

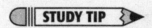
STUDY TIP

Practice After every page,
stop reading and think
about what you've read. Try
to think of examples from
everyday life. Don't go on
to the next section until you
think you understand.

✓ **READING CHECK**

1. Describe What did
the people watching the
cannonballs see that told
them the cannonballs fell at
the same rate?

_____

_____

_____

TAKE A LOOK

2. Predict The golf ball
is heavier than the table
tennis ball. On the figure,
draw three circles to show
where the golf ball will be at
times 2, 3, and 4.

How Much Acceleration Does Gravity Cause?

Because of gravity, all objects accelerate, or speed up, toward Earth at a rate of 9.8 meters per second per second. This is written as 9.8 m/s/s or 9.8 m/s^2. So, for every second an object falls, its velocity (speed) increases by 9.8 m/s. This is shown in the figure below.

Math Focus
3. Calculate How fast is the ball moving at the end of the third second? Explain your answer.

_____

_____

_____

_____

_____

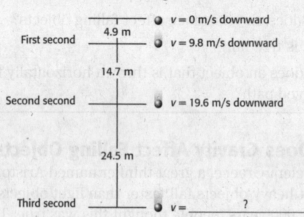

A falling object accelerates at a constant rate. The object falls faster and farther each second than it did the second before.

What Is the Velocity of a Falling Object?

Suppose you drop a rock from a cliff. How fast is it going when it reaches the bottom? If you have a stopwatch, you can calculate its final velocity.

If an object starts from rest and you know how long it falls, you can calculate its final velocity by using this equation:

$$v_{final} = g \times t$$

In the equation, v_{final} stands for final velocity in meters per second, g stands for the acceleration due to gravity (9.8 m/s^2), and t stands for the time the object has been falling (in seconds).

If the rock took 4 s to hit the ground, how fast was it falling when it hit the ground?

Step 1: Write the equation.

$$v_{final} = g \times t$$

Math Focus
4. Calculate A penny is dropped from the top of a tall stairwell. What is the velocity of the penny after it has fallen for 2 s? Show your work.

Step 2: Place values into the equation, and solve for the answer.

$$v_{final} = 9.8 \, \frac{m/s}{s} \times 4 \, s = 39.2 \, m/s$$

The velocity of the rock was 39.2 m/s when it hit the ground.

How Can You Calculate How Long an Object Was Falling?

Suppose some workers are building a bridge. One of them drops a metal bolt from the top of the bridge. When the bolt hits the ground, it is moving 49 m/s. How long does it take the bolt to fall to the ground?

Step 1: Write the equation.

$$t = \frac{v_{final}}{g}$$

Step 2: Place values into the equation, and solve for the answer.

$$t = \frac{49 \text{ m/s}}{9.8 \frac{\text{m/s}}{\text{s}}} = 5 \text{ s}$$

The bolt fell for 5 s before it hit the ground.

How Does Air Resistance Affect Falling Objects?

Try dropping a pencil and a piece of paper from the same height. What happens? Does this simple experiment show what you just learned about falling objects? Now crumple the paper into a tight ball. Drop the crumpled paper and the pencil from the same height.

What happens? The flat paper falls more slowly than the crumpled paper because of air resistance. *Air resistance* is the force that opposes the motion of falling objects. ☑

How much air resistance will affect an object depends on the size, shape, and speed of the object. The flat paper has more surface area than the crumpled sheet. This causes the flat paper to fall more slowly.

How Air Resistance Affects Velocity

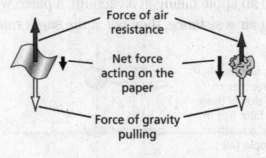

Force of air resistance

Net force acting on the paper

Force of gravity pulling

Math Focus

5. Calculate A rock falls from a cliff and hits the ground with a velocity of 98 m/s. How long does the rock fall? Show your work.

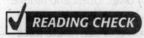
READING CHECK

6. Identify Which has more air resistance, the flat paper or the crumpled paper?

_____

TAKE A LOOK

7. Explain Why does the crumpled paper fall faster than the flat paper?

_____

_____

_____

What Is Terminal Velocity?

As the speed of a falling body increases, air resistance also increases. The upward force of air resistance keeps increasing until it is equal to the downward force of gravity. At this point, the total force on the object is zero, so the object stops accelerating.

When the object stops accelerating, it does not stop moving. It falls without speeding up or slowing down. It falls at a constant velocity called the terminal velocity. **Terminal velocity** is the speed of an object when the force of air resistance equals the force of gravity. ☑

Air resistance causes the terminal velocity of hailstones to be between 5 m/s and 40 m/s. Without air resistance, they could reach the ground at a velocity of 350 m/s! Air resistance also slows sky divers to a safe landing velocity.

The parachute increases the air resistance of this sky diver and slows him to a safe terminal velocity.

What Is Free Fall?

Free fall is the motion of an object when gravity is the only force acting on the object. The figure below shows a feather and an apple falling in a vacuum, a place without any air. Without air resistance, they fall at the same rate.

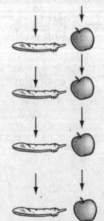

Air resistance usually causes a feather to fall more slowly than an apple falls. But in a vacuum, a feather and an apple fall with the same acceleration because both are in free fall.

READING CHECK

8. Describe When does an object reach its terminal velocity?

_____

_____

TAKE A LOOK
9. Identify A sky diver is falling at terminal velocity. Draw and label an arrow showing the direction and size of the force due to gravity on the sky diver. Draw and label a second arrow showing the direction and size of the force of air resistance on the sky diver.

TAKE A LOOK
10. Predict When air resistance acts on the apple and the feather, which falls faster?

_____

_____

Orbiting Objects Are in Free Fall

Satellites and the space shuttle orbit Earth. You may have seen that astronauts inside the shuttle float unless they are belted to a seat. They seem weightless. In fact, they are not weightless, because they still have mass.

Weight is a measure of the pull of gravity on an object. Gravity acts between any two objects in the universe. Every object in the universe pulls on every other object. Every object with mass has weight. ☑

The force of gravity between two objects depends on the masses of the objects and how far apart they are. The more massive the objects are, the greater the force is. The closer the objects, the greater the force.

Your weight is determined mostly by the mass of Earth because it is so big and so close to you. If you were to move away from Earth, you would weigh less. However, you would always be attracted to Earth and to other objects, so you would always have weight.

Astronauts float in the shuttle because the shuttle is in free fall. That's right—the shuttle is always falling. Because the astronauts are in the shuttle, they are also falling. The astronauts and the shuttle are falling at the same rate. That is why the astronauts seem to float inside the shuttle. ☑

Isaac Newton first predicted this kind of free fall in the late 17th century. He reasoned that if a cannon were placed on a mountain and fired, the cannon ball would fall to Earth. Yet, if the cannon ball were shot with enough force, it would fall at the same rate that Earth's surface curves away. The cannon ball would never hit the ground, so it would orbit Earth. The figure below shows this "thought experiment."

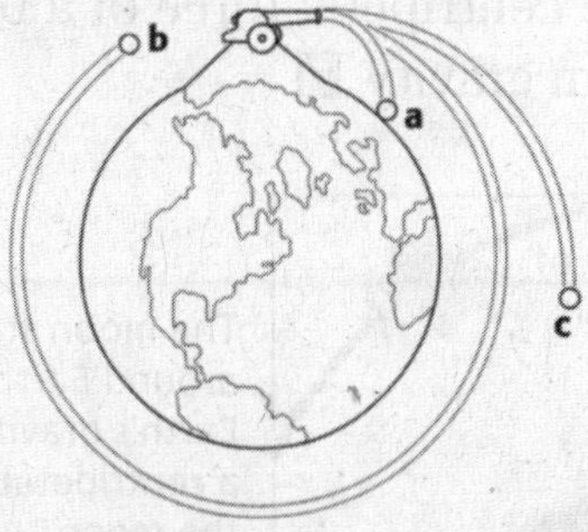

Newton's cannon is a "thought experiment." Newton reasoned that a cannon ball shot hard enough from a mountain top would orbit Earth.

☑ READING CHECK

11. Explain When will an object have no weight? Explain your answer.

_____

_____

☑ READING CHECK

12. Explain Why don't the astronauts in the orbiting shuttle fall to the floor?

_____

_____

_____

Critical Thinking

13. Infer Compared with cannon ball **b**, what do you think cannon ball **c** would do?

_____

_____

SECTION 1 Gravity and Motion *continued*

What Motions Combine to Make an Object Orbit?

An object is in orbit (is orbiting) when it is going around another object in space. When the space shuttle orbits Earth, it is moving forward. Yet, the shuttle is also in free fall. The figure below shows how these two motions combine to cause orbiting.

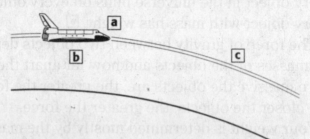

a. The space shuttle moves forward at a constant speed. If there were no gravity, the space shuttle would continue to move in a straight line.

b. The space shuttle is in free fall because gravity pulls it toward Earth. The space shuttle would move straight down if it were not traveling forward.

c. The path of the space shuttle follows the curve of Earth's surface. This path is known as an orbit.

TAKE A LOOK

14. Identify On the figure, draw a line showing the path that the space shuttle would take if gravity were not acting on it.

What Force Keeps an Object in Orbit?

Many objects in space are orbiting other objects. The moon orbits Earth, while Earth and the other planets orbit the sun. These objects all follow nearly circular paths. An object that travels in a circle is always changing direction.

If all the forces acting on an object balance each other out, the object will move in the same direction at the same speed forever. So, objects cannot orbit unless there is an unbalanced force acting on them. *Centripetal force* is the force that keeps an object moving in a circular path. Centripetal force pulls the object toward the center of the circle. The centripetal force of a body orbiting in space comes from gravity. ☑

READING CHECK

15. Identify What must be applied to an object to change its direction?

_____

_____

TAKE A LOOK

16. Identify Draw an arrow on the figure to show the direction that centripetal force acts on the moon.

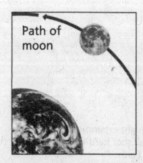

Path of moon

The moon stays in orbit around Earth because Earth's gravity provides a centripetal force on the moon.

What Is Projectile Motion?

Projectile motion is the curved path an object follows when it is thrown near the Earth's surface. The motion of a ball that has been thrown forward is an example of projectile motion.

Projectile motion is made up of two parts: horizontal motion and vertical motion. Horizontal motion is motion that is parallel to the ground. Vertical motion is motion that is perpendicular to the ground. The two motions do not affect each other. Instead, they combine to form the curved path we call projectile motion. ☑

When you throw a ball forward, your hand pushes the ball to make it move forward. This force gives the ball its horizontal motion. After the ball leaves your hand, no horizontal forces act on the ball (if we forget air resistance for now). So the ball's horizontal velocity does not change after it leaves your hand.

However, gravity affects the vertical part of projectile motion. Gravity pulls the ball straight down. All objects that are thrown accelerate downward because of gravity.

READING CHECK

17. List What two motions combine to make projectile motion?

_____

_____

Critical Thinking

18. Infer If you are playing darts and you want to hit the bulls-eye, where should you aim?

_____

_____

TAKE A LOOK

19. List On the figure, fill in the three blanks with the correct words.

20. Apply Concepts If there were no air resistance, how fast would the ball's downward velocity be changing? Explain your answer.

_____

_____

_____

_____

a After the ball leaves the pitcher's hand, the ball's _____ velocity is constant.

b The ball's vertical velocity increases because _____ causes it to accelerate downward.

c The two motions combine to form a _____ path.

Section 1 Review

SECTION VOCABULARY

free fall the motion of a body when only the force of gravity is acting on the body	**terminal velocity** the constant velocity of a falling object when the force of air resistance is equal in magnitude and opposite in direction to the force of gravity
projectile motion the curved path that an object follows when thrown, launched, or otherwise projected near the surface of Earth	

1. Explain Is a parachutist in free fall? Why or why not?

_____

_____

_____

2. Identify Cause and Effect Complete the table below to show how forces affect objects.

Cause	Effect
Gravity acts on a falling object.	
	The falling object reaches terminal velocity.

3. Calculate A rock at rest falls off a cliff and hits the ground after 3.5 s. What is the rock's velocity just before it hits the ground? Show your work.

4. Identify What force must be applied to an object to keep it moving in a circular path?

_____

5. Explain Which part of projectile motion is affected by gravity? Explain how it is affected.

_____

_____

_____

CHAPTER 20 Forces and Motion

SECTION
2 Newton's Laws of Motion

BEFORE YOU READ

After you read this section, you should, be able to answer these questions:

• What is net force?

• What happens to objects that have no net force acting on them?

• How are mass, force, and acceleration related?

• How are force pairs related by Newton's third law of motion?

National Science Education Standards
PS 2b, 2c

What Is a Net Force?

A *force* is a push or a pull. It is something that causes an object to change speed or direction. There are forces acting on all objects every second of the day. They are acting in all directions.

At first, this might not make sense. After all, there are many objects that are not moving. Are forces acting on an apple sitting on a desk? The answer is yes. Gravity is pulling the apple down. The desk is pushing the apple up.

In the figure below, the arrows represent the size and direction of the forces on the apple.

STUDY TIP ✏

Summarize in Pairs Read each of Newton's laws silently to yourself. After reading each law, talk about what you read with a partner. Together, try to figure out any ideas that you didn't understand.

Forces Acting on an Apple

Table pushing up

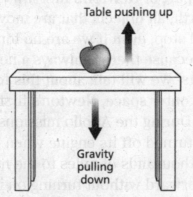

Gravity pulling down

So, why doesn't the apple move? The apple is staying where it is because all the forces balance out. There are no unbalanced forces. That is, there is no net force on the apple. *Net force* is the total force acting on an object. If the net force on an object is zero, the object will not change speed or direction. ☑

TAKE A LOOK
1. Compare What is the size of the force pulling the apple down compared with the size of the force pushing it up?

_____

2. Explain What will not happen to an object if the net force acting on it is zero?

_____

_____

What Is Newton's First Law of Motion?

Newton's first law of motion describes objects that have no unbalanced forces, or no net force, acting on them. It has two parts:

1. An object at rest will remain at rest.
2. An object moving at a constant velocity will continue to move at a constant velocity. ☑

PART 1: OBJECTS AT REST

An object that is not moving is said to be at rest. A golf ball on a tee is an example of an object at rest. An object at rest will not move unless an unbalanced force is applied to it. The golf ball will keep sitting on the tee until it is struck by a golf club.

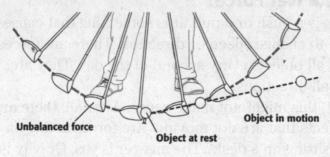

Unbalanced force **Object in motion** **Object at rest**

A golfball will remain at rest on a tee until it is acted on by the unbalanced force of a moving club.

PART 2: OBJECTS IN MOTION

The second part of Newton's first law can be hard to picture. On Earth, all objects that are moving eventually slow down and stop, even if we are no longer touching them. This is because there is always a net force acting on these objects. We will talk about this force later.

However, in outer space, Newton's first law can be seen easily. During the Apollo missions to the moon, the spacecraft turned off its engine when it was in space. It then drifted thousands of miles to the moon. It could keep moving forward without turning on its engines because there was no unbalanced force to slow it down.

Spacecraft traveling to the moon at constant velocity

READING CHECK

3. Identify What will happen to an object at rest if no unbalanced forces act on it?

_____

_____

4. Identify What will happen to an object moving at constant velocity?

_____

_____

TAKE A LOOK

5. Predict What would happen to the distances between the moving-ball images if the unbalanced force were greater?

_____

_____

TAKE A LOOK

6. Apply Concepts What must the spacecraft do to land softly on the moon?

_____

_____

How Does Friction Affect Newton's First Law?

On Earth, friction makes observing Newton's first law difficult. If there were no friction, a ball would roll forever until something got in its way. Instead, it stops quickly because of friction.

Friction is a force that is produced whenever two surfaces touch each other. Friction always works against motion. ☑

Friction makes a rolling ball slow down and stop. It also makes a car slow down when its driver lets up on the gas pedal.

What Is Inertia?

Newton's first law is often called the *law of inertia*. **Inertia** is the ability of an object to resist any change in motion. In order to change an object's motion, a force has to overcome the object's inertia. So, in order to move an object that is not moving, you have to apply a force to it. Likewise, in order to change the motion of an object that is moving, you have to apply a force to it. The greater the object's inertia, the harder it is to change its motion.

How Are Mass and Inertia Related?

An object that has a small mass has less inertia than an object with a large mass. Imagine a golf ball and a bowling ball. Which one is easier to move?

The golf ball has much less mass than the bowling ball. The golf ball also has much less inertia. This means that a golf ball will be much easier to move than a bowling ball. ☑

Inertia makes it harder to accelerate a car than to accelerate a bicycle. Inertia also makes it easier to stop a moving bicycle than a car moving at the same speed.

✔ **READING CHECK**

7. Describe How does friction affect the forward motion of an object?

_____

_____

STANDARDS CHECK

PS 2b An object that is not subjected to a force will continue to move at constant speed and in a straight line.

Word Help: constant

a quantity whose value does not change

8. Explain When a moving car stops suddenly, why does a bag of groceries on the passenger seat fly forward into the dashboard?

_____

_____

_____

_____

✔ **READING CHECK**

9. Explain Why is a golf ball easier to throw than a bowling ball?

_____

_____

What Is Newton's Second Law of Motion?

Newton's second law of motion describes how an object moves when an unbalanced force acts on it. The second law has two parts:

1. The acceleration of an object depends on the mass of the object. If two objects are pushed or pulled by the same force, the object with the smaller mass will accelerate more. ☑

READING CHECK

10. Apply Concepts Which object will accelerate more if the same force is applied to both: a pickup truck or a tractor-trailer truck?

2. The acceleration of an object depends on the force applied to the object. If two objects have the same mass, the one you push harder will accelerate more.

Acceleration   Acceleration   Acceleration

If the force applied to the carts is the same, the acceleration of the empty cart is greater than the acceleration of the loaded cart.

Acceleration increases when a larger force is exerted.

TAKE A LOOK

11. Compare On the figure, draw arrows showing the size and direction of the force that the person is applying to the cart in each picture.

How Is Newton's Second Law Written as an Equation?

Newton's second law can be written as an equation. The equation shows how acceleration, mass, and net force are related to each other:

$$a = \frac{F}{m}, \text{ or } F = m \times a$$

In the equation, a is acceleration (in meters per second squared), m is mass (in kilograms), and F is net force (in newtons, N). One newton is equal to one kilogram multiplied by one meter per second squared. ☑

Newton's second law explains why all objects fall to Earth with the same acceleration. In the figure on the top of the next page, you can see how the larger force of gravity on the watermelon is balanced by its large mass.

READING CHECK

12. Identify What are the units of force?

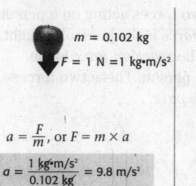

$a = \dfrac{F}{m}$, or $F = m \times a$

$a = \dfrac{1 \text{ kg·m/s}^2}{0.102 \text{ kg}} = 9.8 \text{ m/s}^2$

$m = 1.02$ kg

$F = 10$ N $= 10$ kg·m/s^2

$a = \dfrac{F}{m}$, or $F = m \times a$

$a = \dfrac{\Box \text{ kg·m/s}^2}{\Box \text{ kg}} = \Box \text{ m/s}^2$

The apple has less mass than the watermelon does. So, less force is needed to give the apple the same acceleration that the watermelon has.

Math Focus
13. Calculate In the figure, fill in the boxes with the correct numbers to calculate the acceleration of the watermelon.

How Can You Solve Problems Using Newton's Second Law?

You can use the equation $F = ma$ to calculate how much force you need to make a certain object accelerate a certain amount. Or you can use the equation $a = \dfrac{F}{m}$ to calculate how much an object will accelerate if a certain net force acts on it.

For example, what is the acceleration of a 3 kg mass if a force of 14.4 N is used to move the mass?

Step 1: Write the equation that you will use.

$$a = \dfrac{F}{m}$$

Step 2: Replace the letters in the equation with the values from the problem.

$$a = \dfrac{14.4 \text{ N}}{3 \text{ kg}} = \dfrac{14.4 \text{ kg·m/s}^2}{3 \text{ kg}} = 4.8 \text{ m/s}^2$$

Math Focus
14. Calculate What force is needed to accelerate a 1,250 kg car at a rate of 40 m/s^2? Show your work in the space below.

What Is Newton's Third Law of Motion?

All forces act in pairs. Whenever one object exerts a force on a second object, the second object exerts a force on the first object. The forces are always equal in size and opposite in direction.

For example, when you sit on a chair, the force of your weight pushes down on the chair. At the same time, the chair pushes up on you with a force equal to your weight.

ACTION AND REACTION FORCES

The figure below shows two forces acting on a person sitting in a chair. The *action force* is the person's weight pushing down on the chair. The *reaction force* is the chair pushing back up on the person. These two forces together are known as a *force pair*.

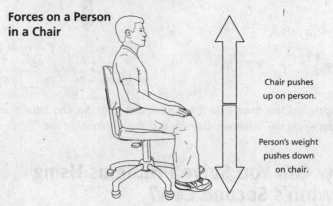

Forces on a Person in a Chair

Chair pushes up on person.

Person's weight pushes down on chair.

Action and reaction forces are also present when there is motion. The figures below show some more examples of action and reaction forces.

The space shuttle's thrusters push gases downward. The gases push the space shuttle upward with equal force.

The bat exerts a force on the ball and sends the ball flying. The ball exerts an equal force on the bat.

The action force and reaction force always act on different objects. For example, when you sit in a chair, the action force (your weight) acts on the chair. However, the reaction force (the chair pushing up on you) acts on you.

SECTION 2 Newton's Laws of Motion *continued*

HARD-TO-SEE REACTION FORCES

In the figure below, a ball is falling toward the Earth's surface. The action force is the Earth's gravity pulling down on the ball. What is the reaction force?

The force of gravity between Earth and a falling object is a force pair.

Believe it or not, the reaction force is the ball pulling up on Earth. Have you ever felt this reaction force when you have dropped a ball? Of course not. However, both forces are present. So, why don't you see or feel Earth rise?

To answer this question, recall Newton's second law. Acceleration depends on the mass and the force on an object. The force acting on Earth is the same as the force acting on the ball. However, Earth has a very, very large mass. Because it has such a large mass, its acceleration is too small to see or feel. ☑

You can easily see the ball's acceleration because its mass is small compared with Earth's mass. Most of the objects that fall toward Earth's surface are much less massive than Earth. This means that you will probably never feel the effects of the reaction force when an object falls to Earth.

TAKE A LOOK

18. Identify On the figure, draw and label arrows showing the size and direction of the action force and the reaction force for the ball falling to the Earth.

✓ **READING CHECK**

19. Explain Why can't you feel the effect of the reaction force when an object falls to Earth?

_____

_____

_____

Name ____________________ Class ______________ Date ______________

Section 2 Review

NSES PS 2b, 2c

SECTION VOCABULARY

inertia the tendency of an object to resist being moved or, if the object is moving, to resist a change in speed or direction until an outside force acts on the object

1. **Explain** How are inertia and mass related?

____

2. **Use Graphics** The hockey puck shown below is moving on the ice at a constant velocity. The arrow represents the constant velocity. In the box, draw the arrow that represents the velocity of the puck if no unbalanced forces act on the puck.

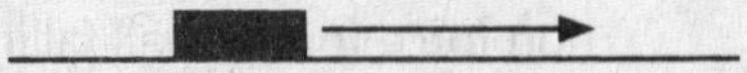

3. **Use Graphics** The ball shown below has two forces acting on it. The arrows represent the size and direction of the forces. In the box, draw the arrow that represents the net force on the ball.

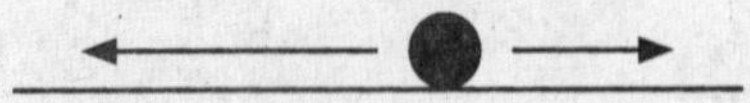

4. **Identify** Describe two things you can do to increase the acceleration of an object.

____

____

5. **Identify** Identify the action and reaction forces when you kick a soccer ball.

____

____

____

6. **Calculate** What force is needed to accelerate a 40 kg person at a rate of 4.5 m/s^2? Show your work.

____

____

____

CHAPTER 20 Forces and Motion

SECTION 3 Momentum

BEFORE YOU READ

After you read this section, you should be able to answer these questions:

• What is momentum?

• How is momentum calculated?

• What is the law of conservation of momentum?

What Is Momentum?

Picture a compact car and a large truck moving at the same velocity. The drivers of both vehicles put on the brakes at the same time. Which vehicle will stop first? You most likely know that it will be the car. But why? The answer is momentum.

The momentum of an object depends on the object's mass and velocity. **Momentum** is the product of the mass and velocity of an object. In the figure below, a car and a truck are shown moving at the same velocity. Because the truck has a larger mass, it has a larger momentum. A greater force will be needed to stop the truck. ☑

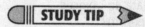

STUDY TIP

Visualize As you read, try to picture in your head the events that are described. If you have trouble imagining, draw a sketch to illustrate the event.

READING CHECK

1. Identify What is the momentum of an object?

_____

_____

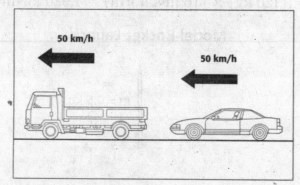

A truck and car traveling with the same velocity do not have the same momentum.

Object	Momentum
A train moving at 30 km/h	
A bird sitting on a branch high in a tree	
A truck moving at 30 km/h	
A rock sitting on a beach	

TAKE A LOOK

2. Predict How could the momentum of the car be increased?

_____

_____

Critical Thinking

3. Apply Concepts Fill in the chart to the left to show which object has the most momentum, which object has a smaller amount of momentum, and which objects have no momentum.

How Can You Calculate Momentum?

If you know what an object's mass is and how fast it is going, you can calculate its momentum. The equation for momentum is

$$p = m \times v$$

In this equation, p is momentum (in kilograms multiplied by meters per second), m is the mass of the object (in kilograms), and v is the velocity of the object (in meters per second). ☑

Like velocity, momentum has direction. The direction of an object's momentum is always the same as the direction of the object's velocity.

Use the following procedure to solve momentum problems:

Step 1: Write the momentum equation.

Step 2: Replace the letters in the equation with the values from the problem.

Let's try a problem. A 120 kg ostrich is running with a velocity of 16 m/s north. What is the momentum of the ostrich?

Step 1: The equation is $p = m \times v$.

Step 2: m is 120 kg and v is 16 m/s north. So,

$$p = (120 \text{ kg}) \times (16 \text{ m/s north}) = 1{,}920 \text{ kg•m/s north}$$

READING CHECK

4. Identify What do you need to know in order to calculate an object's momentum?

_____

_____

Math Focus

5. Calculate A 6 kg bowling ball is moving at 10 m/s down the alley toward the pins. What is the momentum of the bowling ball? Show your work.

_____

TAKE A LOOK

6. Describe Use a metric ruler to draw an arrow next to the rocket to show its momentum. The size of the rocket's momentum is 15 kg•m/s. The scale for the arrow should be 1 cm = 10 kg•m/s.

Model Rocket Launch

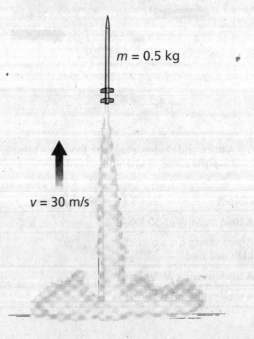

$m = 0.5$ kg

$v = 30$ m/s

What Is the Law of Conservation of Momentum?

When a moving object hits an object at rest, some or all of the momentum of the first object is transferred to the second object. This means that the object at rest gains all or some of the moving object's momentum. During a collision, the total momentum of the two objects remains the same. Total momentum doesn't change. This is called the *law of conservation of momentum.*

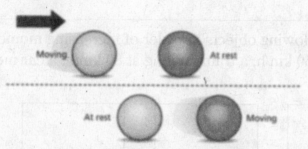

Momentum

Moving   At rest

At rest   Moving

The momentum before a collision is equal to the momentum after the collision.

TAKE A LOOK
7. Identify Draw an arrow showing the size and direction of the darker ball's momentum after its collision with the lighter ball.

The law of conservation of momentum is true for any colliding objects as long as there are no outside forces. For example, if someone holds down the darker ball in the collision shown above, it will not move. In that case, the momentum of the lighter ball would be transferred to the person holding the ball. The person is exerting an outside force. ☑

The law of conservation of momentum is true for objects that either stick together or bounce off each other during a collision. In both cases, the velocities of the objects will change so that their total momentum stays the same.

READING CHECK
8. Identify When is the law of conservation of momentum not true for two objects that are interacting?

_____
_____

When football players tackle another player, they stick together. The velocity of each player changes after the collision because of conservation of momentum.

Although the bowling ball and bowling pins bounce off each other and move in different directions after a collision, momentum is neither gained nor lost.

Say It
Explain Words In a group, discuss how the everyday use of the word *momentum* differs from its use in science.

Section 3 Review

SECTION VOCABULARY

momentum a quantity defined as the product of the mass and velocity of an object	

1. Explain A car and a train are moving at the same velocity. Do the two objects have the same momentum? Explain your answer.

_____

_____

2. Show Relationships Put the following objects in order of increasing momentum: a parked car, a train moving at 50 km/h, a train moving at 80 km/h, a car moving at 50 km/h.

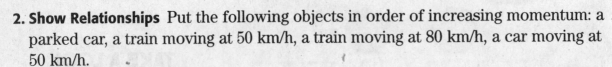

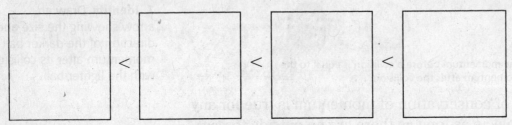

3. Calculate A 2.5 kg puppy is running with a velocity of 4.8 m/s south. What is the momentum of the puppy? Show your work.

4. Explain What is the law of conservation of momentum?

_____

_____

5. Calculate A ball has a momentum of 1 kg•m/s north. It hits another ball of equal mass that is at rest. If the first ball stops, what is the momentum of the other ball after the collision? (Assume there are no outside forces.) Explain your answer.

_____

_____

_____

SECTION 1 Fluids and Pressure

BEFORE YOU READ

After you read this section, you should be able to answer these questions:

- What are fluids?
- What is atmospheric pressure?
- What is water pressure?
- What causes fluids to flow?

What Are Fluids?

You have something in common with a dog, a sea gull, and a dolphin. You and all these other animals spend a lifetime moving through fluids. A **fluid** is any material that can flow and that takes the shape of its container. Fluids have these properties because their particles can easily move past each other. Liquids and gases are fluids. ☑

Fluids produce pressure. **Pressure** is the force exerted on a given area. The motions of the particles in a fluid are what produce pressure. For example, when you pump up a bicycle tire, you push air into the tire. Air is made up of tiny particles that are always moving. When air particles bump into the inside surface of the tire, the particles produce a force on the tire. The force exerted on the area of the tire creates air pressure inside the tire.

The air particles inside the tire hit the walls of the tire with a force. This force produces a pressure inside the tire. The pressure keeps the tire inflated.

PRESSURE AND BUBBLES

Why are bubbles round? It's because fluids (such as the gas inside the bubbles) exert the same pressure in all directions. This gives the bubbles their round shape.

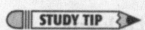
STUDY TIP

Explain As you read this section, study each figure. In your notebook, describe what each figure tells you about pressure.

READING CHECK

1. Identify What is a fluid?

_____

_____

_____

TAKE A LOOK

2. Define What is pressure?

_____

_____

SECTION 1 Fluids and Pressure *continued*

CALCULATING PRESSURE

Remember that pressure is a force exerted on an area. You can use this equation to calculate pressure:

$$pressure = \frac{force}{area}$$

The SI unit of force is the pascal. One **pascal** (Pa) is equal to a force of one newton pushing on an area of one square meter (1 N/m²). 1 Pa of pressure is very small. A stack of 120 sheets of notebook paper exerts a pressure of about 1 Pa on a table top. Therefore, scientists usually give pressure in kilopascals (kPa). 1 kPa equals 1,000 Pa.

Let's calculate a pressure. What is the pressure produced by a book that has an area of 0.2 m² and a weight of 10 N? Solve pressure problems using the following procedure:

Math Focus
3. Calculate What pressure is exerted by a crate with a weight of 3,000 N on an area of 2 m²? Show your work.

Step 1: Write the equation. $pressure = \dfrac{force}{area}$

Step 2: Substitute and solve. $= \dfrac{10 \text{ N}}{0.2 \text{ m}^2} = 50 \dfrac{\text{N}}{\text{m}^2} = 50 \text{ Pa}$

What Is Atmospheric Pressure?

The *atmosphere* is the layer of gases that surrounds Earth. Gravity holds the atmosphere in place. The pull of gravity gives air weight. The pressure caused by the weight of the atmosphere is called **atmospheric pressure**.

Atmospheric pressure is exerted on everything on Earth, including you. At sea level, the pressure is about 101,300 Pa (101.3 kPa). This means that every square centimeter of your body has about 10 N (2 lbs) of force pushing on it.

Why doesn't your body collapse under this pressure? Like the air in a balloon, the fluids inside your body exert pressure. This pressure inside your body acts against the atmospheric pressure.

TAKE A LOOK
4. Describe What would be the length of the arrows if the balloon were inflated more? Explain your answer.

_____

_____

_____

_____

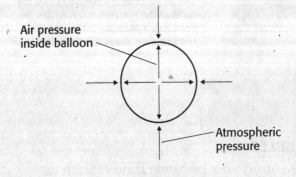

Air pressure inside balloon

Atmospheric pressure

The air inside the balloon produces a pressure inside the balloon. The pressure inside the balloon equals the atmospheric pressure outside the balloon. Therefore, the balloon stays inflated.

SECTION 1 **Fluids and Pressure** *continued*

PRESSURE, ALTITUDE, AND DEPTH

It is very difficult to climb Mount Everest. One reason is that there is not very much air at the top of Mount Everest. The atmospheric pressure on top of Mount Everest is only about one-third of that at sea level. As you climb higher, the pressure gets lower and lower. At the top of the atmosphere, the pressure is almost 0 Pa. ☑

Increasing altitude

12,000 m above sea level	Airplanes fly at about 12,000 m above sea level. Atmospheric pressure there is about 20 kPa.
9,000 m above sea level	The top of Mount Everest is about 9,000 m above sea level. Atmospheric pressure there is about 30 kPa.
4,000 m above sea level	La Paz, the capital of Bolivia, is about 4,000 m above sea level. Atmospheric pressure in La Paz is about 51 kPa.
0 m above sea level	At sea level, atmospheric pressure is about 101 kPa.

Air pressure is greatest at Earth's surface because the entire weight of the atmosphere is pushing down there. This is true for all fluids. As you get deeper in a fluid, the pressure gets higher. You can think of being at sea level as being "deep" in the atmosphere. ☑

PRESSURE CHANGES AND YOUR BODY

What happens to your body when atmospheric pressure changes? You may have felt your ears "popping" when you were in an airplane or in a car climbing a mountain. Air chambers behind your ears help to keep the pressure in your ears equal to air pressure. The "pop" happens because the pressure inside your ears changes as air pressure changes.

✔ READING CHECK

5. Describe As altitude increases, what happens to atmospheric pressure?

_____

_____

Math Focus
6. Calculate About what fraction of atmospheric pressure at sea level is atmospheric pressure at La Paz?

_____

_____

✔ READING CHECK

7. Explain Why is atmospheric pressure greatest at the surface of Earth?

_____

_____

_____

What Affects Water Pressure?

Water is a fluid. Therefore, it exerts a pressure. Like air pressure, water pressure increases as depth increases, as shown in the figure below. The pressure increases as the diver gets deeper because more and more water is pushing on her. In addition, the atmosphere pushes down on the water. Therefore, the total pressure on the diver is the sum of the water pressure and the atmospheric pressure. ☑

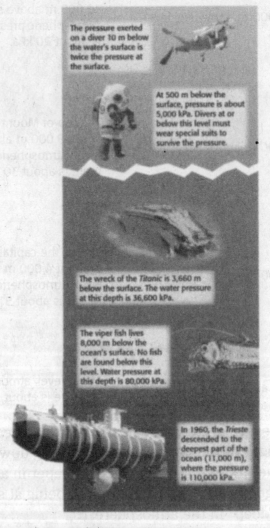

The pressure exerted on a diver 10 m below the water's surface is twice the pressure at the surface.

At 500 m below the surface, pressure is about 5,000 kPa. Divers at or below this level must wear special suits to survive the pressure.

The wreck of the *Titanic* is 3,660 m below the surface. The water pressure at this depth is 36,600 kPa.

The viper fish lives 8,000 m below the ocean's surface. No fish are found below this level. Water pressure at this depth is 80,000 kPa.

In 1960, the *Trieste* descended to the deepest part of the ocean (11,000 m), where the pressure is 110,000 kPa.

READING CHECK

8. Explain Why does pressure increase as depth increases?

_____

_____

_____

Say It

Discuss In a small group, talk about the kinds of adaptations that deep-water organisms, such as the viper fish, may have to help them survive at very high water pressures.

Critical Thinking

9. Infer What is the total pressure in kPA 10 m below the water? Hint: the total pressure is the sum of the atmospheric pressure and the water pressure.

_____

DENSITY EFFECTS ON WATER PRESSURE

Density is a measure of how closely packed the particles in a substance are. It is a ratio of the mass of an object to its volume. Water is about 1,000 times denser than air. Water has more mass (and weighs more) than the same volume of air. Therefore, water exerts more pressure than air. The pressure exerted by 10 m of water is 100 kPa. This is almost the same as the pressure exerted by the whole atmosphere.

What Causes Fluids to Flow?

All fluids flow from areas of high pressure to areas of low pressure. Imagine a straw in a glass of water. Before you suck on the straw, the air pressure inside the straw is equal to the air pressure on the water. When you suck on the straw, the air pressure inside the straw decreases. However, the pressure on the water outside the straw stays the same. The pressure difference forces water up the straw and into your mouth.

PRESSURE DIFFERENCE AND BREATHING

The flow of air from high pressure to low pressure is also what allows you to breathe. In order to inhale, a muscle in your chest moves down. This makes the volume of your chest bigger, so your lungs have more room to expand. As your lungs expand, the pressure inside them goes down. Atmospheric pressure is now higher than the pressure inside your lungs, so air flows into your lungs. The reverse of this process happens when you exhale, as shown in the figure below.

Critical Thinking

10. Apply Concepts Why does the air pressure inside a straw go down when you suck on the straw?

_____

_____

_____

_____

_____

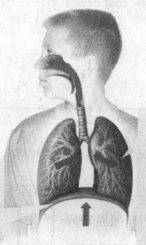

When you exhale, a muscle in your chest moves upward. The volume of your chest decreases.

As the volume of your chest decreases, the pressure in your lungs increases. The pressure in your lungs becomes greater than the pressure outside your lungs. Therefore, the air flows out of your lungs (higher pressure) into the air (lower pressure).

TAKE A LOOK
11. Explain Why does air flow out of your lungs when you exhale?

_____

_____

_____

PRESSURE DIFFERENCES AND TORNADOES

During a tornado, wind speeds can reach 300 miles per hour or more! Some of the damaging winds caused by a tornado are due to pressure differences. The air pressure inside a tornado is very low. Because the air pressure outside the tornado is high, the air rushes into the tornado and produces strong winds. The winds cause the tornado to act as a giant vacuum cleaner. Objects are pulled in and lifted up by these winds.

Section 1 Review

SECTION VOCABULARY

atmospheric pressure the pressure caused by the weight of the atmosphere **fluid** a nonsolid state of matter in which the atoms or molecules are free to move past each other, as in a gas or liquid	**pascal** the SI unit of pressure (symbol, Pa) **pressure** the amount of force exerted per unit area of a surface

1. **Describe** How do fluids exert pressure on a container?

_____

_____

2. **Evaluate** Define density in terms of mass and volume. How does density affect pressure?

_____

_____

_____

3. **Calculate** The water in a glass has a weight of 2.5 N. The bottom of the glass has an area of 0.012 m². What is the pressure exerted by the water on the bottom of the glass? Show your work.

4. **Describe** Fill in the blank spaces in the chart below to show how air moves in and out of your lungs when you breathe.

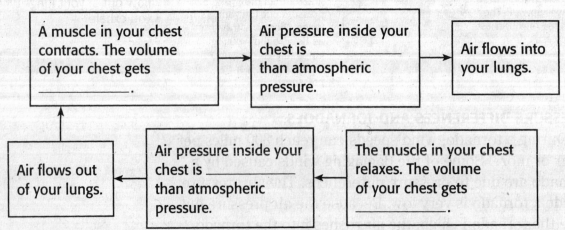

CHAPTER 21 Forces in Fluids

SECTION
2 Buoyant Force

National Science
Education Standards
PS 1a, 2c

BEFORE YOU READ

After you read this section, you should be able to answer
these questions:

- What is buoyant force?
- What makes objects sink or float?
- How can we change an object's density?

What Is Buoyant Force and Fluid Pressure?

Why does an ice cube that has been pushed under the
water pop back up? A force called buoyant force pushes
the ice cube up to the water's surface. **Buoyant force** is the
upward force that a fluid exerts on all objects in the fluid. If
an object is buoyant, that means it will float on water like a
raft. Or rise in the air like a helium-filled balloon.

Look at the figure below. Water exerts a pressure on all
sides of the object in the water. The water produces the same
amount of horizontal force on both sides of the object. These
equal forces balance one another.

However, the vertical forces are not equal. Remember that
fluid pressure increases with depth. There is more pressure
on the bottom of the object than on the top. ☑

The longer arrows in the figure below show the larger
pressures. You can see that the arrows are longest under-
neath the object. This shows that the water applies a net
upward force on the object. This upward force is buoyant
force. It is what makes the object float.

There is more pressure at the bottom of an object because
pressure increases with depth. The differences in pressure
produce an upward buoyant force on the object.

Buoyant force is what makes you feel lighter when you
float in a pool of water. The buoyant force of the water
pushes up on your body and reduces your weight.

STUDY TIP

Learn New Words As you
read, underline words you
don't understand. When you
figure out what they mean,
write the words and their
definitions in your notebook.

☑ **READING CHECK**

1. Identify Why is the force
on the bottom of an object in
a fluid larger than the force
on the top?

_____

_____

TAKE A LOOK
2. Identify What produces
buoyant force?

_____

_____

_____

_____

SECTION 2 Buoyant Force *continued*

DETERMINING BUOYANT FORCE

Archimedes, a Greek mathematician who lived in the third century BCE, discovered how to find buoyant force. Archimedes found that objects in water *displace*, or take the place of, water. The weight of the displaced water equals the buoyant force of the water. This is now known as **Archimedes' principle**.

You can find buoyant force by measuring the weight of the water that an object displaces. Suppose a block of ice displaces 250 mL of water. The weight of 250 mL of water is about 2.5 N. The weight of the displaced water equals the buoyant force. Therefore, the buoyant force on the block is 2.5 N.

Notice that only the weight of the displaced fluid determines the buoyant force on an object. The weight of the object does not affect buoyant force.

What Makes Objects Float or Sink?

An object in a fluid will sink if its weight is greater than the buoyant force. An object floats only when the buoyant force is equal to or less than the object's weight. ☑

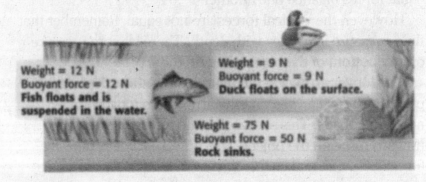

Weight = 12 N
Buoyant force = 12 N
Fish floats and is suspended in the water.

Weight = 9 N
Buoyant force = 9 N
Duck floats on the surface.

Weight = 75 N
Buoyant force = 50 N
Rock sinks.

SINKING

The rock in the figure above weighs 75 N. It displaces 5 L, or about 50 N, of water. According to Archimedes' principle, the buoyant force is about 50 N. Since the weight of the rock is greater than the buoyant force, the rock sinks.

FLOATING

The fish in the figure weighs 12 N. It displaces a volume of water that weighs 12 N. The buoyant force and the fish's weight are equal, so the fish floats in the water. It does not sink to the bottom or rise to the surface—it is *suspended* in the water.

Math Focus

3. Calculate A can of soda displaces about 360 mL of water when it is put in a tank of water. The weight of 360 mL of water is about 3.6 N. What is the buoyant force on the can of soda?

_____

_____

✓ **READING CHECK**

4. Identify Will an object sink or float if its weight is less than the buoyant force?

_____

_____

TAKE A LOOK

5. Explain Why does the fish float in the middle of the water?

_____

_____

_____

Critical Thinking

6. Apply Concepts If the duck in the figure weighed 10 N, would more or less of the duck be underwater?

_____

_____

BUOYING UP

If the duck dove underwater, it would displace more than 9 N of water. As a result, the buoyant force on the duck would be greater than the duck's weight. When the buoyant force on an object is greater than the object's weight, the object is *buoyed up*, or pushed up in the water.

An object is buoyed up until the part underwater displaces an amount of water that equals the object's weight. Therefore, the part of the duck that is underwater displaces 9 N of water.

How Does Density Affect Floating?

Remember that *density* is the mass of an object divided by its volume. How does the density of the rock compare to the density of water? The volume of the rock is 5 L, and it displaces 5 L of water. The weight of the rock is 75 N, and the weight of 5 L of water is 50 N. The weight of an object is a measure of its mass. In the same volume, the rock has more mass than water. Therefore, the rock is more dense than water. ☑

The rock sinks because it is denser than water. The duck floats because it is less dense than water. The fish floats suspended in the water because it has the same density as the water. ☑

MORE DENSE OR LESS DENSE THAN AIR

Why does an ice cube float on water but not in air? An ice cube floats in water because it is less dense than water. However, most substances are more dense than air. The ice cube is more dense than air, so it does not float in air.

One substance that is less dense than air is helium, a gas. When a balloon is filled with helium, the filled balloon becomes less dense than air. Therefore, the balloon floats in air, like the one in the picture below.

This balloon floats because the helium in it is less dense than air.

STANDARDS CHECK

PS 2c If more than one force acts on an object along a straight line, then the forces will reinforce or cancel one another, depending on their direction and magnitude. Unbalanced forces will cause changes in the speed or direction of an object's motion.

7. Explain If a log floating on the water is pushed underwater, why will it pop back up?

_____

_____

_____

✓ **READING CHECK**

8. Define What is density?

_____

_____

✓ **READING CHECK**

9. Explain How does an object's density determine whether it floats or sinks?

_____

_____

_____

TAKE A LOOK

10. Infer The balloon in the picture is filled with about 420,000 L of helium. Does 420,000 L of air have a greater or smaller mass than the 420,000 L of helium in the balloon? Explain your answer.

_____

_____

_____

What Affects an Object's Density?

The total density of an object can change if its mass or volume changes. If volume increases and mass stays the same, density decreases. If mass increases and volume stays the same, density increases.

TAKE A LOOK
11. Compare What is the volume of the ship compared to the volume of the steel used to make the ship?

_____

_____

_____

_____

A block of steel is denser than water, so it sinks. If that block is shaped into a hollow form, the overall density of the form is less than water. Therefore, the ship floats.

CHANGING SHAPE

Steel is almost eight times denser than water. Yet huge steel ships cruise the oceans with ease. If steel is more dense than water, how can these ships float? The reason a steel ship floats has to do with its shape. If the ship were just a big block of steel, it would sink very quickly. However, ships are built with a hollow shape. The hollow shape increases the volume that the steel takes up without increasing the mass of the steel.

Increasing the volume of the steel produces a decrease in its density. When the volume of the ship becomes large enough, the overall density of the ship becomes less than water. Therefore, the ship floats. ☑

Most ships are built to displace more water than is necessary for the ship to float. These ships are made this way so that they won't sink when people and cargo are loaded onto the ship.

READING CHECK
12. Explain How can changing the shape of an object lower its overall density?

_____

_____

_____

CHANGING MASS

A submarine is a ship that can travel both on the surface of the water and underwater. Submarines have *ballast tanks* that can open to let seawater flow in. When seawater flows in, the mass of the submarine increases. Therefore, its overall density increases. When seawater is pushed out, the overall density of the submarine decreases and it rises to the surface.

SECTION 2 Buoyant Force *continued*

Water flows into the ballast tanks. The submarine becomes more dense and sinks.

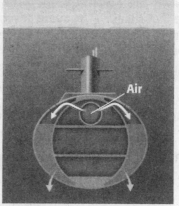

Air

Compressed air forces water out of the ballast tanks. The submarine becomes less dense and floats to the surface.

TAKE A LOOK
13. **Describe** How does a submarine increase its density?

_____

_____

CHANGING VOLUME

Some fish can change their overall density by changing their volume. Most bony fish have an organ called a *swim bladder*. This swim bladder can fill with gases or release gases. The gases are less dense than the rest of the fish. When gases go into the swim bladder, the overall volume of the fish increases, but the mass of the fish does not change as much. This lowers the overall density of the fish and keeps it from sinking in the water. ☑

The fish's nervous system controls the amount of gas in the bladder. Some fish, such as sharks, do not have a swim bladder. These fish must swim constantly to keep from sinking.

✓ **READING CHECK**

14. **Explain** How do most bony fish change their overall density?

_____

_____

_____

Swim bladder

Most bony fish have a swim bladder, an organ that allows them to adjust their overall density.

Section 2 Review

SECTION VOCABULARY

Archimedes' Principle the principle that states that the buoyant force on an object in a fluid is an upward force equal to the weight of the volume of fluid that the object displaces	**buoyant force** the upward force that keeps an object immersed in or floating on a liquid

1. Predict In Figure 1, a block of wood is floating on the surface of some water. In Figure 2, the same block of wood is pushed beneath the surface of the water. In the space below, predict what will happen to the wood when the downward force in Figure 2 is removed. Use the term *buoyant force* in your answer.

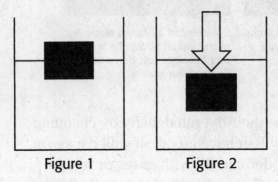

Figure 1   Figure 2

_____

_____

_____

2. Calculate A container that is filled with mercury has a mass of 4810 g. If the volume of the container is 355 mL, what is its overall density? Show your work. Round your answer to the nearest tenth.

3. Identify Give two ways that an object's overall density can change.

_____

_____

4. Explain How can knowing an object's density help you to predict whether the object will float or sink in a fluid?

_____

_____

_____

CHAPTER 21 Forces in Fluids
SECTION 3 **Fluids and Motion**

National Science
Education Standards
PS 1a

BEFORE YOU READ

After you read this section, you should be able to answer these questions:

• How does fluid speed affect pressure?

• How do lift, thrust, and wing size affect flight?

• What is drag?

• What is Pascal's principle?

What Are Fluid Speed and Pressure?

Usually, when you think of something flowing, it is a liquid such as water. But remember that a fluid is any material that can flow and that takes the shape of its container. So, gases such as air are fluids, too. Both liquids and gases flow when forces act on them.

An 18th century Swiss mathematician Daniel Bernoulli, found that fast-moving fluids have a lower pressure than slow-moving fluids. **Bernoulli's principle** states that as the speed of a moving fluid increases, the fluid's pressure decreases. ☑

Have you ever watched an airplane take off and wondered how it could stay in the air? Look at the picture of the airplane below and you'll see Bernoulli's principle at work.

Wing Design and Lift

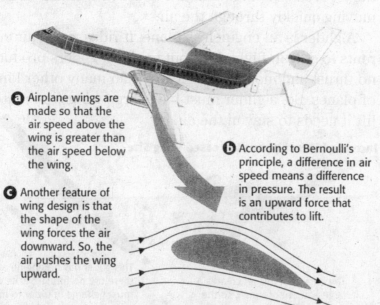

ⓐ Airplane wings are made so that the air speed above the wing is greater than the air speed below the wing.

ⓒ Another feature of wing design is that the shape of the wing forces the air downward. So, the air pushes the wing upward.

ⓑ According to Bernoulli's principle, a difference in air speed means a difference in pressure. The result is an upward force that contributes to lift.

STUDY TIP

Reading Organizer As you read this section, create an outline of the section. Use headings from the section in your outline.

READING CHECK

1. Describe What is Bernoulli's principle?

_____

_____

_____

_____

Critical Thinking

2. Infer Why do you need a windy day to fly a kite?

_____

_____

_____

_____

What Factors Affect Flight?

A common airplane in the skies today is the Boeing 737 jet. Even without passengers, the plane weighs 350,000 N (about 79,000 lbs). How can something so big and heavy get off the ground? Wing shape plays a role in helping these big planes, as well as smaller planes, fly.

According to Bernoulli's principle, the fast-moving air above the wing exerts less pressure than the slow-moving air below the wing. As a result, the greater pressure below the wing exerts an upward force. This upward force, known as **lift**, pushes the wings (and the rest of the airplane or bird) upward against the downward pull of gravity. ☑

THRUST AND LIFT

The amount of lift caused by a plane's wing is determined partly by the speed of the air around the wing. Thrust determines the speed of a plane. **Thrust** is the forward force that the plane's engine produces. Usually, a plane with a large amount of thrust moves faster than a plane that has less thrust. This faster speed means greater lift.

WING SIZE, SPEED, AND LIFT

The size of a plane's wings also affects the plane's lift. The jet plane in the picture below has small wings, but its engine gives a large amount of thrust. This thrust pushes the plane through the sky at great speeds. As a result, the jet creates a large amount of lift with small wings by moving quickly through the air.

A glider is an engineless plane. It rides rising air currents to stay in flight. Without engines, gliders produce no thrust and move more slowly than many other kinds of planes. So, a glider must have large wings to create the lift it needs to stay in the air. ☑

Increased Thrust Versus Increased Wing Size

The engine of this jet creates a large amount of thrust, so the wings don't have to be very big.

This glider has no engine and therefore no thrust. So, its wings must be large in order to maximize the amount of lift achieved.

✓ **READING CHECK**

3. Explain What is lift?

_____

_____

_____

✓ **READING CHECK**

4. Identify What are two factors that affect lift?

_____

_____

SECTION 3 Fluids and Motion *continued*

BERNOULLI, BIRDS, AND BASEBALL

Birds don't have engines, so they must flap their wings in order to supply thrust and lift. A small bird must flap its wings at a fast pace to stay in the air. But a hawk flaps its wings only occasionally. It flies with little effort because it has larger wings. Fully extended, a hawk's wings allow it to glide on wind currents and still have enough lift to stay in the air. ☑

Bernoulli's Principle and the Screwball

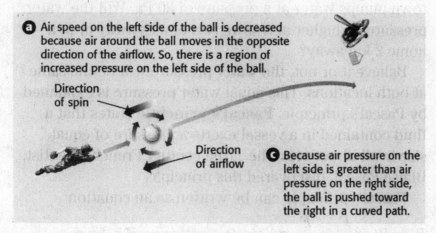

a Air speed on the left side of the ball is decreased because air around the ball moves in the opposite direction of the airflow. So, there is a region of increased pressure on the left side of the ball.

Direction of spin

Direction of airflow

c Because air pressure on the left side is greater than air pressure on the right side, the ball is pushed toward the right in a curved path.

The picture above shows how a baseball pitcher can use Bernoulli's principle to throw a confusing screwball. It is hard for a batter to hit.

DRAG AND MOTION IN FLUIDS

Have you ever walked into a strong wind and noticed that the wind seemed to slow you down or push you backward? Fluids exert a force that opposes the motion of objects moving through them. The force in a fluid that opposes or restricts motion is called **drag**. ☑

In a strong wind, air "drags" on your body and makes it difficult for you to move into the wind. Drag also works against the forward motion of a plane or bird in flight. Drag is usually caused by an irregular flow of air. An irregular or unpredictable flow of fluids is known as *turbulence*.

Copyright © by Holt, Rinehart and Winston. All rights reserved.

✔ READING CHECK

5. Identify What does a bird supply by flapping its wings?

_____

_____

TAKE A LOOK

6. Identify Suppose the pitcher threw the ball and it curved toward the batter. Would the ball be spinning in a clockwise or counterclockwise direction?

_____

_____

✔ READING CHECK

7. Decribe What is drag?

_____

_____

_____

SECTION 3 Fluids and Motion *continued*

TURBULENCE AND LIFT

Turbulence causes drag and that reduces lift. Drag can be a serious problem for airplanes moving at high speeds. As a result, airplanes have ways to reduce turbulence. For example, airplanes have flaps on their wings. When the flaps move, it changes the shape or area of a wing. This change can reduce drag and increase lift. ☑

What Is Pascal's Principle?

Imagine that the water-pumping station in your town pumps water at a pressure of 20 Pa. Will the water pressure be higher at a store two blocks away or at a home 2 km away?

Believe it or not, the water pressure will be the same at both locations. This equal water pressure is explained by Pascal's principle. **Pascal's principle** states that a fluid contained in a vessel exerts a pressure of equal size in all directions. The 17the century French scientist, Blaise Pascal, discovered this principle.

Pascal's principle can be written as an equation:

$$P_1 = P_2 \text{ or } \frac{F_1}{A_1} = \frac{F_2}{A_2},$$ where P is pressure, F is force,

and A is area. $P_1 = P_2$ means the pressure in the fluid is the same everywhere in the fluid. However, if the areas pushed on by a fluid are different, the forces will be different. In the figure below, Area 2 is larger than Area 1.

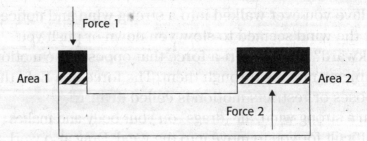

Suppose that force 1 = 10 N, area 1 = 2 cm², and area 2 = 100 cm², what is force 2? Rearrange the equation and put in the values.

$$F_2 = A_2 \times \frac{F_1}{A_1} = 100 \text{ cm}^2 \times \frac{10 \text{ N}}{2 \text{ cm}^2} = 500 \text{ N}$$

Pushing on one end of the fluid with a 10 N force caused a 500 N force on the other end. This will be used on the next page to explain how car brakes work.

Critical Thinking

9. Infer Would Pascal's principle still apply if there is a leak in the town's water system? Explain your answer.

_____

_____

_____

_____

Math Focus

10. Determine If force 1 = 20 N, area 1 = 5 cm², and area 2 = 200 cm², what is force 2? Show your work.

SECTION 3 Fluids and Motion *continued*

PASCAL'S PRINCIPLE AND MOTION

Hydraulic devices use Pascal's principle to move or lift objects. Hydraulic means the devices operate using fluids, usually oil. In hydraulic devices liquids cannot be easily compressed, or squeezed, into a smaller space. Cranes, forklifts, and bulldozers have hydraulic devices that help them lift heavy objects.

Hydraulic machines can multiply forces. Car brakes are a good example of this. In the picture below, a driver's foot exerts pressure on a cylinder of liquid. This pressure is transmitted to all parts of the liquid-filled brake system. The liquid moves the brake pads. The pads press against the wheels and friction stops the car. ☑

The force is multiplied. This is because the pistons that push the brake pads are larger than the piston pushed by the brake pedal.

READING CHECK

11. Explain How can forklifts and bulldozers lift such heavy loads?

_____

_____

_____

Because of Pascal's principle, the touch of a foot can stop tons of moving metal.

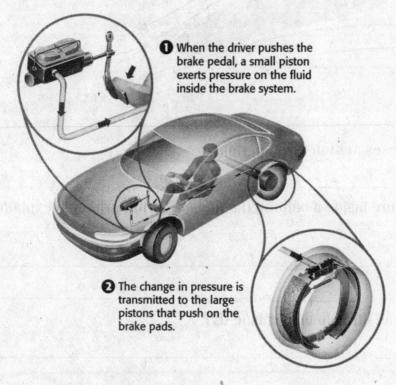

❶ When the driver pushes the brake pedal, a small piston exerts pressure on the fluid inside the brake system.

❷ The change in pressure is transmitted to the large pistons that push on the brake pads.

Section 3 Review

SECTION VOCABULARY

Bernoulli's principle the principle that states that the pressure in a fluid decreases as the fluid's velocity increases	**lift** a upward force on an object that moves in a fluid
drag a force parallel to the velocity of the flow; it opposes the direction of an aircraft and, in combination with thrust, determines the speed of the aircraft	**Pascal's principle** the principle that states that a fluid in equilibrium contained in a vessel exerts a pressure of equal intensity in all direction
	thrust the pushing or pulling force exerted by the engine of an aircraft or rocket

1. Explain What is the relationship between pressure and fluid speed?

_____

_____

2. Label Write "fastest air speed" and "highest pressure" in the correct places on the wing.

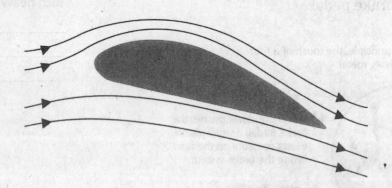

3. Identify What force opposes motion through fluid?

_____

4. Infer Where is the pressure inside a balloon the highest? What principle explains your answer?

_____

_____

5. Explain How do thrust and lift help an airplane fly?

_____

_____

_____

CHAPTER 22 Work and Machines

SECTION 1 Work and Power

National Science
Education Standards
PS 3a

BEFORE YOU READ

After you read this section, you should be able to answer these questions:

- What is work?
- How do we measure work?
- What is power and how is it calculated?

What Is Work?

You may think of work as a large homework assignment. You have to read a whole chapter by tomorrow. That sounds like a lot of work, but in science, work has a different meaning. **Work** is done when a force causes an object to move in the direction of the force. You might have to do a lot of thinking, but you are not using a force to move anything. You are doing work when you to turn the pages of the book or move your pen when writing. ☑

The student in the figure below is bowling. She is doing work. She applies a force to the bowling ball and the ball moves. When she lets go of the ball she stops doing work. The ball keeps rolling but she is not putting any more force on the ball.

STUDY TIP

As you read this section, write the questions in Before You Read in your science notebook and answer each one.

READING CHECK

1. Describe When is work done on an object?

_____
_____
_____

You might be surprised to find out that bowling is work!

The direction of the force

The direction of the bowling ball

STANDARDS CHECK

PS 3a Energy is a property of many substances and is associated with heat, light, electricity, mechanical energy, motion, sound, nuclei, and the nature of a chemical. Energy is transferred in many ways.

2. Identify How is energy transmitted to a bowling ball to make it move down the alley?

_____
_____
_____

How Is Energy Transferred When Work Is Done?

The bowler in the figure above has done work on the bowling ball. Since the ball is moving, it now has *kinetic* energy. The bowler has transferred energy to the ball.

SECTION 1 Work and Power *continued*

When Is Work Done On An Object?

Applying a force does not always mean that work is done. If you push a car but the car does not move, no work is done on the car. Pushing the car may have made you tired. If the car has not moved, no work is done on the car. When the car moves, work is done. If you apply a force to an object and it moves in that direction, then work is done.

How Are Work and Force Different?

You can apply a force to an object, but not do work on the object. Suppose you are carrying a heavy suitcase through an airport. The direction of the force you apply to hold the suitcase is up. The suitcase moves in the direction you are walking. The direction the suitcase is moving is not the same as the direction of the applied force. So when you carry the suitcase, no work is done on the suitcase. Work is done when you lift the suitcase off the ground.

Work is done on an object if two things happen.

1. An object moves when a force is applied.

2. The object moves in the direction of the force. ☑

In the figure below, you can see how a force can cause work to be done.

3. Describe What two things must happen to do work on an object?

_____
_____
_____
_____

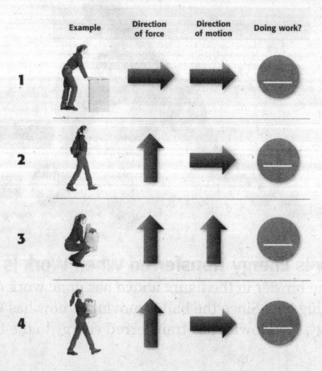

Example	Direction of force	Direction of motion	Doing work?
1			
2			
3			
4			

TAKE A LOOK
4. Identify In the figure, there are four examples. For each example, decide if work is being done and write your answers in the figure.

How Is Work Calculated?

An equation can be written to calculate the work (W) it takes to move an object. The equation shows how work, force, and distance are related to each other:

$$W = F \times d$$

In the equation, F is the force applied to an object (in newtons). d is the distance the object moves in the direction of the force (in meters). The unit of work is the newton-meter (N×m). This is also called a **joule** (J). When work is done on an object, energy is transferred to the object. The joule is a unit of energy.

Let's try a problem. How much work is done if you push a chair that weighs 60 N across a room for 5 m?

Step 1: Write the equation.

$$W = F \times d$$

Step 2: Place values into the equation, and solve for the answer.

$$W = 60 \text{ N} \times 5 \text{ m} = 300 \text{ J}$$

The work done on the chair is 300 joules.

Math Focus
5. Calculate How much work is done pushing a car 20 m with a force of 300 N? Show your work.

W = 80 N × 1 m = 80 J
The force to lift an object is the same as the force of gravity on the object. In other words, the object's weight is the force.

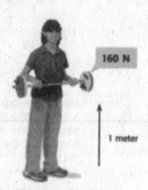

W = 160 N × 1 m = 160 J
The amount of work increases when the weight of an object increases. More force is needed to lift the object.

80 N

2 meters

W = _____

The amount of work also increases when the distance increases.

Math Focus
6. Calculate In the last figure, a barbell weighing 80 N is lifted 2 m off of the ground. How much work is done? Show your work.

Two Paths, Same Work?

A car is pushed to the top of a hill using two different paths. The first path is a long road that has a low, gradual slope. The second path is a steep cliff. Pushing the car up the long road doesn't need as much force as pulling it up the steep cliff. But, believe it or not, the same amount of work is done either way. It is clear that you would need a different amount of force for each path.

Look at the figure below. Pushing the car up the long road uses a smaller force over a larger distance. Pulling it up the steep cliff uses a larger force over a smaller distance. When work is calculated for both paths, you get the same amount of work for each path. ☑

7. Explain Suppose it takes the same amount of work for two people to move an object. One person applies less force in moving the object than the other. How can they both do the same amount of work?

_____

_____

_____

_____

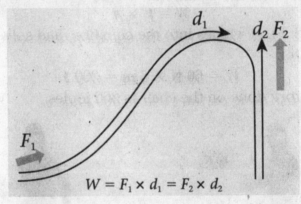

$$W = F_1 \times d_1 = F_2 \times d_2$$

There are two paths to move the car to the top of the hill. d_1 follows the shape of the hill. d_2 is straight from the bottom to the top of the hill. The same work is done in both paths. The distance and force for each of the paths are different.

Let's do a calculation. Suppose path 1 needs a force of 200 N to push the car up the hill for 30 m. Path 2 needs a force of 600 N to pull the car up the hill 10 m. Show that the work is the same for both paths.

Step 1: Write the equation.

$$W_1 = F_1 \times d_1 \text{ and } W_2 = F_2 \times d_2$$

Step 2: Place values into the equation, and solve for the answer.

$$W_1 = 200 \text{ N} \times 30 \text{ m} \times 600 \text{ N and } W_2 =$$
$$600 \text{ N} \times 10 \text{ m} = 600 \text{ J}$$

The amount of work done is the same for both paths.

What Is Power?

The word *power* has a different meaning in science than how we often use the word. **Power** is how fast energy moves from one object to another.

Power measures how fast work is done. The *power output* of something is another way to say how much work can be done quickly. For example, a more powerful weightlifter can lift a barbell more quickly than a less powerful weightlifter.

How Is Power Calculated?

To calculate power (P), divide the work (W) by the time (t) it takes to do the work. This is shown in the following equation: $P = \dfrac{W}{t}$

Power is written in the units joules per second (J/s). This is called a **watt**. One *watt* (W) is the same as 1 J/s.

Let's do a problem. A stage manager at a play raises the curtain by doing 5,976 J of work on the curtain in 12 s. What is the power output of the stage manager?

Step 1: Write the equation.

$$P = \dfrac{W}{t}$$

Step 2: Place values into the equation, and solve for the answer.

$$P = \dfrac{5,976 \text{ J}}{12 \text{ s}} = 498 \text{ W}$$

Wood can be sanded by hand or with an electric sander. The electric sander does the same amount of work faster.

Critical Thinking

8. Apply Concepts An escalator and a elevator can transport a person from one floor to the next. The escalator does it in 15 s and the elevator takes 10 s.

Which does more work on the person? Which has the greater power output?

_____

_____

_____

_____

Math Focus

9. Calculate A light bulb is on for 12 s, and during that time it uses 1,200 J of electrical energy. What is the wattage (power output) of the light bulb?

TAKE A LOOK

10. Explain Why does the electric sander have a higher power output than sanding by hand?

_____

_____

_____

Section 1 Review

SECTION VOCABULARY

joule the unit used to express energy; equivalent to the amount of work done by a force of 1 N acting through a distance of 1 m in the direction of the force (symbol, J)	**watt** the unit used to express power; equivalent to a joule per second (symbol, W)
power the rate at which work is done or energy is transformed	**work** the transfer of energy to an object by using a force that causes the object to move in the direction of the force

1. Explain Is work always done on an object when a force is applied to the object? Why or why not?

_____

_____

2. Analyze Work is done on a ball when a pitcher throws it. Is the pitcher still doing work on the ball as it flies through the air? Explain your answer.

_____

_____

3. Calculate A force of 10 N is used to push a shopping cart 10 m. How much work is done? Show how you got your answer.

4. Compare How is the term work different than the term power?

_____

_____

_____

5. Identify How can you increase your power output by changing the amount of work that you do? How can you increase your power output by changing the time it takes you to do the work?

_____

_____

6. Calculate You did 120 J of work in 3 s. How much power did you use? Show how you got your answer.

CHAPTER 22 Work and Machines

SECTION
2 **What Is a Machine?**

BEFORE YOU READ

After you read this section, you should be able to answer these questions:

- What is a machine?
- How does a machine make work easier?
- What is mechanical advantage?
- What is mechanical efficiency?

What Is a Machine?

Imagine changing a flat tire without a jack to lift the car or a tire iron to remove the bolts. Would it be easy? No, you would need several people just to lift the car! Sometimes you need the help of machines to do work. A **machine** is something that makes work easier. It does this by lowering the size or direction of the force you apply. ☑

When you hear the word machine, what kind of objects do you think of? Not all machines are hard to use. You use many simple machines every day. Think about some of these machines. The following table lists some jobs you use a machine to do.

Work	Machine you could use
Removing the snow in your driveway	
Getting you to school in the morning	
Painting a room	
Picking up the leaves from your front yard	
Drying your hair	

Two Examples of Everyday Machines

STUDY TIP
Brainstorm Think of ways that machines make your work easier and write them down in your science notebook.

READING CHECK
1. Describe How does a machine make work easier?

_____

_____

_____

TAKE A LOOK
2. Identify Complete the table by filling in the last column.

How Do Machines Make Work Easier?

You can use a simple machine, such as a screwdriver, to remove the lid from a paint can. An example of this is shown in the figure below. The screwdriver is a type of *lever*. The tip of the screwdriver is put under the lid and you push down on the screwdriver. The tip of the screwdriver lifts the lid as you push down. In other words, you do work on the screwdriver, and the screwdriver does work on the lid.

WORK IN, WORK OUT

When you use a machine, you do work and the machine does work. The work you do on a machine is called the **work input**. The force you apply to the machine to do the work is the *input force*. The work done by the machine on another object is called the **work output**. The force the machine applies to do this work is the *output force*. ☑

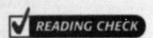
3. Identify What is work input? What is work output?

_____

_____

_____

_____

You do work on the machine and the machine does work on something else.

Output force

Input force

TAKE A LOOK

4. Identify What is the force you put on a screwdriver called?

_____

What is the force the screwdriver puts on the lid called?

_____

5. Identify To make work easier, what force is lowered?

_____

HOW MACHINES HELP

Machines do not decrease the amount of work you do. Remember that work equals the force applied times the distance ($W = F \times d$). Machines lower the force that is needed to do the same work by increasing the distance the force is applied. This means less force is needed to do the same work.

In the figure above, you apply a force to the screwdriver and the screwdriver applies a force to the lid. The force the screwdriver puts on the lid is greater than the force you apply. Since you apply this force over a greater distance, your work is easier. ☑

SECTION 2 What Is a Machine? *continued*

SAME WORK, DIFFERENT FORCE

Machines make work easier by lowering the size or direction (or both) of the input force. A machine doesn't change the amount of work done. A ramp can be used as a simple machine shown in the figure below. In this example a ramp makes work easier because the box is pushed with less force over a longer distance.

Input Force and Distance

The boy lifts the box. The input force is the same as the weight of the box.

The girl uses a ramp to lift the box. The input force is the less than the weight of the box. She applies this force for a longer distance.

Look at the boy lifting a box in the figure above. Suppose the box weighs 450 N and is lifted 1 m. How much work is done to move the box?

Step 1: Write the equation.
$$W = F \times d$$

Step 2: Place values into the equation, and solve.
$$W = 450 \text{ N} \times 1 \text{ m} = 450 \text{ N} \times \text{m, or } 450 \text{ J}$$

Look at the girl using a ramp. Suppose the force to push the box is 150 N. It is pushed 3 m. How much work is done to move the box?

Step 1: Write the equation.
$$W = F \times d$$

Step 2: Place values into the equation, and solve.
$$W = 150 \text{ N} \times 3 \text{ m} = 450 \text{ N} \times \text{m, or } 450 \text{ J}$$

Work done to move box is 450 J.

The same amount of work is done with or without the ramp. The boy uses more force and a shorter distance to lift the box. The girl uses less force and a longer distance to move the box. They each use a different force and a different distance to do the same work.

TAKE A LOOK
6. Describe Notice that the box is lifted the same distance by the boy and the girl. Which does more work on the box?

_____

_____

Math Focus
7. Calculate How much work is done when a 50 N force is applied to a 0.30 m screwdriver to lift a paint can lid?

FORCE AND DISTANCE CHANGE TOGETHER

When a machine changes the size of the output force, the distance must change. When the output force increases, the distance the object moves must decrease. This is shown in the figure of the nutcracker below. The handle is squeezed with a smaller force than the output force that breaks the nut. So, the output force is applied over a smaller distance.

Machines Change the Size and/or Direction of a Force

A nutcracker increases the force but applies it over a shorter distance.

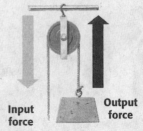

A simple pulley changes the direction of the input force, but the size of the output force is the same as the input force.

What Is Mechanical Advantage?

Some machines can increase the size of the force more than others. A machine's **mechanical advantage** tells you how much the force increases. The mechanical advantage compares the input force with the output force.

CALCULATING MECHANICAL ADVANTAGE

A machine's mechanical advantage can be calculated by using the following equation:

$$mechanical\ advantage\ (MA) = \frac{output\ force}{input\ force}$$

Look at this example. You push a box weighing 500 N up a ramp (output force) by applying 50 N of force. What is the mechanical advantage of the ramp?

Step 1: Write the equation.

$$mechanical\ advantage\ (MA) = \frac{output\ force}{input\ force}$$

Step 2: Place values into the equation, and solve.

$$MA = \frac{500\ N}{50\ N} = 10$$

The mechanical advantage of the ramp is 10.

Math Focus
8. Explain Why is the input arrow shorter than the output arrow in the photo of the nutcracker? Why are the arrows the same length in the figure of the pulley?

_____

_____

_____

_____

Math Focus
9. Calculate What is the mechanical advantage of a nutcracker if the input force is 65 N and the output force is 130 N?

What Is a Machine's Mechanical Efficiency?

No machine changes all of the input work into output work. Some of the work done by the machine is lost to *friction*. Friction is always present when two objects touch. The work done by the machine plus the work lost to friction is equal to the work input. This is known as the *Law of Conservation of Energy*.

The **mechanical efficiency** of a machine compares a machine's work output with the work input. A machine is said to be efficient if it doesn't lose much work to friction.

CALCULATING MECHANICAL EFFICIENCY

A machine's mechanical efficiency is calculated using the following equation:

$$\text{mechanical advantage (MA)} = \frac{output\ force}{input\ force} \times 100$$

The 100 in the equation means that mechanical efficiency is written as a percentage. It tells you the percentage of work input that gets done as work output.

Let's try a problem. You do 100 J of work on a machine and the work output is 40 J. What is the mechanical efficiency of the machine?

Step 1: Write the equation.

$$\text{mechanical effeciency (ME)} = \frac{work\ output}{work\ input} \times 100$$

Step 2: Place values into the equation, and solve.

$$ME = \frac{40\ J}{100\ J} \times 100 = 40\%$$

Process Chart

You apply an input force to a machine.

↓

The machine changes the size and/or direction of the force.

↓

The machine applies an _____ _____ on the object.

↓

The mechanical efficiency and/or advantage of the machine can be determined.

Math Focus

10. Calculate What is the mechanical efficiency of a simple pulley if the input work is 100 N and the output work is 90 N?

Math Focus

11. Identify Fill in the missing words on the process chart.

_____

Section 2 Review

SECTION VOCABULARY

machine a device that helps do work by either overcoming a force or changing the direction of the applied force	**work input** the work done on a machine; the product of the input force and the distance through which the force is exerted
mechanical advantage a number that tells how many times a machine multiplies force	**work output** the work done by a machine; the product of the output force and the distance through which the force is exerted
mechanical efficiency a quantity, usually expressed as a percentage, that measures the ratio of work output to work input in a machine	

1. Explain Why is it easier to move a heavy box up a ramp than it is to lift the box off the ground?

_____

_____

2. Identify What are the two ways that a machine can make work easier?

_____

3. Compare What is the difference between work input and work output?

_____

_____

4. Calculate You apply an input force of 20 N to a hammer that applies an output force of 120 N to a nail. What is the mechanical advantage of the hammer? Show your work.

5. Explain Why is a machine's work output always less than the work input?

_____

6. Calculate What is the mechanical efficiency of a machine with a work input of 75 J and a work output of 25 J? Show your work.

CHAPTER 22 Work and Machines

SECTION 3 Types of Machines

After you read this section, you should be able to answer these questions:

- What are the six simple machines?
- What is a compound machine?

What Are the Six Types of Simple Machines?

All machines are made from one or more of the six simple machines. They are the lever, the pulley, the wheel and axle, the inclined plane, the wedge, and the screw. They each work differently to change the size or direction of the input force.

What Is a Lever?

A commonly used simple machine is the **lever**. A *lever* has a bar that rotates at a fixed point, called a *fulcrum*. The force that is applied to the lever is the *input force*. The object that is being lifted by the lever is called the *load*. A lever is used to apply a force to move a load. There are three classes of levers. They all have a different location for the fulcrum, the load, and the input force on the bar. ☑

FIRST-CLASS LEVERS

In first-class levers, the fulcrum is between the input force and the load as shown in the figure below. The direction of the input force always changes in this type of lever. They can also be used to increase either the force or the distance of the work.

Examples of First-Class Levers

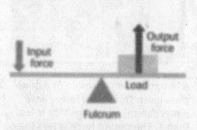

The fulcrum can be located closer to the load than to the input force. This lever has a mechanical advantage that is greater than 1. The output force is larger than the input force.

The fulcrum can be located exactly in the middle. This lever has a mechanical advantage that is equal to 1. The output force is the same as the input force.

STUDY TIP

As you read through the section, study the figures of the types of machines. Make a list of the six simple machines and a sentence describing how each works.

READING CHECK

1. Describe How does a lever do work?

_____

_____

_____

_____

Critical Thinking

2. Predict Suppose the fulcrum in the figure to the far left is located closer to the input force. How will this change the mechanical advantage of the lever? Explain.

_____

_____

_____

_____

_____

SECOND-CLASS LEVERS

In second-class levers, the load is between the fulcrum and the input force as shown in the figure below. They do not change the direction of the input force. Second-class levers are often used to increase the force of the work. You apply less force to the lever than the force it puts on the load. This happens because the force is applied over a larger distance.

Critical Thinking

3. Explain How does a second-class lever differ from a first-class lever?

_____
_____
_____
_____

Examples of Second-Class Levers

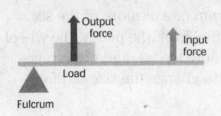

In a second-class lever, the output force, or load, is between the input force and the fulcrum.

A wheelbarrow is an example of a second-class lever. Second-class levers have a mechanical advantage that is greater than 1.

THIRD-CLASS LEVERS

In third-class levers, the input force is between the fulcrum and the load as shown in the figure below. The direction of the input force does not change and the input force does not increase. This means the output force is always less than the input force. Third-class levers do increase the distance that the output force works.

Critical Thinking

4. Explain Why can't a third-class lever have a mechanical advantage of 1 or more?

_____
_____
_____

Examples of Third-Class Levers

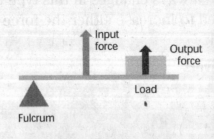

In a third-class lever, the input force is between the fulcrum and the load.

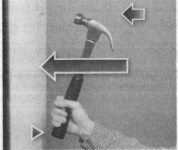

A hammer is an example of a third-class lever. Third-class levers have a mechanical advantage that is less than 1. The output force is less than the input force. Third-class levers increase the distance that the output force acts on.

What Is a Pulley?

When you open window blinds by pulling on a cord, you are using a pulley. A **pulley** is a simple machine with a grooved wheel that holds a rope or a cable. An input force is applied to one end of the cable. The object being lifted is called the load. The load is attached to the other end. The different types of pulleys are shown in the figure at the bottom of the page. ☑

FIXED PULLEYS

A *fixed pulley* is connected to something that does not move, such as a ceiling. To use a fixed pulley, you pull down on the rope to lift the load. The direction of the force changes. Since the size of the output force is the same as the input force, the mechanical advantage (*MA*) is 1. An elevator is an example of a fixed pulley. ☑

MOVABLE PULLEYS

Moveable pulleys are connected directly to the object that is being moved, which is the load. The direction does not change, but the size of the force does. The mechanical advantage (*MA*) of a movable pulley is 2. This means that less force is needed to move a heavier load. Large construction cranes often use movable pulleys.

BLOCK AND TACKLES

If you use a fixed pulley and a movable pulley together, you form a pulley system. This is a *block and tackle*. The mechanical advantage (*MA*) of a block and tackle is equal to the number of sections of rope in the system.

✓ READING CHECK

5. Describe What is a pulley?

_____

_____

_____

✓ READING CHECK

6. Describe Why can't a fixed pulley have a mechanical advantage greater than 1?

_____

_____

_____

Types of Pulleys

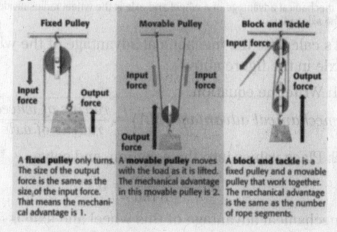

Fixed Pulley	Movable Pulley	Block and Tackle

A **fixed pulley** only turns. The size of the output force is the same as the size of the input force. That means the mechanical advantage is 1.

A **movable pulley** moves with the load as it is lifted. The mechanical advantage in this movable pulley is 2.

A **block and tackle** is a fixed pulley and a movable pulley that work together. The mechanical advantage is the same as the number of rope segments.

TAKE A LOOK
7. Identify The section of rope labeled Input force for the block and tackle is not counted as a rope segment. There are four rope segments in this block and tackle. What is the mechanical advantage of the block and tackle?

_____

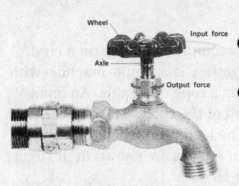

a When a small input force is applied to the wheel, it turns in a circular distance.

b When the wheel turns, so does the axle. The axle is smaller than the wheel. Since the axle turns a smaller distance, the output force is larger than the input force.

Critical Thinking

8. Explain If the input force remains constant and the wheel is made smaller, what happens to the output force?

_____

_____

What Is a Wheel and Axle?

Did you know that a faucet is a machine? The faucet in the figure above is an example of a **wheel and axle**. It is a simple machine that is made up of two round objects that move together. The larger object is the *wheel* and the smaller object is the *axle*. Some examples of a wheel and axle are doorknobs, wrenches, and steering wheels.

MECHANICAL ADVANTAGE OF A WHEEL AND AXLE

The mechanical advantage (*MA*) of a wheel and axle can be calculated. To do this you need to know the *radius* of both the wheel and the axle. Remember, the radius is the distance from the center to the edge of the round object. The equation to find the mechanical advantage (*MA*) of a wheel and axle is:

$$mechanical\ advantage\ (MA) = \frac{radius\ of\ wheel}{radius\ of\ axle}$$

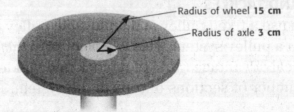

Radius of wheel **15 cm**

Radius of axle **3 cm**

The mechanical advantage of a wheel and axle is the wheel radius divided by the axle radius.

Math Focus

9. Calculate A car has a wheel and axle. If the radius of the axle is 7.5 cm and the radius of the wheel is 75 cm, what is the mechanical advantage? Show your work.

Let's calculate the mechanical advantage of the wheel and axle in the figure above.

Step 1: Write the equation.

$$mechanical\ advantage\ (MA) = \frac{radius\ of\ wheel}{radius\ of\ axle}$$

Step 2: Place values into the equation, and solve.

$$MA = \frac{15\ cm}{3\ cm} = 5$$

The mechanical advantage of this wheel and axle is 5.

What Is an Inclined Plane?

The Egyptians built the Great Pyramid thousands of years ago using the **inclined plane**. An *inclined plane* is a simple machine that is a flat, slanted surface. A ramp is an example of an inclined plane.

Using an inclined plane to move a heavy object into a truck is easier than lifting the object. The input force is smaller than the object's weight. The same work is done, but it happens over a longer distance. ☑

You do work to push a piano up a ramp. This is the same amount of work you would do to lift it straight up. An inclined plane lets you apply a smaller force over a greater distance.

READING CHECK

10. Explain How does an incline plane make lifting an object easier?

_____

_____

_____

MECHANICAL ADVANTAGE OF INCLINED PLANES

The mechanical advantage (*MA*) of an inclined plane can also be calculated. The length of the inclined plane and the height the object that is lifted must be known. The equation to find the mechanical advantage (*MA*) of an inclined plane is:

$$\text{mechanical advantage } (MA) = \frac{\text{length of inclined plane}}{\text{height load raised}}$$

We can calculate the mechanical advantage (*MA*) of the inclined plane shown in the figure above.

Step 1: Write the equation.

$$\text{mechanical advantage } (MA) = \frac{\text{length of inclined plane}}{\text{height load raised}}$$

Step 2: Place values into the equation, and solve for the answer.

$$MA = \frac{3 \text{ m}}{0.6 \text{m}} = 5$$

The mechanical advantage (*MA*) is 5.

If the length of the inclined plane is much greater than the height, the mechanical advantage is large. That means an inclined plane with a gradual slope needs less force to move objects than a steep-sloped one.

Math Focus

11. Determine An inclined plane is 10 m and lifts a piano 2.5 m. What is the mechanical advantage of the inclined plane? Show your work.

What Is a Wedge?

A knife is often used to cut because it is a **wedge**. A *wedge* is made of two inclined planes that move. Like an inclined plane, a wedge needs a small input force over a large distance. The output force of the wedge is much greater than the input force. Some useful wedges are doorstops, plows, ax heads, and chisels.

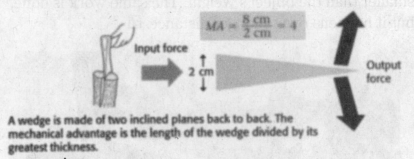

$$MA = \frac{8 \text{ cm}}{2 \text{ cm}} = 4$$

Input force

2 cm

Output force

A wedge is made of two inclined planes back to back. The mechanical advantage is the length of the wedge divided by its greatest thickness.

MECHANICAL ADVANTAGE OF WEDGES

The mechanical advantage of a wedge can be found by dividing the length of the wedge by its greatest thickness. The equation to find a wedge's mechanical advantage is:

$$mechanical\ advantage\ (MA) = \frac{length\ of\ wedge}{largest\ thickness\ of\ wedge}$$

A wedge has a greater mechanical advantage if it is long and thin. When you sharpen a knife you are making the wedge thinner. This needs a smaller input force.

What Is a Screw?

A **screw** is an inclined plane that is wrapped around a cylinder. To turn a screw, a small force over a long distance is needed. The screw applies a large output force over a short distance. Screws are often used as fasteners.

Threads

If you could unwind a screw, you would have a very long inclined plane.

Say It

MECHANICAL ADVANTAGE OF SCREWS

To find the mechanical advantage of a screw you need to first unwind the inclined plane. Then, if you compare the length of the inclined plane with its height you can calculate the mechanical advantage. This is the same as calculating the mechanical advantage of an inclined plane. The longer the spiral on a screw and the closer the threads, the greater the screw's mechanical advantage.

SECTION 3 Types of Machines *continued*

What Is a Compound Machine?

There are machines all around you. Many machines do not look like the six simple machines that you have read about. That is because most of the machines in the world are **compound machines**. These are machines that are made of two or more simple machines. A block and tackle is one example of a compound machine that you have already seen. It is made of two or more pulleys. ☑

A common example of a compound machine is a can opener. A can opener may look simple, but it is made of three simple machines. They are the second-class lever, the wheel and axle, and the wedge. When you squeeze the handle, you are using a second-class lever. The blade is a wedge that cuts the can. When you turn the knob to open the can, you are using a wheel and axle.

READING CHECK

13. Describe What is a compound machine?

_____

_____

_____

A can opener is a compound machine. The handle is a second-class lever, the knob is a wheel and axle, and a wedge is used to open the can.

TAKE A LOOK

14. Describe Describe the process of using a can opener. Tell the order in which each simple machine is used and what it does to open the can.

_____

_____

_____

_____

_____

_____

MECHANICAL EFFICIENCY OF COMPOUND MACHINES

The *mechanical efficiency* of most compound machines is low. Remember that mechanical efficiency tells you what percentage of work input gets done as work output. This is different than the mechanical advantage. The efficiency of compound machines is low because they usually have many moving parts. This means that there are more parts that contact each other and more friction. Recall that friction lowers output work. ☑

Cars and airplanes are compound machines that are made of many simple machines. It is important to lower the amount of friction in these compound machines. Friction can often damage machines. Grease is usually added to cars because it lowers the friction between the moving parts.

READING CHECK

15. Identify Why do most compound machines have low mechanical efficiency?

_____

_____

_____

_____

Section 3 Review

SECTION VOCABULARY

compound machine a machine made of more than one simple machine	**screw** a simple machine that consists of an inclined plane wrapped around a cylinder
inclined plane a simple machine that is a straight, slanted surface, which facilitates the raising of loads; a ramp	**wedge** a simple machine that is made up of two inclined planes and that moves; often used for cutting
lever a simple machine that consists of a bar that pivots at a fixed point called a fulcrum	**wheel and axle** a simple machine consisting of two circular objects of different sizes; the wheel is the larger of the two circular objects
pulley a simple machine that consists of a wheel over which a rope, chain, or wire passes	

1. Compare Use a Venn Diagram to compare a first-class lever and a second-class lever.

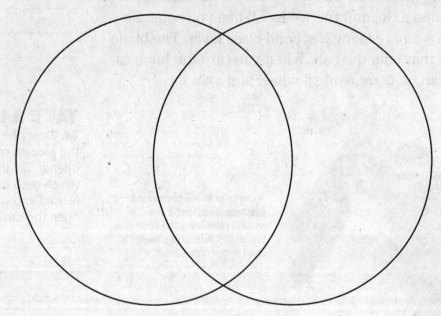

2. Calculate A screwdriver is used to put a screw into a piece of wood. The radius of the handle is 1.8 cm and the radius of shaft is 0.6 cm. What is the mechanical advantage of using the screwdriver? Show your work.

3. Compare What is the difference between a wedge and a screw?

_____

_____

4. Analyze When there is a lot of friction in a machine, what is lowered and causes mechanical efficiency to be lowered?

_____

_____

CHAPTER 23 Energy and Energy Resources

SECTION 1

What Is Energy?

National Science
Education Standards
PS 3a, 3d, 3e, 3f

BEFORE YOU READ

After you read this section, you should be able to answer
these questions:

• How are energy and work related?

• How is kinetic energy different from potential energy?

• What are some of the other forms of energy?

What Is Energy?

A tennis player needs energy to hit a ball with her
racket. The ball has energy as it flies through the air.
Energy is all around you, but what is energy?

In science, **energy** is the ability to do work. *Work* is
done when a force makes an object move in the direction
of the applied force. How do energy and work help you
play tennis? The tennis player does work on her racket
by applying a force to it. The racket does work on the
ball to make it fly into the air.

When the racket does work on the ball, energy moves
from the racket to the ball. Energy is the reason the racket
can do work. So, work is the transfer of energy. Both work
and energy are written in the units joules (J).

STUDY TIP

Make a Venn Diagram to
compare and contrast kinetic
energy and potential energy.
Make a list of other forms
of energy and tell if they are
kinetic energy or potential
energy.

✔ READING CHECK

1. Identify When work is done
by one object on another, what
is transferred?

_____

_____

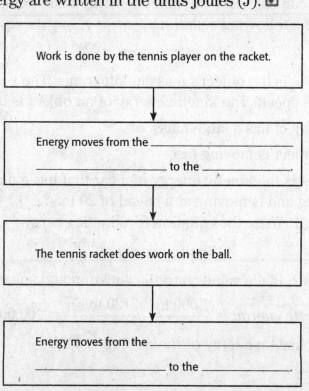

TAKE A LOOK

2. Identify Fill in the
process chart to show how
energy moves when work is
done on an object.

What Is Kinetic Energy?

When the tennis player hits the ball with the racket, energy moves from the racket to the ball. The tennis ball has kinetic energy. **Kinetic energy** is the energy of motion. All moving objects have kinetic energy. Like all other forms of energy, kinetic energy can be used to do work. The kinetic energy of a hammer does work on a nail. This is seen in the figure below. ☑

3. Identify For an object to have kinetic energy, what must it be doing?

_____
_____

The hammer has kinetic energy.

When you swing a hammer, it is moving. It has kinetic energy. This energy does work on the nail driving it into the wood.

CALCULATING KINETIC ENERGY

You can calculate the kinetic energy of an object by using the following equation.

$$kinetic\ energy = \frac{mv^2}{2}$$

The m is the object's mass in kilograms. The v is the object's speed. The kinetic energy of an object is large if:

• the object has a large mass, or

• the object is moving fast

What is the kinetic energy of a car that has a mass of 1,000 kg and is moving at a speed of 20 m/s?

Step 1: Write the equation for kinetic energy.

$$kinetic\ energy = \frac{mv^2}{2}$$

Step 2: Place values into the equation and solve.

$$kinetic\ energy = \frac{1,000\ kg \times (20\ m/s)^2}{2} = 200,000\ J$$

The kinetic energy of the car is 200,000 J.

Math Focus

4. Calculate What is the kinetic energy of a 0.50 kg hammer that hits the floor at a speed of 10 m/s?

SECTION 1 What Is Energy? *continued*

What Is Potential Energy?

An object does not have to be moving to have energy. **Potential energy** is the energy an object has because of its position. This kind of energy is harder to see because we do not see the energy at work. In the figure below, when the bow is pulled back, it has potential energy. Work has been done on it, and that work has been turned into potential energy. ☑

The bow and the string have energy that is stored as potential energy. When the man lets go of the string, the potential energy does work on the arrow.

GRAVITATIONAL POTENTIAL ENERGY

When you lift an object, you do work on it. You move it in an opposite direction from the force of gravity. As you lift the object, you transfer energy to the object and give it gravitational potential energy. The amount of *gravitational potential energy* of an object depends on the object's weight and its distance from the ground.

CALCULATING GRAVITATIONAL POTENTIAL ENERGY

The gravitational potential energy of an object can be determined by using the following equation:

gravitational potential energy = weight × height

The weight is in newtons (N) and the height is in meters (m). Gravitational potential energy is written in newton × meters (N × m). This is the same as a joule (J).

Let's do a calculation. What is the gravitational potential energy of a book with a weight of 13 N at a height of 1.5 m off the ground?

Step 1: *gravitational potential energy = weight × height*

Step 2: *gravitational potential energy* = 13 N × 1.5 m = 19.5 J

The book now has 19.5 J of potential energy.

Interactive Textbook

429

Energy and Energy Resources

✓ **READING CHECK**

5. Identify What causes an object to have potential energy?

_____

_____

Critical Thinking

6. Infer What would the man need to do to give the arrow more potential energy?

_____

_____

Math Focus

7. Calculate What is the potential energy of a 300 N rock climber standing 100 m from the base of a rock wall?

What Is Mechanical Energy?

Look at the figure below. All the energy in the juggler's pins is in the form of mechanical energy. **Mechanical energy** is the total energy of motion and position of an object. In other words, it is the kinetic energy plus the potential energy of an object. ☑

READING CHECK

8. Identify Adding what two energies gives the mechanical energy of an object?

_____

_____

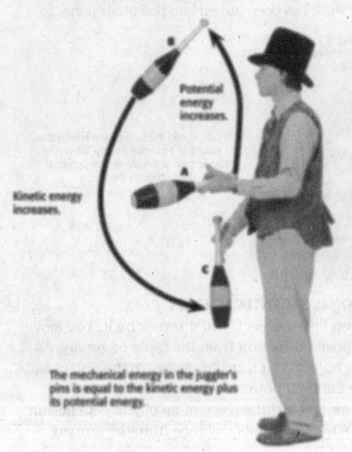

Potential energy increases.

Kinetic energy increases.

The mechanical energy in the juggler's pins is equal to the kinetic energy plus its potential energy.

TAKE A LOOK

9. Identify In the figure, circle the pin with the most potential energy.

 **Say It**

Discuss Suppose a batter hits a pop-up in baseball that goes straight up. With a partner, discuss the changes in kinetic and potential energy as the ball leaves the bat and rises to its highest height.

MECHANICAL ENERGY IN A JUGGLER'S PIN

The mechanical energy of an object doesn't change unless energy is transferred to or from another object.

Look again at the figure of the juggler. The juggler moves the pin by doing work on it. He gives the pin kinetic energy. When he lets go of the pin, the pin's kinetic energy changes into potential energy. As the pin goes up, it slows down. When all of the pin's kinetic energy is turned into potential energy, it stops going up.

When the pin starts to fall, its energy is mostly potential energy. As it falls, the potential energy is changed back into kinetic energy. At different times, the pin may have more kinetic energy or more potential energy. The total mechanical energy at any point is always the same.

What Are the Other Forms of Energy?

Energy can be in a form other than mechanical energy. The other energy forms are thermal, chemical, electrical, sound, light, and nuclear energy. All of these energy forms are connected in some way to kinetic energy and potential energy.

THERMAL ENERGY

Matter is made of particles that are moving. These particles have kinetic energy. *Thermal energy* is all the kinetic energy from the movement of the particles in an object. ☑

The figure below shows the thermal energy of particles at different temperatures. Particles move faster at higher temperatures than at lower temperatures. The faster the particles move, the greater their kinetic energy and thermal energy are.

☑ **READING CHECK**

10. Describe What is thermal energy?

_____

_____

_____

The Thermal Energy in Water

The particles in an ice cube vibrate in fixed positions and do not have a lot of kinetic energy.

The particles in water in a lake can move more freely and have more kinetic energy than water particles in ice do.

The particles in water in steam move rapidly, so they have more energy than particles in liquid water do.

CHEMICAL ENERGY

Chemical compounds such as sugar, salt, and water store energy. These compounds are made of many atoms that are held together by chemical bonds. Work is done to join the atoms together to form these bonds. *Chemical energy* is the energy stored in the chemical bonds that hold the compounds together. Chemical energy is a type of potential energy because it depends on the position of the atoms in the compound. ☑

☑ **READING CHECK**

11. Identify Chemical energy is a form of what type of energy?

_____

SECTION 1 What Is Energy? *continued*

ELECTRICAL ENERGY

You use electrical energy every day. Electrical outlets in your home allow you to use this energy. *Electrical energy* is the energy of moving particles called electrons. Electrons are the negatively charged particles of atoms.

What happens when you plug an electrical device, such as an amplifier shown in the figure below, into an outlet? You use electrical energy. The electrons in the wires move to the amplifier. The moving electrons do work on the speaker in the amplifer. This makes the sound that you hear from the amplifier.

Electrical energy has both kinetic energy and potential energy. When electrical energy runs through a wire, it uses its kinetic energy. Electrical energy that is waiting to be used is potential energy. This potential energy is in the wire before you plug in an electrical appliance.

STANDARDS CHECK

PS 3d Electrical circuits provide a means of transferring electrical energy when heat, light, sound, and chemical changes are produced.

12. Describe How does an amplifier make sound?

_____

_____

_____

_____

TAKE A LOOK
13. Identify What is the energy source for the amplifier? What kind of energy is transmitted by the guitar?

_____

_____

_____

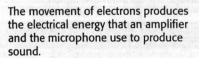

The movement of electrons produces the electrical energy that an amplifier and the microphone use to produce sound.

As the guitar strings vibrate, they cause particles in the air to vibrate. These vibrations transmit sound energy.

SOUND ENERGY

Sound energy is the energy from a vibrating object. *Vibrations* are small movements of particles of an object. In the figure above, the guitar player pulls on the guitar string. This gives the string potential energy. When she lets go of the string, the potential energy turns into kinetic energy. This makes the string vibrate.

When the guitar string vibrates, some of its kinetic energy moves to nearby air particles. These vibrating air particles cause sound energy to travel. When the sound energy reaches your ear, you hear the sound of the guitar.

SECTION 1 What Is Energy? *continued*

LIGHT ENERGY

Light helps you see, but not all light can be seen. We use light in microwaves, but we do not see it. *Light energy* is made from vibrations of electrically charged particles. Light energy is like sound energy. They both happen because particles vibrate. However, light energy doesn't need particles to travel. This makes it different than sound energy. Light energy can move through a *vacuum*, which is an area where there is no matter.

Microwave Oven

The energy used to cook food in a microwave is a form of

_____ _____.

NUCLEAR ENERGY

Another kind of energy is stored in the nucleus of an atom. This energy is *nuclear energy*. This energy is stored as potential energy.

There are two ways nuclear energy can be given off by a nucleus. When two or more small nuclei join together, they give off energy in a reaction called *fusion*. The sun's light and heat come from fusion reactions.

The second way nuclear energy is given off is when a nucleus splits apart. This process is known as *fission*. Large nuclei, like uranium, can be broken apart with fission. Fission is used to create electrical energy at nuclear power plants. ☑

Our Sun

Without _____

_____ that gives the sun its energy, life on Earth would not be possible.

TAKE A LOOK
14. Identify Complete the sentence found in the figure.

✓ **READING CHECK**
15. Compare How do nuclear fusion and fission differ?

_____

_____

_____

TAKE A LOOK
16. Identify Complete the sentence found in the figure.

Section 1 Review

SECTION VOCABULARY

energy the capacity to do work **kinetic energy** the energy of an object that is due to the object's motion	**mechanical energy** the amount of work an object can do because of the object's kinetic and potential energies **potential energy** the energy that an object has because of the position, shape, or condition of the object

1. Identify An object's kinetic energy depends on two things. What are they?

_____

2. Calculate A book weighs 16 N and is placed on a shelf that is 2.5 m from the ground. What is the gravitational potential energy of the book? Show your work.

3. Calculate What is the kinetic energy of a 2,000 kg bus that is moving at 25 m/s? Show your work.

4. Explain A girl is jumping on a trampoline. When she is at the top of her jump, her mechanical energy is in what form? Explain why.

_____

_____

5. Identify How are sound energy and light energy similar?

_____

_____

6. Conclude Which type of nuclear reaction is more important for life on Earth? Explain why.

_____

_____

CHAPTER 23 Energy and Energy Resources

SECTION
2   **Energy Conversions**

National Science
Education Standards
PS 3a, 3f

BEFORE YOU READ

After you read this section, you should be able to answer
these questions:

• What are energy conversions?

• How do machines use energy conversions?

What Is an Energy Conversion?

Think of a book sitting on a shelf. The book has gravitational potential energy when it is on the shelf. What happens
if the book falls off the shelf? Its potential energy changes into
kinetic energy. This is an example of an energy conversion.

An **energy conversion** is a change from one form of
energy to another. Any form of energy can change into
any other form of energy. One form of energy can sometimes change into more than one other energy form.

KINETIC AND POTENTIAL ENERGY CONVERSION

A common energy conversion happens between potential energy and kinetic energy. When the skateboarder
in the figure below moves down the half-pipe, he has a
lot of kinetic energy. As he travels up the half-pipe, his
kinetic energy changes into potential energy.

Potential Energy and Kinetic Energy

At the top of his jump, the skateboarder looks like he has almost
stopped moving. His

_____

is at a maximum.

At the bottom of the half-pipe, the skateboarder has a lot of
speed. His _____ _____ is at a maximum.

STUDY TIP

Make a list in your science
notebook of the energy
conversions discussed in this
section.

TAKE A LOOK
1. Identify Complete the
two blanks in the figure.

ELASTIC POTENTIAL ENERGY

Another example of potential energy changing into kinetic energy is shown in a rubber band. When you stretch a rubber band, you give it potential energy. This energy is called *elastic potential energy*. When you let go of the rubber band, it flies across the room. The stored potential energy in the stretched rubber band is turned into kinetic energy. ☑

How Do Chemical Energy Conversions Happen?

You may have heard that breakfast is the most important meal of the day. Why is eating breakfast so important? *Chemical energy* comes from the food you eat. Your body changes the chemical energy from food into several different energy forms that it can use. Some of these are:

• mechanical energy, to move your muscles

• thermal energy, to keep your body temperature constant

• electrical energy, to help your brain think

So eating breakfast helps your body do all of your daily activities.

TAKE A LOOK
3. Identify Fill in the process chart to show how energy is converted in your body.

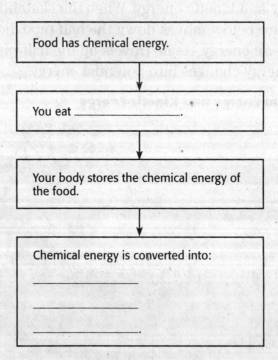

Food has chemical energy.

You eat _____.

Your body stores the chemical energy of the food.

Chemical energy is converted into:

_____

_____

_____

SECTION 2 Energy Conversions *continued*

From Light Energy to Chemical Energy

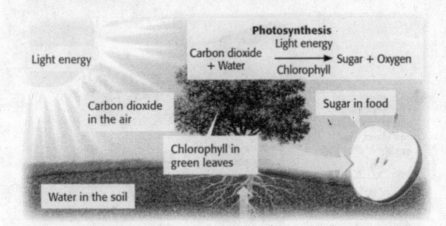

STANDARDS CHECK

PS 3f The sun is a major source of energy for changes on the earth's surface. The sun loses energy by emitting light. A tiny fraction of that light reaches the earth, transferring energy from the sun to the earth. The sun's energy arrives as light with a range of wave lengths, consisting of visible light, infrared, and ultraviolet radiation.

4. Identify Light energy from the sun is converted into what kind of energy by green leaves?

_____

_____

ENERGY CONVERSIONS IN PLANTS

Did you know the chemical energy in the food you eat comes from the sun's energy? When you eat fruits, vegetables, or grains, you are taking in chemical energy. Energy from the sun was used to make the chemical energy in the food. Many animals also eat plants. If you eat meat from these animals, you are also taking in energy that comes from the sun.

The figure above shows how light energy is used to make new material that has chemical energy. When photosynthesis happens in plants, light energy from the sun is changed into chemical energy. *Photosynthesis* is a chemical reaction in plants that changes light energy into chemical energy. When we eat fruits, vegetables, or grains, we are eating the chemical energy that is stored in the plants.

The chemical energy from a tree can also be changed into thermal energy. This change happens when you burn a tree's wood. If you go back far enough, you would see that the energy from a wood fire comes from the sun.

Discuss With a partner, describe how light energy from the sun becomes chemical energy in meat from animals.

Energy Conversions in a Hair Dryer

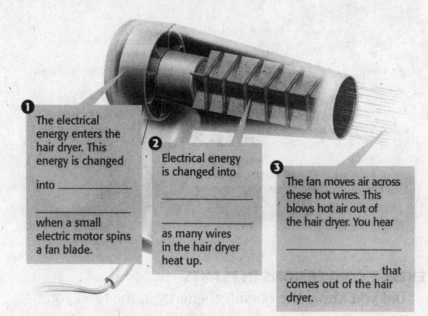

① The electrical energy enters the hair dryer. This energy is changed

into _____

_____

when a small electric motor spins a fan blade.

② Electrical energy is changed into

_____

_____

as many wires in the hair dryer heat up.

③ The fan moves air across these hot wires. This blows hot air out of the hair dryer. You hear

_____

_____ that comes out of the hair dryer.

TAKE A LOOK

5. Identify Fill in the blanks in the figure.

Why Are Energy Conversions Important?

Energy conversions happen everywhere. Heating our homes and getting energy from food are just a few examples of how we use energy conversions. Machines also convert and use energy. This can be seen from the figure of the hair dryer above. Electrical energy by itself won't dry your hair. The hair dryer changes electrical energy into thermal energy. This heat helps dry your hair.

ELECTRICAL ENERGY CONVERSIONS

You use electrical energy all of the time. Some examples of when you use electrical energy are listening to the radio, making toast, and taking a picture. Electrical energy changes into other kinds of energy easily.

TAKE A LOOK

6. Identify Look at the table. In each example, think about what kind of energy conversion happens. Then, fill in the last column of the table with the kind (or kinds) of energy that is formed.

Some Conversions of Electrical Energy		
Alarm Clock	Electrical energy →	
Hair dryer	Electrical energy →	
Light bulb	Electrical energy →	
Blender	Electrical energy →	

SECTION 2 Energy Conversions *continued*

How Do Machines Use Energy?

You have seen how energy can change into different forms. Another way to look at energy is to see how machines use energy. Remember that a machine can make work easier. It does this by changing the size or direction (or both) of the force that does the work. The machine converts the energy you put into the machine into work.

There is another way that machines can help you use energy. They can convert energy into a form that you need, such as a stove using electrical energy to cook food. ☑

An example of how energy is used by a machine is shown in the figure below. The biker puts a force on the pedals. This transfers kinetic energy from the biker into kinetic energy to move the pedals. This energy moves to other parts of the bike. The bike lets the biker use less force over a greater distance. This makes his work easier.

✓ READING CHECK

7. Describe What are two ways a machine converts energy?

_____

_____

_____

_____

Energy Conversions in a Bicycle

For your bike to start and keep moving, energy must be transferred and converted.

❶ Your body stores
_____ energy. This energy is changed into kinetic energy when your muscles contract and relax.

❷ Your legs transfer
_____ energy to the pedals by pushing them in a circle.

❹ The chain moves and transfers kinetic energy to the back wheel.
This gets you _____ and keeps you moving.

❸ The pedals transfer _____ energy to the wheel and then to the chain.

TAKE A LOOK
8. Identify Complete the blanks in the figure.

_____

_____

_____

_____

Section 2 Review

NSES PS 3a, 3f

SECTION VOCABULARY

energy conversion a change from one form of energy to another	

1. **Identify** Suppose a skier is standing still at the top of a hill. What kind of energy does the skier have?

_____

2. **Explain** What happens to the energy of the skier in Question 1 if he goes down the hill?

_____

3. **Describe** How does your body get the energy that it needs to do all of its daily activities? Describe the energy conversions that take place.

_____

_____

_____

4. **Explain** What energy conversion happens when green plants use energy from the sun to make sugar?

_____

_____

5. **Analyze** A vacuum cleaner is a machine that uses electrical energy. What are three forms of energy that the electrical energy changes into in the vacuum cleaner?

_____

_____

6. **Compare** How are the energy conversions of a machine similar to the energy conversions in your body?

_____

_____

_____

CHAPTER 23 Energy and Energy Resources
SECTION
3 **Conservation of Energy**

National Science
Education Standards
PS 3a

BEFORE YOU READ

After you read this section, you should be able to answer these questions:

• How does friction affect energy conversions?

• What is the law of conservation of energy?

• What happens to energy in a closed system?

Where Does Energy Go?

Most roller coasters use a chain to pull the cars up to the top of the first hill. As the cars go up and down the rest of the hills, their potential energy and kinetic energy keep changing. Since no additional energy is put into the cars, they never go as high as the first hill. Is energy lost? No, it just changes into other forms of energy.

When a roller coaster's energy is changing, some of its energy is lowered by friction. **Friction** is a force that is present when two objects touch each other. There is friction between the cars' wheels and the track. There is also friction between the cars and the air around them. Friction slows the motion of a roller coaster. ☑

STUDY TIP

In your science notebook, list the energy conversions in this section and note why energy is not conserved in the conversion.

READING CHECK

1. Identify What causes the energy of a roller coaster to be lowered when the coaster's energy changes form?

_____

_____

Energy Conversions in a Roller Coaster

ⓐ The PE is

_____
at the top of the first hill.

ⓑ Because of friction, the KE at the bottom of the first

hill is _____ than the PE at the top of the hill.

ⓒ The PE at the top of the second hill is

_____ than the KE and PE from the first hill.

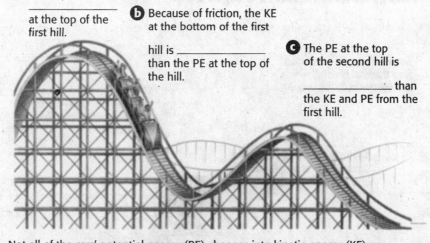

Not all of the cars' potential energy (PE) changes into kinetic energy (KE) as the cars go down the first hill. When the cars move up hill 2, not all of its KE is converted into PE at the top of hill 2. Some of it is changed into thermal energy because of friction.

TAKE A LOOK

2. Complete Fill in the blanks in the figure with the following terms (they can be used more than once): greater, less, largest, smallest.

3. Identify Into what form
of energy does friction
convert mechanical energy?

_____

What Happens to Energy in a Closed System?

A *closed system* is a group of objects that move energy
only to each other. On a roller coaster, a closed system
would be the track, the cars, and the air around the cars.

Some of the mechanical energy (kinetic energy plus
potential energy) of the cars changes into thermal energy.
This happens because of friction. Mechanical energy is
also converted into sound energy. You hear this energy
when you are near the roller coaster. The rest of the
potential energy changes into kinetic energy. This is seen
by the roller coaster racing down the track.

If you add up all of this energy, it equals the cars'
potential energy at the top of the first hill. The energy at
the top of the hill is the same as the energy that is converted to other forms. So, energy is conserved and not
lost in a closed system.

LAW OF CONSERVATION OF ENERGY

Energy is always conserved. This is always true, so it
is called a law. In the **law of conservation of energy**,
energy cannot be created or destroyed. In a closed system,
energy can change from one form to another, but the total
energy is always the same. It doesn't matter how many or
what kinds of energy conversions take place. The energy
conversions that happen in a light bulb are shown in the
figure below. ☑

Energy Conservation in a Light Bulb

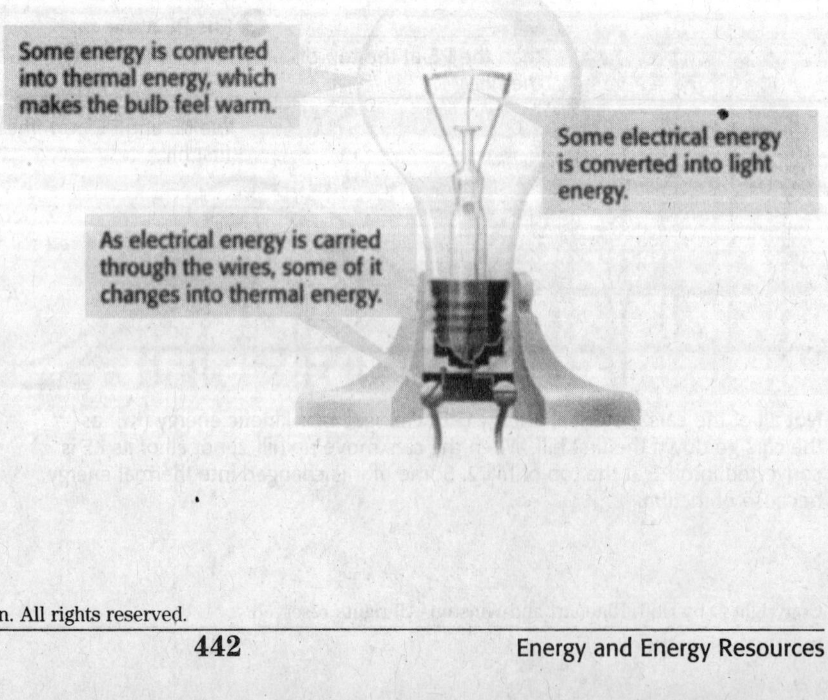

☑ READING CHECK

4. Describe What is the law
of conservation of energy?

_____

_____

Critical Thinking

5. Explain Why is the light
bulb *not* a closed system?

_____

_____

_____

_____

Some energy is converted
into thermal energy, which
makes the bulb feel warm.

Some electrical energy
is converted into light
energy.

As electrical energy is carried
through the wires, some of it
changes into thermal energy.

Why Can't a Machine Run Forever Without Adding Energy?

Any time an energy conversion happens, some of the energy always changes into thermal energy. This thermal energy is in the form of friction. The energy from friction is not useful energy. It is not used to do work.

Think about a car. You put gas into a car. Not all of the chemical energy from the gas makes the car move. Some energy is wasted as thermal energy. This energy leaves through the radiator and the exhaust pipe.

It is impossible to create a machine that runs forever without adding more energy to it. This kind of machine is called a *perpetual motion machine*. A machine has to have a constant supply of energy because energy conversions always produce wasteful thermal energy. ☑

EFFICIENT ENERGY CONVERSIONS

If a car gets good gas mileage, it is energy efficient. The *energy efficiency* is found by comparing the amount of starting energy with the amount of useful energy produced. A car with high energy efficiency goes farther than other cars with the same amount of gasoline.

More efficient energy conversions waste less energy. For example, smooth, *aerodynamic* cars have less friction between the car and the air around it. They use less energy, so they are more efficient. If less energy is wasted, then less energy is needed to run the car.

Car A

The shape of newer cars lowers the friction between the car and the air passing over it.

Car B

6. Describe Why can't there be a perpetual motion machine?

_____

_____

7. Describe Which car is more aerodynamic in shape?

_____

Section 3 Review

SECTION VOCABULARY

friction a force that opposes motion between two surfaces that are in contact	**law of conservation of energy** the law that states that energy cannot be created or destroyed but can be changed from one form to another

1. Explain Suppose you drop a ball. It bounces a few times and then stops. Does the energy disappear? Explain your answer.

_____

_____

_____

2. Identify Fill in the following concept map for a closed system of a roller coaster. Include the parts of the closed system and the energy that is produced.

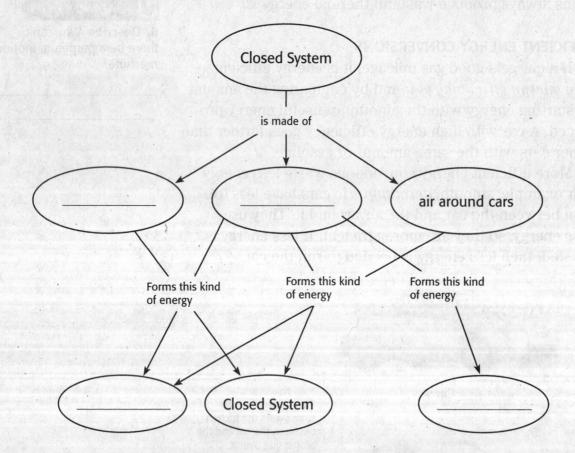

3. Identify What causes thermal energy always formed in an energy conversion?

_____

_____

**National Science
Education Standards**
PS 3a, 3e, 3f

BEFORE YOU READ

**After you read this section, you should be able to answer
these questions:**

- What is an energy resource?
- How do we use nonrenewable energy resources?
- What are renewable energy resources?

What Is an Energy Resource?

Energy is used for many things. It is used to light our
homes, to make food and clothing, and to move people
from place to place. An *energy resource* is a natural product that can be changed into other energy forms to do
work. There are many types of energy resources.

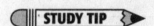

STUDY TIP

In your science notebook,
make a table that lists nonrenewable and renewable
energy resources.

What Are Nonrenewable Energy Resources?

Some energy resources are **nonrenewable resources**.
These are resources that can never be replaced or are
replaced more slowly than they are used. ☑

Oil, natural gas, and coal are nonrenewable resources
called fossil fuels. **Fossil fuels** are energy resources that
formed from buried plants and animals that lived a very long
time ago. Millions of years ago, the plants stored energy
from the sun by photosynthesis. The animals stored and ate
the energy from the plants. When we burn fossil fuels today,
we are using the sun's energy from millions of years ago. ☑

READING CHECK

1. Identify What are nonrenewable resources?

_____

_____

_____

Formation of Fossil Fuels

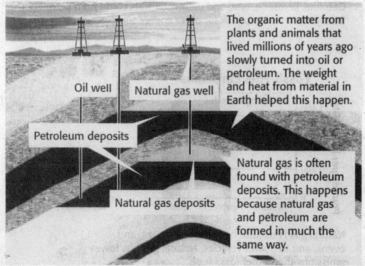

The organic matter from
plants and animals that
lived millions of years ago
slowly turned into oil or
petroleum. The weight
and heat from material in
Earth helped this happen.

Oil well   Natural gas well

Petroleum deposits

Natural gas deposits

Natural gas is often
found with petroleum
deposits. This happens
because natural gas
and petroleum are
formed in much the
same way.

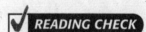

READING CHECK

2. Identify Where did the
energy contained in fossil
fuels come from?

_____

_____

USES OF FOSSIL FUELS

All fossil fuels have stored energy from the sun. This can be changed into other kinds of energy. The figure below shows how we use fossil fuels.

The three most common fossil fuels are coal, natural gas, and oil (petroleum). Burning coal is a way to produce electrical energy. Gasoline, wax, and plastics are made from petroleum. Natural gas is often used to heat homes. ☑

3. Identify What are the three most common fossil fuels?

_____

_____

Everyday Uses of Some Fossil Fuels

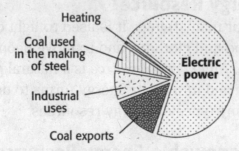

Most coal used in the United States is burned. This produces steam that runs electrical generators.

Math Focus

4. Analyze Graph What is the annual oil production trend after the year 2010?

_____

_____

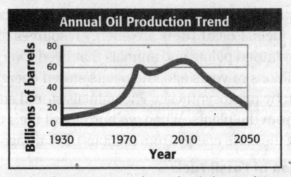

Gasoline, kerosene, wax, and petrochemicals come from petroleum. Scientists continue to look for other energy sources.

Fossil-Fuel Emissions

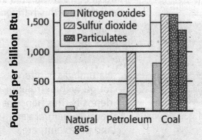

Natural gas is used to heat homes, stoves, and ovens, and to power vehicles. Natural gas has lower emissions than other fossil fuels.

ELECTRICAL ENERGY FROM FOSSIL FUELS

Electrical energy can be produced when fossil fuels are burned. Most of the electrical energy produced in the United States is from fossil fuels. *Electric generators* change the chemical energy from the fossil fuels into electrical energy. This is shown in the figure below.

Converting Fossil Fuels into Electrical Energy

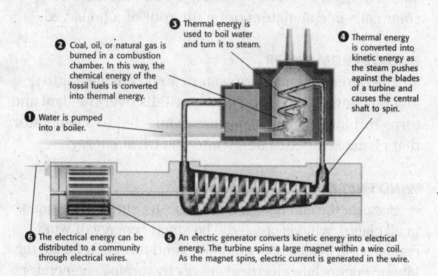

❸ Thermal energy is used to boil water and turn it to steam.

❷ Coal, oil, or natural gas is burned in a combustion chamber. In this way, the chemical energy of the fossil fuels is converted into thermal energy.

❹ Thermal energy is converted into kinetic energy as the steam pushes against the blades of a turbine and causes the central shaft to spin.

❶ Water is pumped into a boiler.

❻ The electrical energy can be distributed to a community through electrical wires.

❺ An electric generator converts kinetic energy into electrical energy. The turbine spins a large magnet within a wire coil. As the magnet spins, electric current is generated in the wire.

NUCLEAR ENERGY

Electrical energy is also produced from nuclear energy. Nuclear energy comes from radioactive elements like uranium. The nucleus of a uranium atom splits into two smaller nuclei in a process called *nuclear fission*. There is not a large supply of radioactive elements, so nuclear energy is a nonrenewable resource. ☑

A nuclear power plant changes the thermal energy from nuclear fission into electrical energy. Splitting uranium atoms creates thermal energy. This thermal energy is changed to electrical energy in a process similar to how fossil fuel power plants work. The figure above shows how this happens.

STANDARDS CHECK

PS 3e In most chemical and nuclear reactions, energy is transferred into or out of a system. Heat, light, mechanical motion, or electricity might all be involved in such transfers.

5. Identify What is most often used to produce electricity in the United States?

_____

_____

✓ **READING CHECK**

6. Identify What nuclear process is used to produce electricity?

_____

_____

What Are Renewable Energy Resource?

Some energy sources are replaced faster than they are used. These are **renewable resources**. Some of these resources can almost produce an endless supply of energy.

SOLAR ENERGY

The energy from the sun can be changed into electrical energy by solar cells. A *solar cell* is a device that changes solar energy into electrical energy. You may have seen solar cells in calculators or on the roof of a house. ☑

ENERGY FROM WATER

The potential energy of water can be changed into kinetic energy in a dam. The water falls over the dam and turns turbines. The turbines are connected to a generator that changes kinetic energy into electrical energy.

WIND ENERGY

Because the sun does not heat Earth's surface the same in all places, wind is created. The kinetic energy of wind can turn the blades of a windmill. Wind turbines change this kinetic energy into electrical energy by turning a generator.

GEOTHERMAL ENERGY

Thermal energy made by the heating of Earth's crust is called *geothermal energy*. Geothermal power plants pump water under the ground near hot rock. The water turns into steam, which is used to turn the turbines of a generator.

BIOMASS

Biomass is organic matter, like plants, wood, or waste. When biomass is burned, it gives off the energy it got from the sun. This can be used to make electrical energy.

READING CHECK

7. Describe What does a solar cell do?

_____

_____

TAKE A LOOK

8. Identify The table lists many renewable resources. Complete the missing boxes in the table.

Renewable energy source	Direct source of energy	Original source of energy
Solar energy	sun	_____
Energy from water	_____	sun
Wind energy	_____	sun
Geothermal energy	heat of earth's crust	_____
Biomass	organic matter	_____

How Do You Decide What Energy Source to Use?

All energy resources have advantages and disadvantages. The table below compares many energy resources. The energy source you choose often depends on where you live, what you need it for, and how much you need. To decide which source to use, advantages and disadvantages must be thought about.

One disadvantage of using fossil fuels you have often heard is that fossil fuels pollute the air. Another disadvantage is that we can run out of fossil fuels if we use them all up.

Some renewable resources have disadvantages, too. It is hard to produce a lot of energy from solar energy. Many renewable resources are limited to places where that resource is available. For example, you need a lot of wind to get power from wind energy. ☑

Energy planning around the world is important. Energy planning means determining your energy needs and your available energy resources, and then using this energy responsibly.

☑ **READING CHECK**

9. Describe What is a disadvantage of solar energy?

_____

_____

Advantages and Disadvantages of Energy Resources		
Resource	**Advantages**	**Disadvantages**
Fossil Fuels	• produces large amounts of energy • easy to get • makes electricity • makes products like plastic	• nonrenewable • produces smog • produces acid precipitation • risk of oil spills
Nuclear	• concentrated energy form • no air pollution	• produces radioactive waste • nonrenewable
Solar	• almost endless source • no pollution	• expensive • works best in sunny areas
Water	• renewable • inexpensive • no pollution	• needs dams, which hurt water ecosystem • needs rivers
Wind	• renewable • inexpensive • no pollution	• needs a lot of wind
Geothermal	• almost endless source • little land needed	• needs a ground hot spot • produces wastewater
Biomass	• renewable • inexpensive	• needs a lot of farmland • produces smoke

TAKE A LOOK

10. Identify What are some advantages to using wind power?

_____

_____

_____

Section 4 Review

SECTION VOCABULARY

fossil fuel a nonrenewable energy resource formed from the remains of organisms that lived long ago	**renewable resource** a natural resource that can be replaced at the same rate at which the resource is consumed
nonrenewable resource a resource that forms at a rate that is much slower than the rate at which the resource is consumed	

1. Compare What is the difference between a nonrenewable energy resource and a renewable energy resource?

_____

_____

_____

2. Analyze Why can it be said that the energy from burning fossil fuels ultimately comes from the sun?

_____

_____

_____

_____

3. Explain How is nuclear energy used to make electrical energy?

_____

_____

_____

_____

4. Identify What is a renewable energy resource that does not depend on the sun?

_____

5. Analyze What are some possible reasons that solar power is not used for all of our energy needs?

_____

_____

_____

CHAPTER 24 | Heat and Heat Technology

SECTION
1 | **Temperature**

National Science Education Standards
PS 3a, 3b

BEFORE YOU READ

After you read this section, you should be able to answer these questions:

• How are temperature and kinetic energy related?
• How is temperature measured?
• What is thermal expansion?

What Is Temperature?

You may think of temperature as how hot or cold something is. But using the words *hot* and *cold* can be confusing. Pretend that you are outside on a hot day. If you step onto a porch with a fan blowing, you might think it feels cool. Then, your friend enters the porch from an air-conditioned house. She thinks the porch is warm!

Using the words cool and warm to describe the porch is confusing. Measuring the temperature on the porch tells you exactly how hot or cold it is. The **temperature** tells you the average kinetic energy of the particles in an object. ☑

TEMPERATURE AND KINETIC ENERGY

All matter is made of atoms or molecules that are always moving. The particles that are moving have kinetic energy. The faster they move, the more kinetic energy they have. Look at the figure below. The more kinetic energy the particles have, the higher the temperature is.

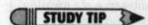

STUDY TIP

Describe Describe how temperature affects the average kinetic energy and thermal expansion. Give several examples of thermal expansion.

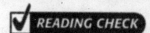

READING CHECK

1. Identify What is used to tell how hot or cold something is?

_____

_____

Gas A   **Gas B**

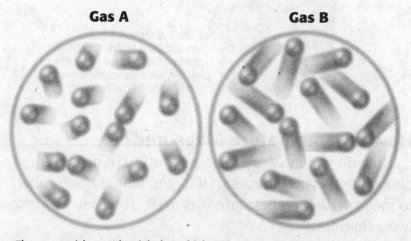

The gas particles on the right have higher kinetic energy than those on the left.

TAKE A LOOK
2. Identify Which gas shows particles at the higher temperature?

_____

_____

AVERAGE KINETIC ENERGY

The particles that make up matter are always moving in different directions and speeds. This movement is random. Since the particles move at different speeds, each particle has its own kinetic energy. The temperature of a substance is a measure of the *average kinetic energy* of all the particles in a substance. A high temperature means more of the particles in the object are moving fast rather than slow.

The temperature of a substance does not depend on how much of it you have. Look at the figure below. A pot of tea and a cup of tea each have a different amount of tea. Their atoms have the same average kinetic energy. There may be more tea in the teapot than in the cup, but they are at the same temperature.

3. Identify The temperature of a substance is a measure of what kind of energy?

_____

_____

_____

Critical Thinking

4. Infer As the tea in the cup cools, what happens to the average kinetic energy of the particles in the tea? What happens to the motion of the particles in the tea?

_____

_____

_____

_____

There is more tea in the teapot than in the cup. But the temperature of the tea in the cup is the same as the temperature of the tea in the teapot.

How Is Temperature Measured?

How do you measure the temperature of a cup of hot chocolate? If you took a sip of it, you would not be able to measure the temperature very well. The best way is to use a thermometer.

USING A THERMOMETER

Many thermometers are thin glass tubes filled with a liquid. Mercury and alcohol are often used in thermometers because they are a liquid over a wide range of temperatures. They also expand at a constant rate.

Thermometers that use liquids can measure temperature because of thermal expansion. **Thermal expansion** is the increase in the volume of a substance when the temperature of the substance increases. When a substance's temperature increases, its particles move faster and farther away from each other. There is more space between the particles, so the substance expands. ☑

If you look at the figure below, all three thermometers are at the same temperature. The alcohol in each thermometer has expanded the same amount. The number reading for each thermometer is different because a different temperature scale is used for each one.

Three Temperature Scales

	Fahrenheit	Celsius	Kelvin
Water boils	212°	100°	373
Body temperature	98.6°	37°	310
Room temperature	68°	20°	293
Water freezes	32°	0°	273

✓ READING CHECK

5. Describe What is thermal expansion?

_____

_____

_____

TEMPERATURE SCALES

There are three different temperature scales that are often used. They are the Fahrenheit scale, the Celsius scale, and the Kelvin scale. When you hear a weather report, you hear the temperature given in degrees Fahrenheit (°F). Scientists often use the Celsius scale. The Kelvin (or absolute) scale is the official SI temperature scale. The Kelvin scale has units called kelvins (K). The Kelvin scale does not use degrees, so 25 K is 25 kelvins. ☑

The lowest temperature on the Kelvin scale is 0 K. This is called **absolute zero**. Absolute zero (−459°F) is the temperature at which all molecules stop moving. It is not possible to reach absolute zero because molecules are always moving.

✓ READING CHECK

6. Identify Which temperature scale is the SI temperature scale?

_____

SECTION 1 Temperature *continued*

TEMPERATURE CONVERSION

For a given temperature, each temperature scale has a different number reading. For example, the freezing point of water is 32°F, 0°C, or 273 K. You can change from one scale to another using the equations in the table below.

Converting Between Temperature Units		
To convert	**Use the equation**	**Example**
Celsius to Fahrenheit °C ⟶ °F	$F = \left(\frac{9}{5}C\right) + 32$	Change 45°C to degrees Fahrenheit.
Fahrenheit to Celsius °F ⟶ °C	$C = \frac{5}{9} \times (F - 32)$	Change 68°F to degrees Celsius.
Celsius to Kelvin °C ⟶ K	$K = C + 273$	Change 45°C to Kelvins.
Kelvin to Celsius K ⟶ °C	$C = K - 273$	Change 32 K to degrees Celsius.

Math Focus

7. Calculate In the last column of the table, calculate the temperature for the temperature scale given.

A change of one Kelvin is the same as a change of one Celsius degree. So a temperature change from 0°C to 1°C is the same as a change from 273 K to 274 K. However, a temperature change of 1°C is higher than a change of 1°F. It's almost two times as large.

What Are Other Examples of Thermal Expansion?

You have learned how thermal expansion is used in thermometers. Thermal expansion has many other uses. Sometimes it is harmful, but other times it is useful.

EXPANSION JOINTS ON HIGHWAYS

Have you ever gone across a bridge in a car? You may have felt bumps every few seconds. The car is going over small spaces called *expansion joints*. If the weather is really hot, the bridge can heat up and expand. When it expands, the bridge can break. Expansion joints separate parts of the bridge so they can expand and not break. ☑

8. Describe What is the purpose of expansion joints in a bridge?

_____

_____

_____

SECTION 1 Temperature *continued*

BIMETALLIC STRIPS IN THERMOSTATS

Another example of thermal expansion happens in a *thermostat*, a device that controls the temperature in your home. The thermostat has a *bimetallic strip* with two different metals coiled together. The strip coils and uncoils as the temperature changes. ☑

Most thermostats have a tube of mercury that is moved by a bimetallic strip. The figure below shows how a mercury thermostat works. The mercury in a tube moves to touch or not touch electrical contacts. Electrical contacts are two pieces of metal that can touch to complete a circuit. This makes the electric circuit in the thermostat close or open. So the thermostat turns on and off the heater in your home.

READING CHECK

9. Explain What does the bimetallic strip do in a thermostat? Why?

_____

_____

_____

How a Thermostat Works

Glass tube with mercury

Electrical contacts

Bimetallic strip

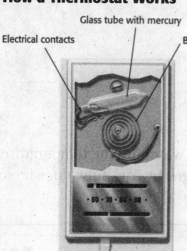

ⓐ When the room temperature is lower than the temperature setting, the bimetallic strip coils. This causes the glass tube above the strip to tilt. Mercury flows and closes the electrical circuit. The result is that the heater turns on.

ⓑ When the room temperature is higher than the temperature setting, the bimetallic strip uncoils. It becomes larger. This causes the glass tube above the strip to flatten out. The mercury moves away and opens the electrical circuit. The result is that the heater turns off.

TAKE A LOOK

10. Explain What happens to the mercury in a thermostat that results in the heater turning on?

_____

_____

THERMAL EXPANSION IN HOT AIR BALLOONS

Thermal expansion is also used in hot air balloons. When air inside a balloon is heated, the air takes up more space. The air particles move faster because they have more kinetic energy. The gas expands to fill the volume of the balloon. The air inside the balloon is less dense than the air outside the balloon. So, the balloon goes up into the air because it is less dense than the air around it.

Name _____ Class _____ Date _____

Section 1 Review

NSES PS 3a, 3b

SECTION VOCABULARY

absolute zero the temperature at which molecular energy is at a minimum (0 K on the Kelvin scale or −273.16°C on the Celsius scale) **thermal expansion** an increase in the size of a substance in response to an increase in the temperature of the substance	**temperature** a measure of how hot (or cold) something is; specifically, a measure of the average kinetic energy of the particles in an object

1. Compare How is the temperature of an object related to its kinetic energy?

_____

_____

_____

2. Calculate The thermometer outside your window reads 77°F. What is the same temperature on the Celsius scale? Show your work.

3. Determine You are doing a science experiment and watching the temperature change in Celsius degrees. If the temperature changes by 5°C, how does it change in Kelvins?

_____

4. Explain How is the liquid in a thermometer used to measure temperature? Why are mercury and alcohol used in thermometers?

_____

_____

_____

5. Analyze How is thermal expansion used to get a hot air balloon off of the ground?

_____

_____

_____

SECTION
2 **What Is Heat?**

National Science Education Standards
PS 3a, 3b

BEFORE YOU READ

After you read this section, you should be able to answer these questions:

- What is heat?
- What is thermal energy?
- How is thermal energy transferred?

What Is Heat?

You might use the word *heat* to describe things that feel hot. However, heat also has to do with things that feel cold. Heat is what makes objects feel hot or cold. **Heat** is the energy that moves between objects that are at different temperatures. ☑

Why do some things feel hot, and other things feel cold? When two objects touch each other, energy moves from one object to the other. This energy is heat. Heat always moves from an object with a higher temperature to an object with a lower temperature. If you touch a cold piece of metal, energy from your hand moves to the metal. So, the metal feels cold when you touch it.

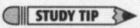
STUDY TIP

Describe Describe one example of energy transfer by each method—conduction, convection, and radiation—that occurs in your science classroom.

READING CHECK

1. Describe What is heat?

_____

_____

_____

The metal stethoscope feels cold because of heat.

TAKE A LOOK
2. Identify In the figure, which way is heat flowing if the stethoscope feels cold?

_____

_____

HEAT AND THERMAL ENERGY

3. Compare When two objects at different temperatures are touching, what will energy do?

_____

_____

_____

_____

_____

If heat is the movement of energy, what kind of energy is moving? The answer is thermal energy. **Thermal energy** is the total kinetic energy of the particles that make up a substance. Thermal energy is measured in joules (J). Thermal energy depends on the substance's temperature and how much of the substance there is. For example, a large lake contains more thermal energy than a smaller lake if both are at the same temperature.

REACHING THE SAME TEMPERATURE

When two objects with different temperatures touch each other, energy moves. Energy moves from the warmer object to the cooler object. This happens until both objects are at the same temperature. When they have the same temperature, the thermal energy of the objects no longer changes. One object might have more thermal energy than the other, but the temperature of both objects is the same.

Transfer of Thermal Energy

❶ Energy is transferred from the particles in the juice to the particles in the bottle. These particles transfer energy to the particles in the ice water, causing the ice to melt.

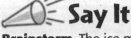 **Say It**

Brainstorm The ice melts in a cooler used to keep a six-pack of cola cold on a hot day. Discuss with a partner energy transfers that occur in the cooler of ice that cause the ice to melt.

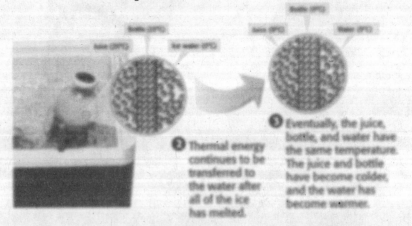

❷ Thermal energy continues to be transferred to the water after all of the ice has melted.

❸ Eventually, the juice, bottle, and water have the same temperature. The juice and bottle have become colder, and the water has become warmer.

SECTION 2 What Is Heat? *continued*

How Is Thermal Energy Transferred?

Every day you see some ways that energy is transferred. Stoves transfer energy to soup in a pot. The temperature of your bath water can change by adding hot or cold water. There are three ways that thermal energy moves from one object to another. They are *conduction*, *convection*, and *radiation*.

CONDUCTION

What happens when you put a cold metal spoon in a bowl of hot soup? The spoon warms up. Even the handle of the spoon gets warm, and it is not touching the soup. The whole spoon gets warm because of conduction. This is shown in the figure below. **Thermal conduction** is the transfer of thermal energy when two objects touch each other. ☑

Conduction also happens within a substance. This is how the handle of the spoon gets warm.

The circles show the energy in the particles of the spoon and the soup. Energy will move from the soup to the spoon until all of the particles have the same energy. Even the handle of the spoon will have the same energy.

When two objects touch, their particles bump into each other. Thermal energy moves from the higher-temperature substance to the lower-temperature substance. When the particles bump into each other, their kinetic energy moves from one particle to another. This makes some particles move faster, and some move slower. This happens until the particles have the same average kinetic energy. Then, both objects will be at the same temperature. ☑

☑ **READING CHECK**

4. Describe What is thermal conduction?

_____

_____

_____

☑ **READING CHECK**

5. Identify What kind of energy is the same for objects at the same temperature?

_____

CONDUCTORS AND INSULATORS

Substances that transfer thermal energy easily are called **thermal conductors**. The cold metal spoon from the figure on the previous page is an example of a conductor. Energy moves easily from the soup to the spoon.

Some substances do not transfer thermal energy very well and are called **thermal insulators**. The bowl from the figure on the previous page does not get hot as quickly as the spoon. Energy does not move easily from the soup to the bowl, so the bowl is an insulator. The table below shows some examples of common conductors and insulators.

Common Conductors and Insulators	
Conductors	**Insulators**
Curling iron	Flannel shirt
Cookie sheet	Oven mitt
Iron pan	Plastic spatula
Copper pipe	Fiberglass insulation
Stove	Ceramic bowl

CONVECTION

The second way thermal energy can be transferred is by convection. **Convection** is the transfer of thermal energy by the movement of a liquid or gas. Look at the figure below. When you boil water in a pot, the water moves in a circular motion because of convection. This circular motion is called a *convection current*. ☑

Critical Thinking

6. Compare What is the difference between thermal conductors and thermal insulators?

_____

_____

_____

_____

☑ READING CHECK

7. Describe What is convection?

_____

_____

_____

TAKE A LOOK

8. Identify Fill in the missing words in the figure.

❶ The cooler water on the surface of the pan is denser, and it

_____

back to the bottom of the pan.

❷ The water moves in the pan because it heats up and cools down. This forms

_____

_____

which are shown by the arrows.

The repeated rising and sinking of water during boiling are due to convection.

SECTION 2 What Is Heat? *continued*

RADIATION

The third way thermal energy is transferred is by radiation. **Radiation** is the transfer of energy by electromagnetic waves. Some examples of electromagnetic waves are visible light and infrared waves. Radiation can transfer energy between particles or through a vacuum, like outer space. Conduction or convection cannot transfer energy through outer space. ☑

All objects radiate electromagnetic waves. The sun gives off visible light that you see. The sun also gives off other waves, like infrared and ultraviolet waves, that you cannot see. When your body takes in infrared waves, you feel warmer.

RADIATION AND THE GREENHOUSE EFFECT

The atmosphere of Earth acts like the windows of a greenhouse. The sun's visible light goes through it. A greenhouse stays warm because it traps energy. The atmosphere traps energy, too. This is called the *greenhouse effect*. You can see how this works in the figure below.

Greenhouse gases absorb infrared light from the sun. This energy is trapped in the atmosphere. Some of these gases are water vapor, carbon dioxide, and methane. Some scientists are concerned about increasing amounts of greenhouse gases in the atmosphere. They think that too much energy will become trapped and make Earth too warm. ☑

READING CHECK

9. Identify Radiation is transferred by what kind of waves?

_____

_____

READING CHECK

10. Describe Why are some scientists concerned about increasing amounts of greenhouse gas in the atmosphere?

_____

_____

_____

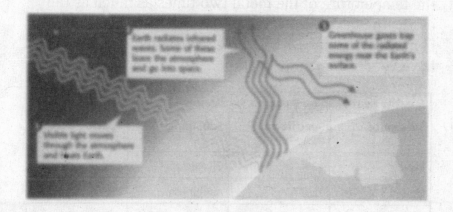

SECTION 2 What Is Heat? *continued*

Why Are Some Substances Warmer Than Others?

Have you ever put on your seat belt on a hot summer day? If so, the metal buckle may have felt hotter than the cloth belt did. Why?

THERMAL CONDUCTIVITY

One reason is because the metal buckle has a higher thermal conductivity than the cloth belt. *Thermal conductivity* is a measure of how fast a substance transfers thermal energy. When you touch the metal, energy moves quickly from the belt to your hand. The cloth and the metal are at the same temperature, but the metal feels hotter. ☑

SPECIFIC HEAT

Another difference between the metal and the cloth is how easily each changes temperature. When the same amount of energy is given to equal masses of metal and cloth, the metal gets hotter. The temperature change depends on the substance's specific heat. **Specific heat** is the amount of energy it takes to change the temperature of 1 kg of a substance by 1°C. ☑

The higher the specific heat of something is, the more energy it takes to raise its temperature. The specific heat of the cloth is more than two times the specific heat of the metal buckle. So, the same thermal energy will raise the temperature of the metal two times as much as the cloth. The table below shows the specific heat of many common substances. The table shows that most metals have very low specific heats.

READING CHECK

11. Identify Which has a higher thermal conductivity, a piece of metal or a piece of cloth?

_____

_____

READING CHECK

12. Define What is the specific heat of a substance?

_____

_____

_____

Math Focus

13. Identify and Explain Suppose that the same mass of each substance in the table gains the same amount of energy. Which substance will have the greatest increase in temperature? Explain why.

_____

_____

_____

Specific Heat of Some Common Substances			
Substance	Specific heat (J/kg·°C)	Substance	Specific heat (J/kg·°C)
Lead	128	Glass	837
Gold	129	Aluminum	899
Copper	387	Cloth of seat belt	1,340
Iron	448	Ice	2,090
Metal of seat belt	500	Water	4,184

CALCULATING HEAT

When a substance changes temperature, the temperature change alone does not tell you how much energy has been transferred. To calculate energy transfer, you must know the substance's mass, its change in temperature, and its specific heat. The equation below is used to calculate the energy, or heat, transferred between objects. ☑

heat = specific heat × mass × change in temperature

How much energy is needed to heat a cup of water to make tea? Using the equation above, you can calculate the heat that is transferred to the water. The temperature of the water increases, so heat is a positive number. You can also use this equation to calculate the heat that leaves an object when it cools down. The value for heat when it cools is negative because the temperature decreases.

✓ **READING CHECK**

14. Describe What data are needed to calculate the amount of energy transferred to a substance?

_____

_____

_____

Mass of water = 0.2 kg
Temperature (before) = 25°C
Temperature (after) = 80°C
Specific heat of
 water = 4,184 J/kg•°C

Information used to calculate heat, the amount of energy transferred to the water, is shown here.

Let's try a problem. You heat 2.0 kg of water to make pasta. The temperature of the water before you heat it is 40°C. The temperature of the water after you heat it is 100°C. How much heat was transferred to the water? (The specific heat of water is 4,184 J/kg•°C).

Math Focus

15. Calculate Use the data in the figure to determine the energy needed to warm the water in the cup.

Step 1: Write the equation.

heat = specific heat × mass × change in temperature

Step 2: Place values into the equation, and solve.

heat = 4,184 J/kg•°C × 2.0 kg × (100°C − 40°C) = 502,080 J.

The heat transferred is 502,080 J.

Section 2 Review

SECTION VOCABULARY

convection the transfer of thermal energy by the circulation or movement of a liquid or gas	**thermal conduction** the transfer of energy as heat through a material
heat the energy transferred between objects that are at different temperatures	**thermal conductor** a material through which energy can be transferred as heat
radiation the transfer of energy as electromagnetic waves	**thermal energy** the kinetic energy of a substance's atoms
specific heat the quantity of heat required to raise a unit mass of homogeneous material 1 K or 1°C in a specified way given constant pressure and volume	**thermal insulator** a material that reduces or prevents the transfer of heat

1. Explain Why can heat describe both hot and cold objects?

_____

_____

2. Identify Use the following Concept Map to describe how thermal energy moves from one object to another.

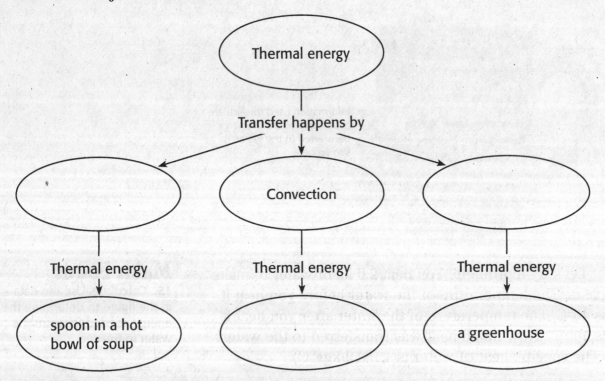

3. Calculate The specific heat of lead is 128 J/kg•°C. How much heat is needed to raise the temperature of a 0.015 kg sample of lead by 10°C? Show your work.

CHAPTER 24 Heat and Heat Technology

SECTION
3 **Matter and Heat**

National Science
Education Standards
PS 3a, 3b

BEFORE YOU READ

After you read this section, you should be able to answer
these questions:

• What are the states of matter?

• How can heat cause a change of state?

• How can heat cause a chemical change?

• What is a calorimeter?

What Are the States of Matter?

Have you ever tried to eat a frozen juice bar outside on a
hot day? The juice bar melts quickly because the sun transfers energy to the bar. This increases the kinetic energy of
the juice bar's molecules and it starts to change to a liquid.

The matter in the frozen juice bar is the same whether
it is frozen or melted. The matter is just in a different form,
or state. The **states of matter** are the physical forms of
a substance. A substance's state depends on three things.
It depends on how fast its particles move, the attractive
forces between the particles, and the atmospheric pressure. The three well-known states of matter are solid, liquid and gas. All three are shown in the figure below. ☑

Suppose you had the same amount of a substance in a
solid, in a liquid, and in a gas. The substance will have the
most thermal energy as a gas and the least as a solid. This
is because the particles in a gas move the fastest.

Particles of a Solid, a Liquid, and a Gas

Particles of a gas
move separately
from one another.
There is almost no
interaction between
the particles.

Particles of a
liquid can slide
past one another.
They move fast
but there is still
some interaction
between particles.

Particles of a solid
are held together
tightly. The particles
vibrate in place.

STUDY TIP

Identify Make a list of three
changes of state and three
chemical changes that you
feel are important. Describe
the role of energy in each.

✓ READING CHECK

1. Identify What are the
physical forms of a substance
called?

_____

_____

TAKE A LOOK
2. Identify Which state of
matter has the least interaction between its particles?

_____

_____

How Does Matter Change State?

A **change of state** is a change of a substance from one state of matter to another. A change of state is a *physical change*. That means a physical property of a substance changes, but the substance is the same. There are four types of changes of state. They are *freezing* (liquid to solid), *melting* (solid to liquid), *boiling* (liquid to gas), and *condensing* (gas to liquid). ☑

3. Identify What are the four types of changes of state?

_____

_____

_____

ENERGY AND CHANGES OF STATE

What would happen if you put an ice cube in a pan and put the pan on a stove burner? The ice will first turn into water and then to steam. You could make a graph of the temperature of the ice as energy is added to it. This graph would look something like the graph below.

Changes of State of Water

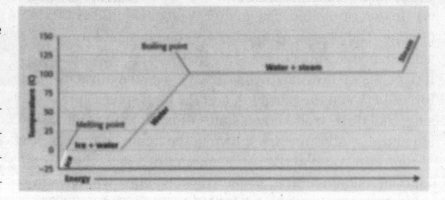

TAKE A LOOK

4. Identify During a change of state, does the temperature increase, decrease, or stay the same? Explain your answer using the graph.

_____

_____

_____

_____

As the ice is heated, the temperature increases from −25°C to 0°C. The ice changes into liquid water at 0°C. Its temperature stays at 0°C until there is only liquid water left in the pan. The energy is being used only to melt the ice, not change its temperature. Then the water temperature increases between 0°C and 100°C. At 100°C the liquid water changes into steam. The water temperature stays at 100°C until there is no more water in the pan.

How Is Heat Involved in Chemical Changes?

We have seen how heat is important in changing the state of matter. Heat is also an important part of chemical changes. A *chemical change* happens when one or more substances combine to make new substances.

To make a new substance, energy is needed to break the old bonds. Energy is released when new bonds are formed. Sometimes, more thermal energy is needed for a reaction to happen than is released when new bonds form. Other times, more energy is given off in the reaction than was needed to break old bonds. ☑

FOOD AND ENERGY

You have probably seen a Nutrition Facts label on some of the food you eat. These labels tell you how much chemical energy the food has. The *Calorie* is the amount of chemical energy in food. One Calorie is the same as 4,184 J. Since the Calorie is a measure of energy, it is also a measure of heat. The amount of energy in food can be measured by a calorimeter.

CALORIMETERS

A *calorimeter* is a device that measures heat. In a *bomb calorimeter*, a food sample is burned in a chamber inside the calorimeter as shown below. The amount of energy (heat) given off equals the energy content of the food.

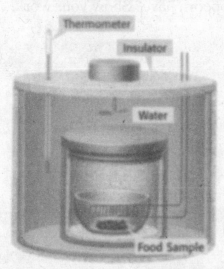

A bomb calorimeter can measure energy content in food by measuring how much heat is given off by a food sample when burned. If a food sample releases 523,000 J of heat, what is the energy content of the food? Since 1 Calorie is 4,184 J,

$$523{,}000 \text{ J} \times \frac{1 \text{ Calorie}}{4{,}184 \text{ J}} = 125 \text{ Calories.}$$

5. Identify When bonds are formed, is energy needed or released?

_____

STANDARDS CHECK

PS 03a Energy is a property of many substances and is associated with heat, light, electricity, mechanical motion, sound, nuclei, and the nature of a chemical. Energy is transferred in many ways.

6. Identify The energy found in food is in what form of energy?

_____

Math Focus

7. Calculate When a peanut is burned in a bomb calorimeter, it releases 8,368 J of energy. What is the energy content of the peanut? Show your work.

Section 3 Review

SECTION VOCABULARY

change of state the change of a substance from one physical state to another	states of matter the physical forms of matter, which include solid, liquid, and gas

1. Analyze What determines if a substance is a solid, a liquid, or a gas?

_____

_____

2. Identify What are the ways that a substance can change its state?

_____

3. Explain During a change of state, why doesn't the temperature of the substance change?

_____

_____

4. Explain What are the ways that heat takes part in a chemical change?

_____

_____

5. Compare How is a physical change different from a chemical change?

_____

_____

6. Calculate A sample of popcorn had 627,600 J of energy when it was measured in a calorimeter. How much energy, in Calories, did the popcorn have? Show your work.

CHAPTER 24 Heat and Heat Technology)

SECTION 4
Heat Technology

National Science Education Standards
PS 3a, 3b

> **BEFORE YOU READ**
>
> **After you read this section, you should be able to answer these questions:**
> - What are the types of heating systems?
> - How does an automobile use heat?
> - How do cooling systems work?
> - How does thermal pollution affect the environment?

What Is Heat Technology?

You may not be surprised to learn that the heater in your home is an example of heat technology. However, cars, refrigerators, and air conditioners are also examples of heat technology. Heat technology heats your home, runs your car, keeps food cold, and keeps you cool.

What Is a Heating System?

The temperature of most homes and buildings is controlled by a *central heating system.* There are different types of central heating systems that you will see on the next few pages.

HOT-WATER HEATING

Water has a high specific heat. This property makes it useful for heating systems because it allows water to hold onto its heat. A hot-water heating system is described in the figure below. ☑

A Hot-Water Heating System

Smoke outlet

Air, heated by hot water in the radiators, circulates in the room by convection currents.

Radiators

Pump

An expansion tank handles the increased volume of heated water.

Hot-water heater

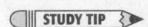
STUDY TIP

Describe As you and a partner read about heating and cooling systems, quiz each other on how they work.

✓ **READING CHECK**

1. Identify What property of water makes it useful for a heating system?

_____

_____

TAKE A LOOK
2. Identify How does the water get from the hot-water heater to the radiators?

_____

_____

Interactive Textbook   **469**   Heat and Heat Technology

SECTION 4 Heat Technology *continued*

WARM-AIR HEATING

The specific heat of air is less than water, so air holds less thermal energy than water. Still, warm-air heating systems are often used in homes. In a warm-air heating system, air is heated by burning fuel (like natural gas) in a furnace. This is shown in the figure below.

The warm air moves through ducts to heat the rooms of the house. The vents are placed on the floor because that is where the cooler air is. A fan pulls the cool air into the furnace. The cool air is heated and returns to the rooms.

TAKE A LOOK

4. Identify How does the air get from the furnace to the vent?

_____

_____

A Warm-Air Heating System

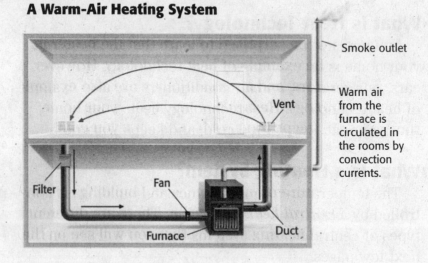

Smoke outlet

Warm air from the furnace is circulated in the rooms by convection currents.

Vent

Filter

Fan

Furnace

Duct

HEATING AND INSULATION

There are many places in a home where heat can leave and enter. It can be wasteful to run a heating or cooling system all of the time. So, insulation is used to lower the amount of energy that leaves or enters a building.

Insulation is a material that prevents the movement of thermal energy. When insulation is used in walls, ceiling, and floors, less heat leaves or enters the building. Insulation helps a house stay warm in winter and cool in summer. ☑

 **READING CHECK**

5. Explain Why is insulation used in buildings and homes?

_____

_____

_____

SOLAR HEATING

The sun gives off a lot of energy. Solar heating systems use some of this energy to heat homes and buildings. A *passive solar heating system* does not have moving parts. It uses the design of the building and special building material to heat with the sun's energy. An *active solar heating system* has moving parts. It uses pumps and fans to move the sun's energy through the building. ☑

The figure of the house below shows how both passive and active solar heating systems work. The large windows and thick concrete walls are part of the passive heating system.

The active heating system is made of several parts. Water is pumped to a solar collector where it is heated by the sun's energy. The hot water moves through pipes. A fan blows over the pipes and thermal energy is transferred to the air. This warm air is used to heat the rooms of the house.

☑ **READING CHECK**

6. Compare How does a passive solar heating system differ from an active solar heating system?

_____

_____

_____

_____

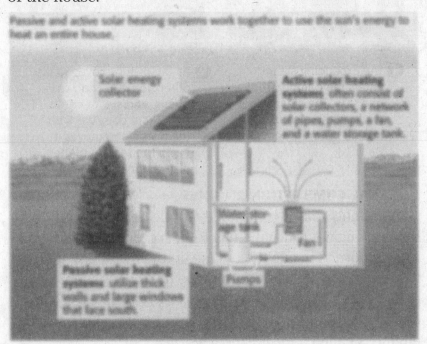

Passive and active solar heating systems work together to use the sun's energy to heat an entire house.

TAKE A LOOK

7. Identify From the figure, list:

1. the parts of the passive solar heating system.
2. the parts of the active solar heating system.

_____

_____

_____

_____

_____

What Are Heat Engines?

Automobiles work because of heat. A car has a **heat engine**, which is a machine that uses heat to do work. Fuel is burned in a heat engine to make thermal energy in a process called *combustion*. If a heat engine burns fuel outside the engine, it is an *external combustion engine*. If a heat engine burns fuel inside the engine, it is an *internal combustion engine*. ☑

☑ **READING CHECK**

8. Identify Where is the fuel burned for an external combustion engine?

_____

_____

EXTERNAL COMBUSTION ENGINES

A simple steam engine is an example of an external combustion engine. This is seen in the figure below. Coal is burned in a boiler (not shown in the figure). The boiler heats water, which turns into steam. The steam comes into the engine and expands, which pushes a piston.

Modern steam engines are used to generate electricity at power plants. The steam turns turbine blades in generators. Generators that use steam to do work convert thermal energy into electrical energy. ☑

READING CHECK

9. Identify Today, what are most steam engines used for?

_____

An External Combustion Engine

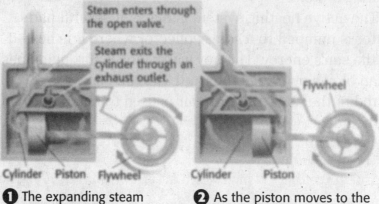

Steam enters through the open valve.

Steam exits the cylinder through an exhaust outlet.

Flywheel

Cylinder   Piston   Flywheel   Cylinder   Piston

1 The expanding steam enters the cylinder from one side. The steam does work on the piston, forcing the piston to move.

2 As the piston moves to the other side, a second valve opens, and steam enters. The steam does work on the piston and moves it back. The motion of the piston turns a flywheel.

INTERNAL COMBUSTION ENGINES

A car engine is a common example of an internal combustion engine. It usually has four, six, or eight cylinders. Fuel is burned inside the engine, in the cylinders. In order for this engine to work, four steps must happen inside each of the cylinders. The cylinders cycle, so each cylinder is at a different step at a different time. This type of engine is called a *four-stroke engine*.

First, gasoline and air enter the cylinder when the piston moves down. This is the *intake stroke*. Then, the *compression stroke* moves the piston up and compresses the gas mixture. Next, in the *power stroke*, a spark plug ignites the fuel mixture. As it expands, it forces the pistons back down. In the last step, the *exhaust stroke*, the piston moves back up. This pushes the exhaust gases out of the cylinder.

Critical Thinking

10. Infer Which stroke coverts chemical energy into mechanical motion?

_____

An Internal Combustion Engine: Four Steps	
Cause	**Effect**
Intake Stroke: The piston in the cylinder moves down.	_____ enter the cylinders
Compression Stroke: The piston in the cylinder moves back up.	the gas mixture gets _____
Power Stroke: A spark plug ignites the fuel mixture.	the gas mixture expands and _____
Exhaust Stroke: The piston in the cylinder moves back up.	_____ are pushed out of the cylinder

TAKE A LOOK
11. Analyze Fill in the Cause-and-Effect Table with the events that happen in an internal combustion engine.

What Is a Cooling System?

An air-conditioned room can feel refreshing on a hot summer day. Cooling systems, like air conditioners, can move thermal energy out of an area so that it feels cooler. Thermal energy normally moves from a high temperature to a lower temperature. An air-conditioning system moves warm air outside. This is against the normal flow of thermal energy. So, it must do work. It's like walking uphill: if you are going against gravity, you must do work.

How a Refrigerator Works

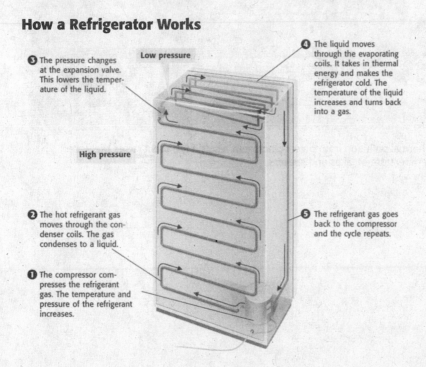

❸ The pressure changes at the expansion valve. This lowers the temperature of the liquid.

Low pressure

❹ The liquid moves through the evaporating coils. It takes in thermal energy and makes the refrigerator cold. The temperature of the liquid increases and turns back into a gas.

High pressure

❷ The hot refrigerant gas moves through the condenser coils. The gas condenses to a liquid.

❶ The compressor compresses the refrigerant gas. The temperature and pressure of the refrigerant increases.

❺ The refrigerant gas goes back to the compressor and the cycle repeats.

TAKE A LOOK
12. Describe When the refrigerator is getting cold, what is happening to the refrigerant?

_____

_____

_____

COOLING AND ENERGY

Most cooling systems need electrical energy to do the work of cooling. Cooling systems have a *compressor* that uses electrical energy to do the work of turning a refrigerant into a liquid. The *refrigerant* is a gas that has a boiling point below room temperature. This allows it to condense (change from a gas to a liquid) easily. ☑

Foods are kept in a refrigerator so they stay fresh. A refrigerator is a cooling system. Thermal energy moves from inside the refrigerator to the coils on the outside of the refrigerator. That is why the lower back of the refrigerator feels warm.

What Is Thermal Pollution?

Heating systems, car engines, and cooling systems all put thermal energy into the environment. However, too much thermal energy can be harmful.

Thermal pollution from power plants can result if the plant raises the water temperature of lakes and streams.

☑ **READING CHECK**

13. Describe What is the purpose of the compressor in a refrigerator?

_____

_____

📢 **Say It**

Investigate Research thermal pollution where you live. See what your local power plant does to minimize thermal pollution. Report your findings to your class.

SECTION 4 **Heat Technology** *continued*

THERMAL POLLUTION

One of the negative effects of too much thermal energy is **thermal pollution**. This happens when the temperature of a body of water rises a lot. Power plants burn fuel, and this gives off thermal energy. Not all of the thermal energy is used to do work. Some is wasted and released into the environment. Many power plants are near a body of water where they can dump the waste thermal energy.

A power plant may use cool water from a body of water to absorb the waste thermal energy. The cool water absorbs the energy and warms up. Then, this warm water may be dumped back into the same water that it came from. This makes the temperature of the water rise. Higher water temperatures in lakes and streams can hurt the animals in it. It can also harm the entire ecosystem of the river or lake. ☑

Some power plants cool the water before putting it back. This helps to lower the amount of thermal pollution.

The flow chart below shows how thermal pollution might occur and its effects.

☑ **READING CHECK**

14. Describe What can thermal pollution do to lakes and streams?

_____

_____

_____

_____

```
┌─────────────────────────────────────────────┐
│ A power plant pumps in cool water from a river. │
└─────────────────────────────────────────────┘
                      │
                      ▼
┌─────────────────────────────────────────────┐
│ The cool water is used to absorb _____ .  │
└─────────────────────────────────────────────┘
                      │
                      ▼
┌─────────────────────────────────────────────┐
│ The cool water heats up.                       │
└─────────────────────────────────────────────┘
                      │
                      ▼
┌─────────────────────────────────────────────┐
│ The warm water is put back into the _____ . │
└─────────────────────────────────────────────┘
                      │
                      ▼
┌─────────────────────────────────────────────┐
│ The water temperature in the river rises.      │
└─────────────────────────────────────────────┘
                      │
                      ▼
┌─────────────────────────────────────────────┐
│ The ecosystem and the animals may be _____ . │
└─────────────────────────────────────────────┘
```

TAKE A LOOK

15. Complete Fill in the Flow Chart with the missing words.

Section 4 Review

NSES   PS 3a, 3b

SECTION VOCABULARY

heat engine a machine that transforms heat into mechanical energy, or work **insulation** a substance that reduces the transfer of electricity, heat, or sound	**thermal pollution** a temperature increase in a body of water that is caused by human activity and that has a harmful effect on water quality and on the ability of that body of water to support life

1. Compare What is the difference between a hot-water heating system and a warm-air heating system?

_____

_____

_____

2. Analyze What happens during the intake and compression strokes of a four stroke engine?

_____

_____

3. Compare What is the difference between an external combustion engine and an internal combustion engine?

_____

_____

4. Explain Why must a cooling system do work to transfer thermal energy?

_____

_____

5. Identify What property must a refrigerant of a cooling system have? Why?

_____

_____

6. Explain Explain how power plants cause thermal pollution and how it affects the environment.

_____

_____

_____